TAXONOMY AND DISTRIBUTION OF THE
CALANOID COPEPOD FAMILY HETERORHABDIDAE

Taxonomy and Distribution of the Calanoid Copepod Family Heterorhabdidae

Taisoo Park

UNIVERSITY OF CALIFORNIA PRESS
Berkeley • Los Angeles • London

BULLETIN OF THE SCRIPPS INSTITUTION OF OCEANOGRAPHY
OF THE UNIVERSITY OF CALIFORNIA, SAN DIEGO
LA JOLLA, CALIFORNIA

Advisory Editors: Charles S. Cox, Gerald L. Kooyman,
Richard H. Rosenblatt (Chairman)

Volume 31

Approved for publication January 1999

UNIVERSITY OF CALIFORNIA PRESS
BERKELEY AND LOS ANGELES, CALIFORNIA

UNIVERSITY OF CALIFORNIA PRESS, LTD.
LONDON, ENGLAND

Library of Congress Cataloging-in-Publication Data

Park, Taisoo.
Taxonomy and distribution of the calanoid copepod family
Heterorhabdidae / Taisoo Park.
p.   cm. — (Bulletin of the Scripps Institution of Oceanography,
University of California, San Diego; v. 31)
Includes bibliographical references (p.   ).
ISBN 0-520-09842-0 (pbk: alk. paper)
1. Heterorhabdidae—Classification. I. Title. II. Series.

QL444.C72 P36 2000
595.3'4—dc21   00-047517

CONTENTS

ABSTRACT

Fifty-nine species, including 25 new species, of the marine calanoid family Heterorhabdidae are described and their geographic ranges defined from specimens found in midwater trawl and plankton net samples collected throughout the Atlantic, Pacific, and Indian Oceans. The family is defined by the specialized characters of the left caudal ramus, the 4th marginal seta of the left caudal ramus, and the basis of the male right P5. Based on these synapomorphies, only *Disseta, Mesorhabdus, Heterostylites, Hemirhabdus, Neorhabdus, Paraheterorhabdus,* and *Heterorhabdus* are referred to the Heterorhabdidae. The genera *Microdisseta* and *Alrhabdus*, previously referred to the family, are removed. The genus *Paraheterorhabdus* is further subdivided into two subgenera—*Paraheterorhabdus* and *Antirhabdus*—and the species of *Heterorhabdus* are placed in four species groups—*spinifrons, papilliger, fistulosus,* and *abyssalis*. Keys are presented and supraspecific taxa diagnosed. Phylogenetic relationships among the genera and the geographic distribution of species are discussed.

Key words: Taxonomy, phylogeny, geographic distribution, worldwide, Copepoda, Calanoida, Heterorhabdidae, new subgenus, new species, keys.

ACKNOWLEDGMENTS

This study is based mainly on midwater trawl and plankton net samples housed in the Planktonic Invertebrates Collections at Scripps Institution of Oceanography. I wish to thank M. Ohman, curator, and A. Townsend, museum scientist, of these collections, who made the samples available for my study. Many additional specimens were obtained from a number of sources through generous help of many individuals. I am particularly indebted to the following persons: F. D. Ferrari of the National Museum of Natural History, Smithsonian Institution, P. Wiebe and J. Craddock of Woods Hole Oceanographic Institution, K. Wishner and J. Saltzman of the University of Rhode Island. I am also indebted to R. H. Rosenblatt of Scripps Institution of Oceanography and F. D. Ferrari of National Museum of Natural History, Smithsonian Institution, for critical reviews which greatly improved the manuscript. This study was supported by National Science Foundation Grant OCE-9402011 and was carried out at Scripps Institution of Oceanography. I wish to acknowledge the generous cooperation of M. Mullin, director of the Marine Life Research Group.

INTRODUCTION

The family Heterorhabdidae was erected by Sars (1902) to accommodate the genera *Heterorhabdus* Giesbrecht 1898 (published in Giesbrecht and Schmeil 1898), *Haloptilus* Giesbrecht 1898, and *Augaptilus* Giesbrecht 1889. Sars (1905) added *Disseta* Giesbrecht 1889 and a new genus, *Mesorhabdus*, to the Heterorhabdidae and transferred *Haloptilus* and *Augaptilus* into a new family, the Augaptilidae. Sars (1925) recognized in the Heterorhabdidae the following 5 genera: *Disseta*, *Heterorhabdus*, *Mesorhabdus*, *Hemirhabdus* Wolfenden 1911, and *Heterostylites* Sars 1920. Brodsky (1950) subdivided the genus *Heterorhabdus* into 2 subgenera—*Euheterorhabdus* (=*Heterorhabdus*) and *Paraheterorhabdus*. In his review of the Heterorhabdidae, Heptner (1972b) described 2 new genera (*Microdisseta* and *Neorhabdus*) for the Heterorhabdidae bringing the number of genera of the family to 7. Grice (1973) described *Alrhabdus johrdeae*, which he referred provisionally to the Heterorhabdidae.

My study is based on worldwide samples mostly from below 1000 m housed at Scripps Institution of Oceanography, the National Museum of Natural History of the Smithsonian Institution, Woods Hole Oceanographic Institution, and the University of Rhode Island. I have identified 59 species, including 25 new species, attributable to the family Heterorhabdidae.

The identification of species was extremely difficult as descriptions in the literature are in most cases either inadequate or too general and thus applicable to more than one species. For certain species, the only useful information available in the literature for its identification is the type locality and body size as the rest of descriptions were devoted mainly to the general features applicable also to the congeners. When only one species occurs in an area, it could be identified by information about its type locality and body size. In this study, however, it was quite common to find several species in the same sample that all agree with the description of a particular species in the literature but can be clearly distinguished from each other by some specialized characters. It was, therefore, essential to redefine all of the previously known species of the family if the new species and inadequately known species were to be properly diagnosed.

Examples of misidentifications in the literature resulting from the insufficient understanding of the previously described species are numerous. For example, *Heterorhabdus abyssalis* is the most confounded species. *Heterorhabdus abyssalis* of Tanaka (1964) and Park (1968) was subsequently referred by Bradford (1971) to *H. proximus* Davis 1949 but found in the present study to be a new species (*H. habrosomus*) clearly distinct from either *H. abyssalis* or *H. proximus*. Ohtsuka et al. (1997) synonymized *Heterorhabdus nigrotinctus* Brady 1918 with *Heterostylites major* (Dahl 1894). *Heterorhabdus nigrotinctus* was found in the present study in the type locality. As figured by Brady (1918), the male 5th pair of legs of *H. nigrotinctus* is distinctly different from that of *H. major.*

When all of the species found in the study were closely examined and compared with one another for their diagnoses, it was necessary to redefine not only all the species but also all the genera and the family itself based on specialized characters shared by all members of each taxon. The family Heterorhabdidae is redefined by the specialized characters of the left caudal ramus, the 4th marginal seta of the left caudal ramus, and the basis of the male right P5. Based on these synapomorphies, only *Disseta, Mesorhabdus, Heterostylites, Hemirhabdus, Neorhabdus, Paraheterorhabdus,* and *Heterorhabdus* are referred to the family and the genera *Microdisseta* and *Alrhabdus*, previously referred to the family, are removed from it. The genus *Paraheterorhabdus* is further divided into 2 subgenera—*Paraheterorhabdus* and *Antirhabdus*—and the species of *Heterorhabdus* are placed in 4 species groups—*spinifrons, papilliger, fistulosus,* and *abyssalis.*

Phylogenetic relationships among the heterorhabdid genera were examined by Heptner (1972b) and Ohtsuka et al. (1997), showing *Disseta* to diverge first, *Microdisseta* to evolve as the second offshoot, and *Hemirhabdus, Neorhabdus,* and *Heterorhabdus* as the terminal branches. However, *Microdisseta* was found in the study not to have synapomorphies that bring heterorhabdid genera together into a monophyletic group and thus it can not be considered to have evolved from the common ancestor of the Heterorhabdid genera.

In this paper I rediagnose the family, redefine its 7 genera, characterize all subgenera and species groups recognized in the family, fully describe, with figures, each of the 59 species found and provide information on its geographic distribution.

MATERIALS AND METHODS

Most of the specimens examined in this study were picked from Isaacs-Kidd midwater trawl (IKMT) samples selected from the collections available at Scripps Institution of Oceanography (SIO). Additional specimens were obtained from IKMT, plankton net and MOCNESS (Multiple Opening/ Closing Net and Environmental Sensing System) samples available at the National Museum of Natural History, Smithsonian Institution (USNM), Woods Hole Oceanographic Institution (WHOI), and the University of Rhode Island (URI). A total of 148 samples (1 plankton net, 126 IKMT and 21 MOCNESS samples) collected throughout the Atlantic, Pacific, and Indian Oceans were examined for specimens. Their sources and the areas they represent are as follows:

1) Ninety-seven IKMT samples from SIO collected obliquely to depths from 500 to 6000 mwo (meter wire out). Of these, 80 were taken in the Pacific, including the Malay Archipelago and the East and South China Seas, 8 in the Indian Ocean, and 9 in the South Atlantic.

2) One plankton net and 9 IKMT samples from USNM collected obliquely to depths from 600 to 3750 mwo. Of these, 3 were taken in the Atlantic, 6 in the Pacific, and 1 in the Indian Ocean.

3) Twenty IKMT and 5 MOCNESS samples from WHOI collected obliquely to depths from 500 to 2800 mwo in the Atlantic.

4) Sixteen MOCNESS samples from URI collected obliquely from various depths between 1293 and 74 m at 13°N, 102°W in the eastern Pacific.

The number of samples examined and the extent of geographic coverage for each ocean are as follows:

1) North Atlantic - 24 samples, including 19 IKMTs to depths of 600 to 4000 mwo and 5 MOCNESS samples from various depths down to 1014 m, throughout the area between 01°10'N (Gulf of Guinea) and 63°25'N (Norwegian Sea).

2) South Atlantic - 13 IKMTs, including 12 collected to depths of 500 to 6000 mwo throughout the area between 10°20'S and 56°37'S and 1 from 2000 mwo at 60°42'S, 59°48'W in the Drake Passage.

3) Northeastern Pacific - 50 samples, including 34 IKMTs to depths from 1000 to 5600 mwo throughout the area between the equator and 40°35'N (off northern California) and from the west coast of America westward to 131°W and 16 MOCNESS samples from various depths between 1293 and 74 m at 13°N, 102°W.

4) Southeastern Pacific - 21 IKMTs to depths from 500 to 5500 mwo throughout the area between the equator and 46°57'S and from the west coast of America westward to 124°W.

5) Central North Pacific - 9 samples, including 8 IKMTs to depths from 2550 to 3000 mwo in the area between the equator and 31°13'N and between 140°W and 180°W, and 1 plankton net sample from 600 - 400 m at 52°20'N, 155°16'W.

6) Central South Pacific - 7 IKMTs to depths from 1700 to 3500 mwo in the area between the equator and 26°S and between 140°W and 180°W.

7) Northwestern Pacific including the Malay Archipelago - 7 IKMTs to depths from 2964 to 3000 mwo in the area between 05°27'S and 35°01'N and between 119°55'E and 146°10'E.

8) Southwestern Pacific including the Tasman Sea - 8 IKMTs to depths from 1500 to 3750 mwo in the area between 09°25'S and 54°30'S and between 150°E and 180°E.

9) Indian Ocean - 9 IKMTs to depths of 1500 to 3000 mwo in the area between 02°28'S and 35°00'S and from the east coast of Africa eastward to 86°23'E.

Citation of type material includes, in order, the number of specimens, collecting gear, sampling depth, source, expedition, cruise number, station number, latitude and longitude, area, and date of collection. The following abbreviations are used in listing type specimens: USNM, National Museum of Natural History; SIO, Scripps Institution of Oceanography; IKMT, Isaacs-Kidd midwater trawl; MOCNESS, Multiple Opening/Closing net and Environmental Sensing System; mwo, meter wire out; MV, Marine Vertebrate cruise; CalCOFI, California Cooperative Oceanic Fisheries Investigations.

With the exception of a few extremely large samples, such as some IKMT samples taken in highly productive coastal waters, the entire sample was examined for adult specimens. For identification, specimens were individually examined and measured, usually in 50:50 glycerine-distilled water. The measurements usually include the body length, as measured from the tip of the forehead to the posterior end of the caudal ramus, and the prosome length (from the tip of the forehead, including the spiniform frontal process, to the posterior end of the prosome) as measured from the lateral side. When the posterior ends of the prosome were produced asymmetrically, the measurement was taken from the side with the longer projection. For detailed morphological observation, specimens were stained with methyl blue in lactic acid and dissected in a drop of clear lactic acid. Drawings were made from stained specimens using a camera lucida. Most of the morphological features important for species identification are illustrated, and specific structures referred to in the description are labeled in the figures. At least one species in each genus, subgenus, and species group is fully described and used as reference in describing the others.

The following abbreviations are used in the descriptions: BL, body length; PL, prosome length; A1, antennule; A2, antenna; Md, mandible; Mx1, maxillule; Mx2, maxilla; Mxp, maxilliped; P1-P5, 1st to 5th legs.

The references cited in the synonymies are only those papers providing descriptions or illustrations by which the identity of the species can be established. Geographic ranges of the species are based only those literature records that can be verified by species descriptions or accompanying illustrations.

Type materials of new species described here have been deposited in the National Museum of Natural History, Washington, DC. All other specimens examined in the study are housed in the Planktonic Invertebrates Collection of Scripps Institution of Oceanography except for those from USNM, which have been deposited in the National Museum of Natural History.

FAMILY HETERORHABDIDAE SARS 1902

Family Heterorhabdidae Sars 1902, p. 117.

Diagnosis. 1) Left caudal ramus normally fused to anal segment, longer than right ramus. 2) Fourth marginal seta (counted from lateral) of left caudal ramus naked and greatly elongated. 3) Basis of male right P5 with a large, plumose inner lobe, which in some taxa is extremely large.

Remarks. The Heterorhabdidae can be distinguished not only from the other families of the superfamily Arietelloidea but also from all other calanoid families by the above three characters. These three characters are highly pronounced in all species examined in the study except for *Paraheterorhabdus (Antirhabdus) compactus*, in which the left caudal ramus is only slightly longer than the right and separated from the anal segment, and the 4th marginal seta of the left caudal ramus is longer than the other marginal setae but not greatly elongated. These character states of this species are believed to be reductive rather than primitive because the species shows an advanced stage of evolution in all other morphological features.

Additional description of female. First pedigerous somite fully separated from cephalosome, 4th and 5th pedigerous somites completely fused. Forehead with a sagittal ridge, visible dorsally in most species as a midanterior tubercular process (Fig.4-c) and in lateral view, extending ventrad beyond sternite into a conical rostrum bearing 2 elongate rostral filaments (Fig.4-b). Urosome 4-segmented. Genital somite with cuticular pores around genital opening (Fig.4-g). Tubule connecting between prosomal gland and spermatheca (Park 1966) is clearly visible (Fig.4-f). Left caudal ramus longer than right and normally fused with anal segment (Fig.4-e). Each caudal ramus with 5 marginal setae and a dorsal appendicular seta. Fourth marginal seta (counted from lateral) naked and greatly elongated. Other marginal setae plumose with setules and in some species, armed with spines (Fig.11-k) in addition to normal setules.

Antennule 25-segmented. In A2 (Fig.1-h), coxa and basis partially fused, with 1 and 2 inner marginal setae, respectively. Endopod 2-segmented; 1st segment with 2 inner marginal setae; 2nd segment divided distally into inner and outer lobes. Inner lobe with 8 marginal setae and normally a posterior appendicular seta. Outer lobe with 6 marginal setae and normally a posterior appendicular seta. Exopod 8-segmented. Primitively, first 7 segments each with an inner marginal seta; however, setae on first 2 segments reduced in most genera. Eighth segment with an inner marginal seta and 3 terminal setae.

Mandibular blade well developed. Masticatory edge (Fig.1-j) armed with an elongate basal spine and a varying number of spiniform teeth; with fewer number of teeth there is an increase in tooth size and in size of gap between teeth (Fig.60-c). Basis with setae ranging in number from 0 to 4. Endopod 2-segmented, with 2-4 inner marginal setae on 1st and 7 or 8 terminal and 1 or 2 anterior appendicular setae on 2nd.

Maxillule (Fig.2-a) with proximal complex, which is clearly divided into 2 segments, plus basis, endopod, and exopod. Proximal complex with 1 outer and 3 inner lobes; first 2 inner lobes and outer lobe on 1st segment and 3rd inner lobe on 2nd. First inner lobe with 6-14 setae; 2nd with 0 or 1 seta; 3rd with 0-3 setae; outer lobe with 3-9 setae. Basis with 0-4 setae. Endopod 1- or 3-segmented, with 1-12 setae. Exopod well developed and elongate, reaching at least distal end of endopod and in most genera, extending far beyond distal end of endopod (Fig.19-d).

Maxilla (Fig.2-b) consisting of 5 segments and endopod. First segment with 2 lobes, 2nd to 5th segments each with a single lobe. First lobe with 2-6 setae in addition to a small spiniform process found in some genera; 2nd with 0-3 setae; 3rd and 4th each with 2 or 3 setae; 5th with 2-4 setae; 6th with 2 or 3 setae. One of setae on 5th and 6th lobes modified into a saberlike or falciform spine (Fig.31-b). Endopod with 2, 6, or 7 setae.

Maxilliped (Fig.2-c) consisting of coxa, basis, and 5-segmented endopod. Coxa with 2-4 lobes. First 2 lobes each with 0 or 1 seta; 3rd lobe with 1, 2, or 3 setae; 4th with 2 or 3 setae; when 3rd lobe has only 1 seta, it may be specialized into a large saberlike or falciform spine (Fig.73-a). Basis with

5

a middle group of 3 setae and a distal group of 2 setae. Five endopodal segments with 3+3+3+3+4 setae, from proximal to distal.

Coxae of P1-4 with an inner marginal seta. Coxa of P5 without setae. Basis of P1 with an inner marginal seta arising from distomedial corner of segment and extending in a distolateral direction over anterior surface of endopod (Fig.2-d), with a small hooklike process on posterior surface (Fig.2-e), and with or without a small outer seta. Bases of P2 and P3 without setae. Bases of P4 and P5 each with a small outer seta.

First 2 endopodal segments with 1 and 2 inner marginal setae, respectively, in P1-P4; in P5 each segment with 1 inner marginal seta. Third endopodal segment with 5 setae in P1, 7 or 8 setae in P2, 8 setae in P3, 7 setae in P4, 6 setae in P5. Exopod with 1+1+2 outer spines, 1 terminal spine and 1+1+4 inner setae in P1, 1+1+3 outer spines, 1 terminal spine and 1+1+5 inner setae in P2-P4, 1+1+2 outer spines, 1 terminal spine and 0+1+4 inner setae in P5. Inner seta of 2nd exopodal segment of P5 spiniform without setules and thicker than inner setae of 3rd exopodal segment (Fig.2-i).

Additional description of male. Similar in morphological details of body to female except that urosome 5-segmented; 1st segment with a gonopore on right side. Left A1 geniculated (Fig.3-c), with knee joint between 18th and 19th segments. First 2 segments of left A1 normally fused; segments 19 and 20 fused in all genera; in some genera, segments 19-21 and segments 22-23 fused. All other cephalosomal appendages and first 4 pairs of legs as in female.

Fifth pair of legs strongly asymmetrical (Fig.5-d), each consisting of 2-segmented basipod and 3-segmented endopod and exopod. Coxa without setae; basis with a small outer seta. Basis of right leg with a large plumose inner lobe, which in some genera is extremely large. Basis of left leg with a plumose inner margin, which in some genera produced into a plumose lobe.

First endopodal segment without setae. Second endopodal segment normally with an inner marginal seta, but which may be missing on left endopod in some genera. Third segment with 6 setae. In right exopod, 1st segment with an outer spine and without inner seta; 2nd with an outer spine and a highly characteristic plumose medial projection (Fig.9-g); 3rd with an outer and a terminal spine. In left exopod, 1st and 2nd segments each with an outer spine and without inner seta; 3rd tapering distally into a long spiniform process, with an outer and inner spine. Distal spiniform process of 3rd segment seems to be terminal spine fused to segment.

Remarks. As members of the superfamily Arietelloidea, the Heterorhabdidae share the following characters with the other families of the superfamily: 1) First pedigerous somite separated from cephalosome; 4th and 5th pedigerous somites fused. 2) Exopod of Mx1 well developed, reaching at least distal end of endopod and in most genera, extending far beyond it. 3) Maxilla greatly elongated. Among the arietelloid families, the Heterorhabdidae, Lucicutiidae, and Augaptilidae share the following characters: 1) Exopod of A2 8-segmented, each of first 7 segments primitively bearing an inner marginal seta, 2) in female P5, 1st exopodal segment without inner marginal seta, 2nd with a spiniform inner marginal seta, 3) in male, left A1 geniculated and gonopore on the right side.

The Heterorhabdidae seems to be most closely related to the Lucicutiidae according to the morphology of Mx2, in which 1 of the setae on the 5th and 6th lobes modified into a saberlike or falciform spine. Within the Heterorhabdidae, however, the large spines of the 5th and 6th lobes of Mx2 show a progressive outgrowth starting from a character state shared with the Lucicutiidae and becoming a highly developed spine in the form of a saber or sickle. The teeth on the masticatory edge of Md show a progressive reduction in number accompanied by a progressive increase in size of individual teeth and the gap between them. Furthermore, the setae of the Mxp coxa show a progressive specialization within the family with a gradual decrease in number accompanied by the extraordinary modification of the middle seta into a large saberlike or falciform spine. These specializations in the Md, Mx2, and Mxp, therefore, serve as key characters not only among the genera of the Heterorhabdidae but also in distinguishing some of the heterorhabdid genera from the other families of the Arietelloidea.

The basis of P1 has a small hooklike process on the posterior surface in all species of the Heterorhabdidae. A similar structure has been described by Ferrari and Saltzman (1998) for *Pleu-*

romamma. However, its significance in systematics of the Calanoida cannot be assessed because of insufficient information on its occurrence in other taxa.

Mauchline (1988) described the pattern of integumental pores of the Heterorhabdidae, which he found to be similar to that of the Arietellidae. However, these two taxa are quite different from each other in morphological details of all cephalosomal appendages and the 5th pair of legs and no synapomorphies have been found. Furthermore, the pore patterns displayed by the genera range from the relatively simple ones found in *Heterorhabdus* and *Mesorhabdus* to the extremely complicated ones of *Hemirhabdus*, which are not consistent with their phylogenetic relationship hypothesized in the present study based on the other anatomical features.

GENERA AND SUBGENERA OF THE FAMILY HETERORHABDIDAE

Disseta Giesbrecht 1889
Mesorhabdus Sars 1905
Heterostylites Sars 1920
Hemirhabdus Wolfenden 1911
Neorhabdus Heptner 1972
Paraheterorhabdus Brodsky 1950
    Subgenus *Paraheterorhabdus* Brodsky 1950
    Subgenus *Antirhabdus*, new subgenus
Heterorhabdus Giesbrecht 1898

Wolfenden (1911) established a new genus, *Alloiorhabdus,* to accommodate *Heterorhabdus austrinus* Giesbrecht 1902, which is the type by page priority, and a new species, *Alloiorhabdus medius*. As pointed out by Vervoort (1951), Wolfenden's putative new genus, based on inaccurate observations on the number of setae on various segments of the swimming legs, is a synonym of *Heterorhabdus*. The specimens described by Wolfenden as *Heterorhabdus austrinus* Giesbrecht 1902 actually belong to *Heterorhabdus farrani* Brady 1918 and the specimens described as *Alloiorhabdus medius* belong to an entirely different genus, *Heterostylites* Sars 1920.

Microdisseta Heptner 1972 and *Alrhabdus* Grice 1973 have been referred to the Heterorhabdidae. However, they could not be referred because the three diagnostic characters of the family described above are not found in these genera.

The genus *Microdisseta* was erected by Heptner (1972b) to accommodate *Disseta minuta* Grice and Hulsemann 1965, which was described from both females (0.68-0.80 mm long) and males (0.73-0.75 mm long) collected from 2000-4750 m at 3 stations in the northeastern Atlantic (between 29°57'N and 40°04'N and between 19°57'W and 23°01'W).

According to Grice and Hulsemann (1965), *Disseta minuta* shares the following features with *Disseta* as well as all other genera of the family Heterorhabdidae: 1st pedigerous somite separate from cephalosome, 4th and 5th pedigerous somites coalesced, urosome 4-segmented in female and 5-segmented in male, caudal rami asymmetrical with left ramus larger, and rostrum with 2 filaments. Of these features, the asymmetrical caudal rami could be singled out as the most important evidence for their possible relationship, and the rest are common not only in the family Heterorhabdidae but also in all other families of the superfamily Arietelloidea. However, both caudal rami in *D. minuta* are fully separated from the anal segment, while only the right caudal ramus is separated from it in the other species of *Disseta*. Each caudal ramus of *D. minuta* has only 4 marginal setae without a lateral marginal seta and without a dorsal appendicular seta, both of which are present in all other species of *Disseta*. The 2nd marginal seta of the left caudal ramus, counted from medial, looks thicker than the other marginal setae as in the other species of *Disseta* but its length is not indicated. Of the cephalosomal appendages, only Md, Mx1, Mx2, and Mxp of the female have been figured for *D. minuta*, of which only Mx2 is similar to that of the other species of *Disseta*. The appendage showing a most significant difference is Mx1, of which the exopod is far short of reaching the distal end of the endopod in *D. minuta*, while it reaches the distal end of the endopod in all other species of *Disseta* and extends far beyond its distal end in all other heterorhabdid genera. The extremely elongate exopod of Mx1 is a major characteristic in distinguishing the superfamily Arietelloidea (=Augatiloidea) including the family Heterorhabdidae from all other calanoid superfamilies (Park 1986). The maxilliped of *D. minuta* is also highly unusual because it has only 2 setae (1 middle and 1 distal) on the basis. Of the swimming legs, only P1 has been figured for *D. minuta*, which is also significantly different both from other species of *Disseta* and from all other genera of the Heterorhabdidae because of the absence of an inner marginal and outer setae of the basis and the presence of an outer seta on the 2nd endopodal segment. In none of the calanoid copepods is P1 known to have an outer seta on the 2nd endopodal segment.

Only the 5th pair of legs and the segments 13-19 of the left A1 of the male have been figured (Grice and Hulsemann 1965). The main difference in the male 5th pair of legs between *D. minuta* and the other species of *Disseta* is the absence in the former of a large plumose inner lobe on the right basis, which is an important diagnostic feature not only for the genus *Disseta* but also for the entire family Heterorhabdidae.

Park (1970) also found 2 females of this species from 1900-980 m in the Caribbean Sea, in which only Md, Mx1, Mx2, and Mxp were intact and found to be the same as figured by Grice and Hulsemann (1965) except: the outer lobe of Mx1 with 7 setae instead of 6; the 1st lobe of Mx2 with 5 setae instead of 4; the 2nd endopodal segment of Mxp with 3 setae instead of 2.

In summary, according to all morphological features so far known for *D. minuta* Grice and Hulsemann 1965, the species must be removed not only from the genus *Disseta* but also from the family Heterorhabdidae because none of the diagnostic characters of these taxa as defined above is found in this species.

It is interesting to note that Ohtsuka et al. (1997) indicated the genus *Microdisseta* as the second offshoot after the genus *Disseta* in the evolution of the family Heterorhabdidae. They derived this conclusion in their cladistic analysis of the heterorhabdid genera based mainly on the number of setae on various appendages without consideration of the more basic question of how the family can be defined as a monophyletic group distinguishable from the other calanoid families. The character states they selected seem inappropriate for analysis of the heterorhabdid genera because if based solely on these character states, any genus of the other families of the Arietelloidea could be accepted as a member of the Heterorhabdidae, making it a polyphyletic group.

The genus *Alrhabdus* was established by Grice (1973) to accommodate *Alrhabdus johrdeae* Grice 1973, which was originally described from female specimens collected near the bottom at 1733 m off New Providence Island, Bahamas (25°10'N, 77°25'W) and provisionally referred to the Heterorhabdidae.

According to Grice (1973), the morphological features allying *Alrhabdus johrdeae* with the Heterorhabdidae were: 1) First pedigerous somite separate from cephalosome and 4th and 5th pedigerous somites coalesced, 2) urosome 4-segmented in female, 3) antennule 25-segmented, 4) first to 5th swimming legs of female each with 3-segmented endopod and exopod, 5) female 5th leg with a spiniform inner seta on 2nd exopodal segment. However, these characters are common not only to the Heterorhabdidae but also to all families of the Arietelloidea.

On the other hand, *A. johrdeae* has a number of morphological characters that differ from the corresponding features of the Heterorhabdidae: 1) Rostrum composed of 2 strong spines, instead of 2 filaments, 2) right caudal ramus longer than left and its 2nd marginal seta (counted from medial) elongated rather than left caudal ramus longer than right and with a greatly elongated seta, 3) exopod of A2 only 1/3 length of endopod rather than about as long as endopod, 4) Mx1 with an exopod far short of reaching distal end of endopod rather than exopod of Mx1 reaching or extending beyond distal end of endopod, 5) first leg without an outer seta on basis, 6) third leg with a well-developed outer seta on basis, 7) outer seta of basis in P5 greatly elongated. The differences shown by these features seem to be much more significant than the similarities exhibited by the other characters described above and the most significant difference is in the caudal rami including their setation and the maxillule, which are the main diagnostic features of the Heterorhabdidae. The known morphology of *A. johrdeae* precludes placement in the Heterorhabdidae and its placement must await the discovery of a male.

KEY TO GENERA AND SUBGENERA OF THE FAMILY HETERORHABDIDAE

Females and Males

1a. Exopod of Mx1 reaching distal end of endopod (Fig.2-a) . . . . *Disseta*
1b. Exopod of Mx1 extending far beyond distal end of endopod (Fig.8-d) 2

2a. First inner lobe of Mx1 with 1 seta transformed into a large spine (Fig.8-d) . . . . . *Mesorhabdus*
2b. First inner lobe of Mx1 with no seta transformed into a large spine 3

3a. Second exopodal segment of female P5 with serrate distal margin (Fig.19-k);
2nd exopodal segment of male right P5 extremely enlarged (Fig.20-f) . . . . *Heterostylites*
3b. Second exopodal segment of female P5 without serrate distal margin;
2nd exopodal segment of male right P5 not enlarged 4

4a. In Mx2, 2nd lobe absent (Fig.31-b) . . . . *Hemirhabdus*
4b. In Mx2, 2nd lobe present 5

5a. In Mx2, 5th and 6th lobes of similar length (Fig.35-b); endopod with 2 setae . . . . . *Neorhabdus*
5b. In Mx2, 5th lobe much longer than 6th (Fig.44-d); endopod with 7 setae 6

6a. In Mx2, 5th lobe with a small seta and a large spine (Fig.44-d) . . . . *Paraheterorhabdus*
6aa. Ventralmost tooth of left Md much longer than that of right Md
(Figs.44-a, b) . . . . *P. Paraheterorhabdus*
6ab. Ventralmost tooth of left Md similar in size to that of right Md
(Figs.57-a, b) . . . . *P. Antirhabdus*, n. subgenus
6b. In Mx2, 5th lobe with 2 large spines (Fig.60-g) . . . . *Heterorhabdus*

GENUS *DISSETA* GIESBRECHT 1889

Disseta Giesbrecht 1889, p. 812.

Type by monotypy: *Disseta palumbii* Giesbrecht 1889

Diagnosis. All principal marginal setae of caudal rami except for 4th on left ramus armed with small spines in addition to normal setules. First segment of A1 with 1+3+3+3 setae/aesthetes (Fig.1-f). In antenna, all 8 exopodal segments each with a well-developed inner marginal seta (Fig.1-h). In mandible, masticatory edge with a basal spine, a group of 6-8 small, contiguous teeth, and 3 or 4 large, spiniform teeth; basis and 1st endopodal segment each with 4 setae, and 2nd endopodal segment with 8 terminal and 2 anterior appendicular setae (Fig.1-i). Maxillule with 2nd and 3rd inner lobes bearing setae; exopod reaching distal end of endopod (Fig.2-a). In maxilla, 5th and 6th lobes each with a spine, which is shorter than the long, normal setae of the lobe (Fig.2-b). Coxa of Mxp with 4 lobes bearing setae (Fig.2-c). Basis of P1 with outer seta (Fig.2-d). Third endopodal segment of P2 with 8 setae. In female 5th leg (Fig.2-i), outer spine of 2nd exopodal segment flanked by 2 large spiniform processes of the segment. In male left, geniculated A1, segments 19-21 fused as 1st postgeniculate complex with 1 proximal and 2 distal setae, which appear to belong to 19th and 21st segments, respectively (Fig.3-c). In male 5th pair of legs (Fig.3-g), 2nd segment of left endopod without seta; 2nd segment of right exopod with spiniform or toothlike medial projection.

Additional description of female. First pedigerous somite fully separate from cephalosome; 4th and 5th pedigerous somites completely coalesced (Fig.1-a). Rostrum consists of 2 filaments borne on a large base, which extends dorsad as a sagittal ridge usually visible dorsally as well as ventrally as midanterior tubercular process of forehead (Fig.1-f). A number of large cuticular pores usually visible on each side of genital field (Fig.1-e). Anal segment partially fused with the following caudal rami; line representing its joint with right ramus is clearly visible in dorsal view but such a line is poorly formed at joint with left ramus (Fig.1-b). Caudal rami strongly asymmetrical with left ramus much larger than right (Fig.1-b). Each caudal ramus bearing 5 principal marginal setae and a small dorsal appendicular seta. Fourth marginal seta (counted from lateral) of left caudal ramus greatly developed, unarmed and usually more than 3 times length of the other principal marginal setae.

Antennule 25-segmented (Fig.1-g), extending beyond posterior end of body. Antenna (Fig.1-h) with well-developed basipod. Coxa and basis with 1 and 2 inner marginal setae, respectively. Endopod 2-segmented, with 2 inner setae on 1st segment, 8 (terminal)+1 (appendicular) setae on inner lobe and 6+1 setae on outer lobe of 2nd segment. Exopod 8-segmented. First 7 segments each with a well-developed inner marginal seta. Eighth segment with an inner marginal and 3 terminal setae.

Mandible (Figs.1-i, j, k) with well-developed masticatory edge armed with a basal spine, a group of 6 or 7 contiguous, relatively small teeth, and 3 evenly spaced large teeth. Left and right masticatory edges basically similar but not exactly symmetrical. Mandibular palp with 4 setae each on basis and 1st endopodal segment, 8 terminal and 2 anterior appendicular setae on 2nd endopodal segment, and 6 setae on exopod. Endopod extending beyond distal end of exopod.

Maxillule (Fig.2-a) with massive 1st inner lobe armed with 14 spiniform setae. Second inner lobe relatively small, fingerlike, with a single apical seta. Third inner lobe extending to the same level as basis, with 2 or 3 terminal setae. Outer lobe with 3 small proximal and 6 large terminal setae. Basis with 3 or 4 setae. Endopod well developed. First and 2nd segments almost fully coalesced, with 3 setae on 1st and 3 or 4 setae on 2nd. Third segment with 5 setae. Exopod massive, extending to the same level as endopod, and with 11 setae—6 large lateral and 5 small distal setae.

Maxilla (Fig.2-b) with 6 lobes followed by 4-segmented endopod. First lobe with 2 proximal, 1 posterior subterminal, and 3 terminal setae in addition to a small spine. Second to 4th lobes each with 1 posterior subterminal and 2 terminal setae. Fifth lobe with 1 posterior subterminal and 2 terminal setae and 1 saberlike spine. Sixth lobe with 1 posterior subterminal and 1 terminal setae and 1 saberlike spine. Terminal setae long, slender, and coarsely spinulose. Saberlike spines of 5th and 6th lobes finely serrated and much shorter, but only slightly thicker, than terminal setae. Endopod with 7 long, slender setae.

Coxa of maxilliped (Fig.2-c) with 4 lobes bearing 1, 1, 3, and 3 setae, respectively, in order from proximal. Basis with a middle group of 3 marginal setae, a distal group of 2 marginal setae, a row of conspicuous short spines extending longitudinally on anterior surface, and a row of long setules along proximal half of inner margin. Endopod 5-segmented. First 4 segments each with 3 setae and a row of small spinules at bases of setae. Last segment with 1 small outer seta and 3 long terminal setae.

First leg (Fig.2-d) with large coxa bearing a well-developed inner marginal seta. Basis with a curved inner marginal seta at distomedial corner of anterior surface, a small outer marginal seta, and a hooklike process on posterior surface close to where exopod is joined (Fig.2-e). Exopod 3-segmented. First and 2nd segments each with an inner seta, an outer spine, and a spiniform process on lateral margin close to outer spine. Third segment with 4 inner setae, 2 outer spines, and a terminal spine. Endopod 3-segmented. First and 2nd segments with 1 and 2 inner setae, respectively, and their distolateral corners produced into a spiniform process. Third segment with 2 inner, 1 outer, and 2 terminal setae. None of endopodal segments armed on anterior surface with setules or spinules that could be referred to von Vaupel-Klein's organ (von Vaupel-Klein 1972; Ferrari and Steinberg 1993).

Second leg (Fig.2-f) with a large coxa bearing an inner marginal seta. Basis without setae. Exopod 3-segmented. First and 2nd segments each with an inner seta and an outer spine. Third segment with 5 inner setae and 3 outer spines in addition to a terminal spine. Endopod 3-segmented, with 1 inner seta on 1st segment, 2 inner setae on 2nd, and 4 inner, 2 outer, and 2 terminal setae on 3rd. Distolateral corner of each endopodal segment produced as a spiniform process. Two outer setae of 3rd endopodal segment each preceded by a spiniform process.

Third leg (Fig.2-g) similar to 2nd except that terminal and outer spines of exopod relatively small.

Fourth leg (Fig.2-h) similar to 3rd except that basis with a small outer seta at distolateral corner of posterior surface, exopod with terminal and outer spines smaller than those of 3rd leg, and 3rd endopodal segment with 7 setae.

Fifth leg (Fig.2-i) without inner seta on coxa. Basis with an outer seta at distolateral corner of posterior surface. Exopod 3-segmented. First segment with an outer spine and without an inner seta. Second segment with an outer spine and a spiniform inner seta. Outer spine of 2nd segment is in a deep pit formed by 2 adjoining spiniform processes at distolateral corner of segment. Third segment with 4 inner setae, 2 small outer spines, and a moderately developed terminal spine. Endopod 3-segmented. First 2 segments each with an inner seta and distolateral corner produced as a spiniform process. Third segment with 2 inner, 2 outer, and 2 terminal setae.

Additional description of male. First pedigerous somite fully separated from cephalosome (Fig.3-a). Fourth and 5th pedigerous somites completely coalesced. Rostrum consists of 2 filaments borne on a large base, which extends dorsad to form midanterior tubercular process of forehead, but the process in dorsal view is not so conspicuously produced as in female.

Urosome 5-segmented including anal segment that is partially fused with the following caudal rami; its joint with right ramus visible dorsally but its joint with left ramus invisible (Fig.3-b). Caudal rami strongly asymmetrical with left ramus much larger than right as in female. Setae of caudal rami also as in female.

Antennules extending beyond posterior end of body. Right antennule similar to that of female. Left antennule (Fig.3-c) geniculated with knee joint between 18th and 19th segments. Proximal 16 segments of antennule similar to those of female except that first 2 segments partially coalesced. First segment after knee joint is a compound segment resulting from the fusion of original segments 19-21 and the following segment is again a compound segment formed by the fusion of original segments 22-23. All the other cephalosomal appendages and 1st to 4th legs similar to those of female.

Fifth pair of legs strongly asymmetrical (Fig.3-g), although they are basically similar to each other, each with 2-segmented basipod followed by 3-segmented endopod and exopod. Left and right exopods in the form of a pair of tongs with 2 distal segments of each turned 90° clockwise or counterclockwise bringing their posterior surfaces to face mediad and then both terminal spines normally turned mediad. Coxa without setae. Basis with a small outer seta and a large inner lobe, which is longer in right leg (Fig.3-d) than the corresponding lobe in left leg (Fig.3-e). First endopodal segment without setae; 2nd with a seta only in right endopod; 3rd with 6 setae in both endopods. In left exopod, 1st and 2nd segments each with an outer spine; 3rd segment with an outer spine, an inner spine, a terminal spine in addition to a large elevated cuticular pore and a spiniform process distally next to terminal spine. In right exopod, each segment with an outer spine in addition to a terminal spine on 3rd segment. Second segment with spiniform medial projection.

Remarks. Disseta is the most primitive genus in the Heterorhabdidae and can be distinguished from the other genera of the family by such primitive features as the inner marginal setae on all exopodal segments of A2, a large number of teeth and setae on Md, a relatively short exopod of Mx1, relatively unspecialized setae and spines of Mx2, a large number of setae on the coxa of Mxp, and an outer seta on the basis of P1.

Nominal species of the genus *Disseta*

Leuckartia scopularis Brady 1883
Disseta palumbii Giesbrecht 1889
Heterorhabdus grandis Wolfenden 1904
       Synonym of *D. palumbii* Giesbrecht 1889 (see Farran 1908, p. 67)
Disseta grandis Esterly 1906
       Synonym of *D. palumbii* Giesbrecht 1889 (see Scott 1909, p. 134)
Disseta atlantica Wolfenden 1911
       Synonym of *D. palumbii* Giesbrecht 1889 (see Sewell 1947, p. 189)
Disseta sp. Esterly 1911
       Synonym of *D. palumbii* Giesbrecht 1889 (see Sewell 1947, p. 189)
Disseta maxima Esterly 1911
       Synonym of *D. scopularis* (Brady 1883) (see Sewell 1947, p. 189)
Disseta minuta Grice and Hulsemann 1965 = *Microdisseta minuta* (Grice and Hulsemann 1965) (see
   Heptner 1972b, p. 60)
Disseta magna Bradford 1971
Disseta coelebs Heptner 1972
       Synonym of *Disseta magna* Bradford 1971
        Remarks. The genus now contains only 3 species (*D. palumbii, D. scopularis,* and *D. magna*).

Key to species of *Disseta*

Females

1a. Dorsally, genital somite (Fig.6-c) relatively short, with smoothly curved
     lateral swellings . . . . *D. magna*
1b. Dorsally, genital somite (Figs.1-b, 4-e) relatively long, with tubercular
     lateral swellings 2

2a. P5, 2nd exopodal segment with a long outer spine (Fig.5-c) . . . . *D. scopularis*
2b. P5, 2nd exopodal segment with a short outer spine (Fig.2-i) . . . . *D. palumbii*

Males

1a. Left P5 (Fig.3-i), 2nd exopodal segment with a large lateral spiniform process . . . . *D. palumbii*
1b. Left P5 (Figs.5-f, 6-h), 2nd exopodal segment without a large lateral spiniform process 2

2a. Left P5 (Fig.6-k), 2nd endopodal segment with an anterodistal lobe; 3rd
    exopodal segment with a small terminal spine . . . . *D. magna*
2b. Left P5 (Fig.5-d), 2nd endopodal segment without an anterodistal lobe; 3rd
    exopodal segment with a large terminal spine . . . . *D. scopularis*

Disseta palumbii Giesbrecht 1889
Figures 1-3

Disseta palumbii Giesbrecht 1889, p. 812—Giesbrecht 1892, p. 369, pl. 29, figs. 2, 8, 14, 19,
    23-25, 27; pl. 38, fig. 44.
Disseta palumboi; Giesbrecht and Schmeil 1898, p. 112.—Farran 1908, p. 67.—Scott 1909, p. 133,
    pl. 41, figs. 11-21.—Sars 1925, p. 221, pl. 60.—Rose 1929, p. 34, fig. 4.—Sewell 1932,
    p. 309, text figs. 102, 103.—Rose 1933, p. 199, fig. 236.—Sewell 1947, p. 185,
    text fig. 48.—Brodsky 1950, p. 343, fig. 240.—Tanaka 1964, p. 32, fig. 190.—Vervoort
    1965, p. 116.
Disseta grandis Esterly 1906, p. 72, pl. 9, fig. 21; pl. 11, figs. 45, 46; pl. 13, fig. 69; pl. 14,
    figs. 88, 94.
Disseta atlantica Wolfenden 1911, p. 313.
Disseta sp. Esterly 1911, p. 331, pl. 28, figs. 40, 41; pl. 30, figs. 76, 80; pl. 31, fig. 100; pl. 32,
    figs. 107, 108.
Heterorhabdus grandis Wolfenden 1904, p. 120, pl. 9, fig. 36.—Wolfenden 1905, p. 8, pl. 4,
    figs. 7, 8.

Description of female. Prosome length, 4.83-5.58 mm; body length, 6.66-7.75 mm. Body rather solidly built (Fig.1-a). Prosome approximately 2.2 times as long as urosome. Dorsally as well as ventrally (Fig.1-f), forehead with a low midanterior tubercular process that extends ventrad to form rostral base. Two rostral filaments frequently of different lengths. Laterally, posterior end of prosome (Figs.1-a, d) extending posteriad into a rather broadly rounded lappet. Dorsally, posterolateral corners of prosome (Figs.1-b, c) more or less asymmetrical, with left side slightly longer than right and each with a rather obtuse posterior end.

First 3 urosomal somites fringed with fine spinules along posterior margin. Genital somite in dorsal view (Figs.1-b, c) asymmetrical with dissimilar lateral tubercular swellings. Tubercular swelling of left side bearing a small rounded ridge or lobular outgrowth, which is somewhat variable in size and shape. Laterally, genital prominence low and rather inconspicuous and close to anterior end of somite (Fig.1-d). Ventral margin of genital somite posterior to genital prominence is more or less concave. In most specimens, tubule connecting spermatheca and prosomal gland was clearly visible. Ventrally (Fig.1-e), genital operculum relatively wide. About 15 cuticular pores clearly visible on each side of genital field.

Genital somite approximately 1.5 times length of 2nd urosomal somite, which is only slightly longer than 3rd (Fig.1-b). Anal segment about 1/2 the length of preceding somite. Left caudal ramus extending beyond posterior end of right by about 1/4 its length as measured along medial margin. Dorsally, anal segment with 2 or 3 cuticular pores on each side and each caudal ramus with 3 cuticular pores close to lateral margin. Fourth marginal seta (counted from lateral) of left caudal ramus nearly as long as body. Fourth marginal seta of right ramus about as long as urosome. All principal marginal caudal setae, except for longest seta of left ramus, are armed with conspicuous spinules in addition to ordinary setules.

Antennule extending beyond posterior end of body by its last 3 or 4 segments (Fig.1-a). Most setae are more or less modified into aesthetes (Fig.1-g). First segment with 4 lobes bearing 1, 3, 3, and 3 setae/aesthetes, respectively; each lobe probably representing a separate segment. First segment, therefore, could be regarded as a composite formed by the fusion of 4 original segments. Segments 2-19 each bearing 3 setae—1 middle and 2 distal. Segment 20 with 1 distal seta, segment 21 with 2 distal setae, segment 22 with 1 posterodistal and 1 anterodistal setae, segment 23 with 1 posterodistal and 2 anterodistal setae, segment 24 with 1 posterodistal and 1 anterodistal setae, segment 25 with 7 terminal setae. Two distalmost setae of segment 25 are borne on a separate base, which could be regarded as a separate segment.

In antenna (Fig.1-h), basipod about as long as endopod, coxa and basis almost completely fused, and endopod only slightly longer than exopod. Inner marginal setae of coxa and basis well developed. Only 1 of 2 marginal setae of 1st endopodal segment well developed. In 2nd endopodal segment, appendicular seta of inner lobe similar in size to lateralmost marginal seta; appendicular seta of outer lobe much better developed than that of inner lobe. All inner marginal and terminal setae of exopod equally well developed.

In mandible (Figs.1-i, j, k), mandibular blade nearly equal in length to combined lengths of basis and exopod. Exopod and endopod similar in length; exopod arising high on lateral margin of basis, while endopod attached to distal end of basis, and therefore, endopod extending beyond distal end of exopod by distal half of its 2nd segment. Left and right masticatory edges asymmetrical only in some minor features. In right masticatory edge (Fig.1-k), group of small, contiguous teeth next to basal spine is flanked at ventral end by a relatively large tooth, which is separated from the group by a deep split; of the 3 large ventral teeth, the middle is monocuspid instead of bicuspid as in the left masticatory edge. Four setae on basis almost as well-developed as those of 1st endopodal segment. Two appendicular setae of 2nd endopodal segment thinner than, and approximately 1/2 length of, terminal setae.

In maxillule (Fig.2-a), setation of 1st inner lobe consists of 3 small anterior submarginal, 9 marginal, and 2 posterior submarginal spiniform setae. Two types of marginal setae are recognizable: 2 proximal setae that are slender and armed with long spinules over their entire length and 7 distal setae that are stout, only proximally armed with long spinules and arranged in 2 rows—anterior row of 3 setae and posterior row of 4 setae. Second inner lobe with a single seta that is nearly as well-developed as marginal setae of 1st inner lobe. Third inner lobe with 3 setae of similar size. Basis with 4 setae. In endopod, 1st and 2nd fused segments with 3 and 4 setae, respectively; 3rd with 5 setae. Setae of 3rd inner lobe, basis, and endopod all well developed. Exopod relatively large, extending to distal end of endopod, with 6 large setae along outer margin and 5 small setae distally.

In maxilla (Fig.2-b), 1st lobe with 2 proximal, 1 posterior subterminal, and 3 terminal setae in addition to a short spiniform process; 2nd to 4th lobes each with 1 posterior subterminal and 2 terminal setae; 5th lobe with 1 posterior subterminal, 2 terminal setae and 1 saberlike spine; 6th lobe with 1 posterior subterminal and 1 terminal setae and 1 saberlike spine. Saberlike spines of 5th and 6th lobes are finely serrated, only a little thicker but much shorter than principal terminal setae. All terminal setae long, slender, and coarsely spinulose with relatively short setules. Posterior subterminal setae of 1st to 4th lobes conspicuously armed with long setules. Two proximal setae of 1st lobe approximately 1/3 length of terminal setae of the same lobe and provided with a few short setules. Four separate segments are recognizable in endopod. Of 7 endopodal setae, a posterior seta of 1st and 2nd segments much shorter than the remaining 5, which are about as long as the principal terminal seta of the 6th lobe. Only posterior setae of first 2 endopodal segments conspicuously armed with setules.

Maxilliped (Fig.2-c) with large coxa about as long as basis, which is a little longer than endopod. All maxillipedal setae similarly plumose. In coxa, seta of 1st lobe approximately 1/2 length of that of 2nd lobe, which is about as long as 2 short setae of 3rd lobe. Long, middle seta of 3rd lobe almost twice length of short setae of same lobe. Fourth lobe with 1 anterior submarginal and 2 marginal setae; 1st marginal seta (counted from proximal) longest, approximately 1.2 and 1.5 times as long as 2nd marginal and anterior submarginal setae, respectively. Distomedial corner of coxa produced into a rounded lobe covered with small spinules. In basis, longitudinal row of short spines on anterior surface extending close to base of 3rd seta; 3 middle setae increasing in length in order from proximal, with last more than twice as long as 1st. All endopodal setae well developed except for outer seta of last segment.

First leg (Fig.2-d) with basipod nearly as long as exopod; first 2 segments of both endopod and exopod reaching the same distance. Outer seta of basis relatively well developed. In exopod, all terminal and outer spines each with an appendicular lateral spinule close to distal end. First outer spine short of reaching base of following outer spine. Second and 3rd outer spines each overreaching base of following outer spine. Terminal spine about as long as combined lengths of 2 distal exopodal segments. Anteriorly, 1st exopodal segment with a short row of setules along distal margin. Lateral sides of all endopodal segments covered with very fine spinules. Hooklike process on posterior surface of basis (Fig.2-e) rather small but clearly visible.

Second leg (Fig.2-f) with basipod about as long as combined lengths of first 2 exopodal segments; endopod reaching distal end of 2nd exopodal segment. Outer spines of exopod well developed, but only 4th overreaching base of its following outer spine. Spiniform process preceding each outer spine well developed. Terminal spine of 3rd exopodal segment approximately 3/4 length of segment. Anteriorly, 1st and 2nd exopodal segments each with 1 cuticular pore close to base of its outer spine; 3rd with 4 cuticular pores on its distal half. In endopod, distolateral spiniform process of 2nd segment better developed than those of 1st and 3rd segments; spiniform processes preceding outer setae of 3rd segment larger than distolateral spiniform process of the same segment.

Third leg (Fig.2-g) similar to 2nd except that terminal and outer spines of exopod relatively small, distolateral spiniform process of 2nd endopodal segment ending in a sharp point, and anterior surface with 2, 3, and 9 cuticular pores, respectively, on 3 exopodal segments.

Fourth leg (Fig.2-h) similar to 3rd except that basis with a small outer seta at distolateral corner of posterior surface, exopod with terminal and outer spines smaller than those of 3rd leg, distolateral spiniform processes of both 1st and 2nd endopodal segments sharply pointed, and with 11 cuticular pores, instead of 9, on anterior surface of 3rd exopodal segment.

Fifth leg (Fig.2-i) with basipod about as long as endopod. Endopod reaching distal end of 2nd exopodal segment excluding spiniform processes. In 2nd exopodal segment, outer spine reaching about halfway to base of 1st outer spine of 3rd segment; inner spiniform seta a little longer than 3rd exopodal segment, thicker in proximal part than endopodal setae but tapering distally like a regular seta, and armed with very short spinules. Third exopodal segment with outer spines shorter than 1/2 length of outer spine of preceding segment, and terminal spine approximately 2/3 length of segment; each outer spine preceded by a spiniform process. Distolateral corners of first 2 endopodal segments each produced into a sharp spiniform process and 2 outer setae of 3rd endopodal segment each preceded by a short spiniform process. Inner setae of first 2 endopodal segments relatively short, with proximal half armed with long setules and distal half with short spinules. Three large cuticular pores found on anterior surface of 3rd exopodal segment.

Description of male. Prosome length, 4.41-5.00 mm; body length, 6.35-7.08 mm. Body (Fig.3-a) relatively slender with prosome about 2 times as long as urosome. First pedigerous somite separated from cephalosome by a clear line of joint. Fourth and 5th pedigerous somites completely coalesced without a trace of segmentation. Laterally, posterior end of prosome broadly rounded, not produced into a lappet as in female. Dorsally, posterolateral corners of prosome (Fig.3-b) more or less symmetrical, each produced into a rather bluntly pointed process.

First 4 urosomal somites (Fig.3-b) fringed with fine spinules along posterior margin. First and 4th urosomal somites similar in length; 2nd and 3rd similar in length and a little longer than 1st. Anal segment approximately 1/2 length of 4th urosomal somite. Combined lengths of anal segment and left caudal ramus only slightly shorter than combined lengths of 2 preceding urosomal somites. Left caudal ramus extending beyond posterior end of right by about 1/4 its length as measured along medial margin. Setae of caudal rami as in female. Dorsally, urosome with 2 cuticular pores on 1st somite, 4 on 2nd and 5 on 3rd. Anal segment and caudal rami with a pore pattern similar to that of female.

Antennule extending beyond posterior end of body by its last 2 or 3 segments. In left antennule (Fig.3-c), 1 of 2 distal setae on segments 10-13 very small. Segment 17 with a spiniform process at anterodistal corner and with 1 middle and 1 distal setae. Segment 18 with 2 distal setae. First postgeniculate segment comprising 3 fused segments (original 19th-21st segments) equipped with 1 middle and 2 distal setae in addition to a spiniform process at anterodistal corner. Segments 22-23 coalesced to form 2nd postgeniculate segment, with 1 anterior seta and 1 posterior seta on segment 22 and 3 anterior setae and 1 posterior seta on segment 23.

In 5th pair of legs, right basis (Fig.3-d) with an elongate inner lobe arising from its distomedial corner and extending distad; left basis (Fig.3-e) with a relatively wide and terminally pointed inner lobe arising from middle of its medial margin and extending in a distomedial direction. Right basis with 2 rows of setules, one along medial margin of segment and another along medial margin of inner lobe. Posteriorly, left basis (Fig.3-f) also with 2 rows of setules, one at base of inner lobe and another along distal margin of inner lobe. Left endopod very characteristic in having a large lobe extending from anterodistal margin of 2nd segment (Fig.3-e). In left exopod (Figs.3-h, i), 2nd segment very characteristic in having an extremely large spiniform process on lateral margin at base of outer spine. Third segment also with a spiniform process at base of outer spine and another at base of terminal spine, but these processes are quite small. In right exopod (Figs.3-j, k), 2nd segment incompletely separated from 3rd and has along medial margin a small, single-pointed process in addition to a large spiniform medial projection. Terminal spine of left exopod moderately developed, approximately 1.5 times length of 3rd segment; that of right exopod slender and only slightly longer than 3rd segment.

Distribution. Since its original discovery at 16°N, 166°E in the central Pacific, *D. palumbii* has been found to be a common member of the circumglobal bathypelagic community, mainly in the tropical and temperate zones of the world oceans (Sewell 1947, Tanaka 1964, Vervoort 1965). *Disseta palumbii* was one of the most common bathypelagic heterorhabdid species found in the present study. A total of 653 specimens including 464 females and 189 males were found. It occurred commonly in samples taken from a depth of 1000 mwo or more in the tropical and temperate zones of all three major oceans. The number of specimens per sample was generally higher in samples taken in equatorial waters, but the largest sample (126 specimens) was a midwater trawl sample collected with 1050 mwo at 26°33'N, 21°27'W in the eastern Atlantic. The northernmost and southernmost latitudes where the species was found in the present study were 40°04'N and 35°59'S respectively in the Atlantic, and 40°35'N (off northern California) and 49°44'S (in the Tasman Sea) respectively in the Indo-Pacific.

Remarks. The female of this species can be readily distinguished from those of the other species of the genus by its characteristic lateral swellings of the genital somite in dorsal view, the lappetlike extension of the posterior end of the prosome in lateral view, and from its close relative, *D. scopularis*, by the relatively short 2nd outer spine of the 5th leg exopod.

The male can easily be recognized by the left 5th leg, which has a large anterodistal lobe on the 2nd endopodal segment and a large spiniform process at the base of the outer spine of the 2nd exopodal segment. A spiniform medial projection and a small spiniform process of the 2nd exopodal segment of the right 5th leg are also useful in identifying the male of this species.

The original description of *D. palumbii* by Giesbrecht (1889) was based on a single female 5.7 mm long. The male was first described by Wolfenden (1905) as *Heterorhabdus grandis*. The body sizes of the specimens found in the present study ranged from 6.66 to 7.75 mm in the female and from 6.35 to 7.08 mm in the male. The largest individuals were found close to American coast in the eastern Pacific and the smallest in equatorial waters of the central Pacific and eastern Atlantic. The female on which Giesbrecht's original description was based is much smaller (5.7 mm long) than the smallest specimen found in the present study (6.66 mm long). However, the morphological details given by Giesbrecht (1892) in descriptions and figures are in full agreement with those of the specimens found in the present study.

Heterorhabdus grandis Wolfenden 1904 was briefly described from 2 female specimens 6.60 mm long taken off the west coast of Ireland. Wolfenden (1905) redescribed the species based on both females and males obtained from off Cape Verde Islands in the equatorial North Atlantic. Wolfenden did not compare his species with *D. palumbii* Giesbrecht 1889. Esterly (1911) considered *Heterorhabdus grandis* Wolfenden 1904 to be identical to a *Disseta* species he found off San Diego. Wolfenden (1911) described *Disseta atlantica* Wolfenden 1911 based on specimens taken at 3 stations in the equatorial and southern Atlantic including the Cape Verde Islands specimens that he had

17

previously identified as *Heterorhabdus grandis* Wolfenden 1904. As Sewell (1947) pointed out, Wolfenden (1911) redescribed *Heterorhabdus grandis* Wolfenden as *D. atlantica*, which is itself a synonym of *D. palumbii* .

Disseta grandis Esterly 1906 based on 3 females and 3 males taken from off San Diego was originally described as closely resembling *D. palumbii* but supposedly distinguished from it mainly by a difference in body size—Esterly's specimens of *D. grandis* were reported to be 8.3 mm long in the female and 7.6 mm in the male, while Giesbrecht's original female specimen of *D. palumbii* was 5.7 mm long. Although Esterly's specimens were much larger than the largest specimen of *D. palumbii* found in the present study and Giesbrecht's specimen was much smaller than the smallest specimen found in the present study, they are all regarded here as belonging to the same species because of their similarity in anatomical details. Esterly's original specimens are no longer available except for some appendages mounted on slides, which are in the Planktonic Invertebrate Collections of Scripps Institution of Oceanography and show no noticeable differences from the specimens found in the present study.

Disseta scopularis (Brady 1883)
Figures 4 and 5

Leuckartia scopularis Brady 1883, p. 51, pl. 14, figs. 1-5.
Disseta scopularis; Scott 1909, p. 134, pl. 42, figs. 1-9.—Wilson 1950, p. 198, pl. 6,
    figs. 47-50.—Tanaka 1964, p. 34, fig. 191.
Disseta maxima Esterly 1911, p. 330, pl. 29, figs. 54, 58; pl. 30, fig. 79.

Description of female. Prosome length, 5.58-6.82 mm; body length, 8.66-10.66 mm. Body elongate (Fig.4-a). Prosome approximately 1.6 times as long as urosome. First pedigerous somite separate from cephalosome by a clear line of joint. Fourth and 5th pedigerous somites coalesced without a visible trace of segmentation. Two rostral filaments of similar size borne on a large base, which in lateral view (Fig.4-b) is produced into a digitiform process. Dorsally, midanterior tubercular process of forehead (Fig.4-c) more or less rectangular. Posterior ends of prosome in lateral view (Fig.4-d) extending posteroventrad into a long, more or less pointed process and in ventral view (Fig.4-g), they are asymmetrical, pointing posteromediad. In dorsal view (Fig.4-e), therefore, they are hidden under genital somite.

Urosome elongate, without spinules along posterior margin of each somite. Genital somite elongate, approximately 1.7 times length of 2nd urosomal somite and asymmetrical in dorsal view with a distinct swelling on left side. Laterally, genital somite (Fig.4-f) with a dorsal hump in anterior half and a distinct ventral indentation about 2/3 its length from anterior end. Genital prominence relatively low, with conspicuous ridges extending posterodorsad from genital opening. Ventrally, genital operculum relatively wide and short, close to anterior end of somite (Fig.4-g); 17-20 cuticular pores clearly visible on each side of genital field. Third urosomal somite only slightly shorter than 2nd and about twice length of anal segment. Dorsally, right caudal ramus (Fig.4-e) separate from anal segment by a line of joint but left caudal ramus fully coalesced with anal segment. Combined lengths of anal segment and left caudal ramus longer than genital somite by 1/5 their length. Left caudal ramus extending beyond posterior end of right by about 1/4 its length as measured along medial margin. Dorsally, 3 or 4 cuticular pores on each side of anal segment and 4 cuticular pores close to lateral margin of each caudal ramus. Setae on caudal rami as in *D. palumbii*.

Antennule extending beyond posterior end of caudal ramus by its last 3 or 4 segments (Fig.4-a). Segmentation and setal arrangement of A1 as in *D. palumbii*. Anatomical details of A2, Md as in *D. palumbii*. Maxillule (Fig.4-h) similar to that of *D. palumbii* except that 3rd inner lobe with 2 setae, instead of 3, basis with 3 setae, instead of 4, and 2nd endopodal segment, which is fused to 1st, bearing 3 setae, instead of 4. Maxilla and maxilliped similar in anatomical details to those of *D. palumbii*.

Swimming legs basically the same as those of *D. palumbii* but more elongated. In exopod of P1 (Fig.4-i), each of first 3 outer spines barely reaching base of following spine. In *D. palumbii*, however, 2nd and 3rd outer spines each overreaching base of following spine. In exopod of P2 (Fig.4-j), all outer spines well developed but none of them overreaching base of following outer spine; 1st and 2nd exopodal segments with 2 and 3 cuticular pores, respectively, on anterior surface, instead of 1 each. Third exopodal segment with 4 cuticular pores, 1 anterior to and 3 posterior to 1st outer spine. In *D. palumbii*, all 4 cuticular pores are posterior to 1st outer spine. Anteriorly, 2nd exopodal segment of P3 (Fig.5-a) with 4, instead of 3, cuticular pores. Third exopodal segment with 9 cuticular pores as in *D. palumbii*, but with a different pattern of distribution. Terminal spine of 3rd exopodal segment approximately 3/7 length of segment. Anteriorly, 2nd and 3rd exopodal segments of P4 (Fig.5-b) with 4 and 9 cuticular pores, respectively, instead of 3 and 11. Terminal spine of 3rd exopodal segment approximately 1/5 length of segment. In exopod of 5th leg (Fig.5-c), 2nd segment including distal spiniform process about 1.8 times length of 3rd segment; its outer spine extending beyond distal end of following outer spine on 3rd segment; inner spiniform seta longer than 3rd segment by 1/4 its length. Third exopodal segment with 2nd outer spine much better developed than 1st, terminal spine only slightly longer than 1/2 length of segment and much shorter than outer spine of 2nd segment, and anteriorly with 2 large cuticular pores posterior to 1st outer spine.

Description of male. Prosome length, 5.33-6.00 mm; body length, 8.50-10.00 mm. Body elongate, otherwise similar in habitus to male of *D. palumbii*. Left A1 geniculate and similar in anatomical details to that of *D. palumbii* male. Other cephalosomal appendages and first 4 pairs of swimming legs similar to those of female.

Fifth pair of legs (Fig.5-d) basically similar to that of *D. palumbii* but in detail shows many remarkable differences by which the species can readily be identified. Basal inner lobe of right leg relatively wide; that of left leg wide and short, with broadly rounded distal end and without a spiniform distal process as found in *D. palumbii*. Anteriorly, 2nd endopodal segment of left P5 with only a curved ridge (Fig.5-d), which may be homologous to the anterodistal lobe found on the same segment of *D. palumbii*. Exopod of left P5 (Figs.5-e, f) is very characteristic, having an enormously developed distal spine and 2nd segment with a large outer spine close to proximal end of segment, which is not preceded by a large spiniform process as in *D. palumbii*. Outer spine of 3rd segment preceded by a spiniform process as in *D. palumbii*. Exopod of right P5 (Figs.5-g-k) with highly modified 2nd and 3rd segments. In 2nd segment, outer spine long, located close to proximal end of segment; medial projection arising from distolateral corner of segment and divided distally into a large toothlike process and a plumose lobe (Fig.5-k). Posteriorly, 3 plumose ridges extending longitudinally from plumose lobe; they seem to represent edges of plumose lobe. Anteriorly (Fig.5-h), 2nd exopodal segment with a ridge extending longitudinally from distal end. Third segment in the form of a wide plate (Figs.5-i, j), wider than long, with a small outer spine on its posterior side and distally with a sigmoid, digitiform process and a terminal spine which is about as long as segment itself excluding the digitiform process.

Distribution. *Disseta scopularis* is known only from the Pacific. Its original description was based on specimens found among surface animals taken between Japan and Honolulu (Brady 1883). The species has since been recorded from the Molucca Sea of the Malay Archipelago (Scott 1909), the eastern tropical Pacific (Wilson 1950), the Izu region of Japan (Tanaka 1964), and the waters off San Diego (Esterly 1911 as *D. maxima*). In the present study, the species is represented by a total of 318 specimens including 199 females and 119 males taken widely in the tropical and temperate waters across the Pacific from the west coasts of both North and South America to the East China Sea and Malay Archipelago. The species was particularly common in the central North Pacific, where as many as 53 specimens were collected in a single midwater trawl sample taken from 3000-0 mwo. In the present study, the northernmost and southernmost latitudes of occurrence in the eastern Pacific were 40°35'N (off California) and 30°46'S (off Chile), respectively. In the central Pacific, the species was found between 31°13'N and 4°57'S; in the western Pacific between 25°37'N and 4°50'S.

Remarks. The species was first described by Brady (1883) from males as *Leuckartia scopularis* and afterward recognized as belonging to the genus *Disseta* by Scott (1909). The female was first described by Wilson (1950) with figures of the habitus in dorsal and lateral views and the 5th leg in anterior view. While his figures of the habitus correctly depict the species, his figure of the 5th leg seems to belong to a different species, as it shows a small outer spine of the 2nd exopodal segment, which in *D. scopularis* is extraordinarily large. *Disseta maxima* Esterly 1911, originally described from a female taken from off San Diego, was found by Sewell (1947) to be synonymous with *D. scopularis.*

Disseta scopularis is the largest species of the genus. The female is readily recognized by its elongated body, a pointed posterior end of the prosome in lateral view, and the extraordinarily long outer spine of the 2nd exopodal segment of P5. The male can easily be recognized by the left P5, in which the terminal spine of the exopod is extremely large and the outer spine of the 2nd exopodal segment is not preceded by a large spiniform process.

Disseta magna Bradford 1971
Figure 6

Disseta magna Bradford 1971, p. 133, fig. 11.
Disseta coelebs Heptner 1972a, p. 1645, figs. 1-8.

Description of female. Prosome length, 5.58-6.00 mm; body length, 7.75-8.66 mm. Body somewhat robust (Fig.6-a). Prosome about twice as long as urosome. Segmentation between cephalosome and 1st pedigerous somite clearly visible; 4th and 5th pedigerous somites completely coalesced without a trace of segmentation. Two rostral filaments of similar length borne on a base, which in lateral view is extended ventrad into a conical process. Dorsally, forehead with a relatively low midanterior tubercular process as in *D. palumbii.* Laterally, posterior end of prosome (Figs.6-b, d) only slightly produced posteriad with a rather broadly rounded margin. Dorsally, posterolateral corners of prosome (Fig.6-c) symmetrical with obtusely angular posterior ends, extending for only a short distance, covering approximately 1/7 length of genital somite.

First 3 urosomal somites (Fig.6-c) fringed with fine spinules along posterior margins as in *D. palumbii.* Dorsally, genital somite symmetrical with smoothly curved lateral margins, widest about 1/3 its length from anterior end, typically with a length:width ratio of 100:83, and approximately 1.5 times as long as 2nd urosomal somite. Laterally, genital prominence (Fig.6-d) more or less truncate, covering about 2/3 length of somite, with genital operculum roughly halfway along ventral side of somite. Ventrally, genital operculum (Fig.6-e) about as long as wide, located roughly at center of somite. Number of cuticular pores found lateral to genital operculum on each side varied from 23 to 34. Surface of genital somite densely covered with fine spinules except for the anterior 1/5 its length and genital prominence.

Third urosomal somite only a little shorter than 2nd. Anal segment approximately 1/2 length of 3rd urosomal somite, partially fused to caudal rami, with the joint visible only with the right caudal ramus. Left caudal ramus extending beyond posterior end of right by approximately 1/4 its length as measured along medial margin. Combined lengths of anal segment plus left caudal ramus about 1.2 times length of genital somite. Number of cuticular pores on dorsal surface of urosome: 3 on each side of genital somite, 1 on each side of 2nd urosomal somite and anal segment, and about 9 on each caudal ramus. Setae on caudal rami as in *D. palumbii* except that dorsal appendicular setae of caudal rami almost 3 times length of the same setae of that species.

Antennule extending beyond posterior end of prosome by its last 3 or 4 segments (Fig.6-a). Setal arrangement of antennule as in *D. palumbii.* Other cephalosomal appendages also as in *D. palumbii.* First to 4th legs similar to those of *D. palumbii* except for numbers of cuticular pores on exopods. Although somewhat variable, 2nd exopodal segment in P3 and P4 with 4-6 pores, while the corresponding segments in *D. palumbii* usually with 3 pores. Fifth leg also similar to that of *D. palumbii* but 3rd exopodal segment more elongated and its 2 outer spines (Fig.6-f) larger than those of *D. palumbii.* Number of cuticular pores found on exopod, 1 each on 1st and 2nd segments and 2 on 3rd segment.

Description of male. Prosome length, 5.16-5.58 mm; body length, 7.66-8.08 mm. Similar in habitus to *D. palumbii* male. All cephalosomal appendages including geniculated left A1 similar in major features to those of *D. palumbii* male. First 4 pairs of swimming legs as in female. Fifth pair of legs basically similar to that of *D. palumbii* male but can easily be distinguished from it by a number of characteristic features described below. Basal inner lobe of right P5 (Fig.6-j) relatively short, pointing straightly distad; that of left P5 (Fig.6-k) also pointing distad, tapering to some degree into a blunt distal end, without a sharp distal point as found in *D. palumbii*. Basis of each leg with 2 rows of setules along medial margin, one proximal to inner lobe and another extending over entire length of lobe itself. Second endopodal segment of left P5 (Fig.6-k) with a large tubercular outgrowth arising from distomedial corner of anterior surface as in *D. palumbii*. Outgrowth is, however, single-lobed distally, while it is more or less bilobed distally in *D. palumbii*. Exopod of left P5 (Figs.6-g, h, l) differing from that of *D. palumbii* in the absence of a large spiniform process adjacent to the outer spine of 2nd segment, which is a prominent feature in that species, and that 3rd segment narrow and more or less quadrate, with outer spine close to distal end of segment. In exopod of right P5 (Figs.6-i, j. m), 2 distal segments broad. Second segment greatly expanded proximomedially and with large toothlike medial projection, which is single-pointed, while it is double-pointed in *D. palumbii*. A small additional spiniform process found on the same segment of *D. palumbii* is missing in this species. Third exopodal segment clearly separate from 2nd, with outer margin smoothly curved; the segment is neither tapering distally as in *D. palumbii* nor expanded widely into a broad plate as in *D. scopularis*. Terminal spine of 3rd exopodal segment about as long as segment itself.

Distribution. Disseta magna seems to be an Indo-Pacific species. Bradford's (1971) original female specimen was from 914-823 m at 41°47'S, 175°01'E off New Zealand and Heptner's (1972a) 3 male specimens (as *D. coelebs*) were collected from the Bougainville Trench (6°13'S, 153°43'E) in the Solomon Sea of the western tropical Pacific. The species was rather rare as compared either with *D. palumbii* or *D. scopularis*. It is represented in the present study by a total of only 65 specimens including 47 females and 18 males found in 12 samples from the Indo-Pacific: 1 from off Chile at 46°57'S, 3 from the eastern tropical Pacific, 3 from off San Diego, 1 from off northern California at 35°29'N, 1 from the central North Pacific at 31°06'N, 155°20'W, 2 from the central South Pacific at 25°17'S, 154°58'W and 24°39'S, 155°00'W, respectively, and 1 from the Indian Ocean at 35°00'S, 60°00'E. The largest number of specimens per sample was 26 in the sample taken at 24°39'S, 155°00'W in the central South Pacific.

Remarks. Disseta magna was originally described from a single female 8.0 mm long taken from off New Zealand. *Disseta coelebs* Heptner 1972 solely based on males seems to be identical to males of *D. magna* found in the present study, and the species is therefore considered a junior synonym of the latter. Heptner's (1972a) specimens (8.15-8.60 mm long) were somewhat larger than the males (7.66-8.08 mm long) found in the present study, but his figure of the 5th pair of legs is in full agreement with the same legs of the present specimens.

The female of *D. magna* is readily distinguished from the other species of the genus by its genital somite, which in dorsal view is perfectly symmetrical and in lateral view has a truncate genital prominence. The male can be distinguished from that of *D. palumbii* by the absence of a large spiniform process guarding the outer spine of the 2nd exopodal segment of the left P5, which is a prominent feature in the male of *D. palumbii,* and from the male of *D. scopularis* by the exopodal terminal spine of the left P5, which is relatively small as compared with the extraordinarily large spine found in *D. scopularis.*

GENUS *MESORHABDUS* SARS 1905

Mesorhabdus Sars 1905, p. 9.

Type by monotypy: *Mesorhabdus annectens* (=*Heterorhabdus brevicaudatus* Wolfenden 1905)

Diagnosis. Third marginal seta of left caudal ramus and 3rd and 4th marginal setae of right caudal ramus armed with small spines in addition to normal setules (Fig.11-k). First segment of A1 with 1+3+3+3 setae/aesthetes (Fig.10-g). In A2, 1st exopodal segment without seta and 2nd with a small marginal seta (Fig.10-h). In mandible (Fig.10-i), masticatory edge with a basal spine, a group of 2 or 3 short, contiguous teeth, and 3 or 4 long spiniform teeth; 2nd endopodal segment with 8 terminal setae plus 2 appendicular setae. Maxillule (Fig.8-d) with 1 of setae on 1st inner lobe developed into a conspicuously large spine; 3rd inner lobe with 1 or 2 setae; exopod extending beyond distal end of endopod by 1/2 its length. In maxilla (Fig.8-f), 5th and 6th lobes each with a large, saberlike spine. Coxa of Mxp with 3 lobes bearing setae (Fig.8-g). Basis of P1 without outer seta (Fig.8-h). Second leg with 8 setae on 3rd endopodal segment (Fig.8-j). Left, geniculated A1 of male with 0 or 1 seta belonging to 20th segment (Figs.9-e, 12-c). Second endopodal segment of male left P5 without seta (Fig.9-g). Second exopodal segment of male right P5 with a large toothlike medial projection.

Additional description of female. First pedigerous somite separate from cephalosome; 4th and 5th pedigerous somites completely coalesced. Laterally, forehead tapering ventrad into a conical rostrum bearing 2 slender distal filaments (Fig.7-e). Dorsally, forehead with low midanterior tubercular process (Fig.10-c), which continues ventrad into rostrum. Dorsally as well as ventrally, genital somite symmetrical, with rather smoothly curved lateral margins. Anal segment partially fused with following caudal rami with a line representing its joint with right ramus usually visible in dorsal view, but such a line is not clearly visible for its joint with left ramus (Fig.10-b). Genital field with large cuticular pores, usually around anterior half of genital operculum (Fig.10-e). Caudal rami strongly asymmetrical with left ramus much longer than right and each ramus carrying 5 principal marginal setae and a small dorsal appendicular seta. Fourth marginal seta, counted from lateral, of left caudal ramus naked, thicker than other marginal setae and about twice length of body. All the other principal marginal setae provided with long setules; 3rd marginal seta of left ramus and 3rd and 4th marginal setae of right ramus armed with short spines in addition to normal setules (Fig.11-k).

Antennule 25-segmented, extending beyond posterior end of body. Antenna (Fig.7-j) with basipod nearly as long as 1st endopodal segment. Coxa and basis with 1 and 2 inner marginal setae, respectively. Endopod 2-segmented, with 2 inner setae on 1st segment, 8 (marginal)+1 (appendicular) setae on inner lobe, and 6+1 setae on outer lobe of 2nd segment. Exopod 8-segmented. First segment without setae; 2nd to 7th each with an inner marginal seta; 8th with a inner marginal and 3 terminal setae.

Mandible (Fig.8-a) with well-developed mandibular blade. Masticatory edges of left and right mandibles asymmetrical (Figs.8-b, c), each armed with a basal spine, a group of 2 or 3 contiguous, relatively short teeth, and 3 or 4 long teeth. Mandibular palp well developed; exopod attached high up along lateral margin of basis and reaching far short of distal end of endopod. All 6 setae of exopod well developed. First endopodal segment with 2 setae and 2nd with 8 terminal and 2 appendicular setae.

In maxillule (Fig.8-d), outer lobe with 5 or 9 setae. First inner lobe well developed with 11-13 setae, one of which is spiniform and much thicker than the others. Second and 3rd inner lobes of similar size, each with 1 or 2 setae. Basis extending mediad like inner lobes, with 2 setae. Endopod small, unsegmented, with 3-5 setae. Exopod large, arising from distal end of appendage and extending along its longitudinal axis, with 5 or 6 long setae along lateral margin and 4 or 5 short setae along medial margin.

Maxilla (Fig.8-f) with 6 well-developed lobes followed by small endopod. First lobe with 1 or 2 relatively small proximal setae, 1 posterior subterminal and 3 terminal setae in addition to a small spiniform process. Second to 4th lobes each with 1 posterior subterminal and 1 or 2 terminal setae. Fifth and 6th lobes each with 2 small setae and a large saberlike spine. Endopod with 6 setae.

Coxa of maxilliped (Fig.8-g) relatively small, approximately 2/3 length of basis, with 3 lobes; 1st lobe with a single seta, 2nd with 1 or 2 setae, 3rd with 2 or 3 setae. Basis with a middle group of 3 setae, a distal group of 2 setae, a band of spinules extending longitudinally on anterior surface, and a row of long setules proximally along inner margin. Endopod nearly as long as basis, with 3 inner marginal setae on each of first 4 segments; 5th segment with 1 small outer and 3 long terminal setae.

Basipod of 1st leg (Fig.8-h) with large coxa and relatively short basis; coxa with a well-developed inner marginal seta; basis with a curved inner marginal seta at distomedial corner of anterior surface and a hooklike process on posterior surface close to joint with exopod. In exopod, each outer spine preceded by a spiniform process. Each endopodal segment with distolateral corner produced into a spiniform process. Anteriorly, endopodal segments without any morphological features referable to von Vaupel-Klein's organ.

Second leg (Fig.8-j) with a large coxa bearing an inner marginal seta and a relatively short unarmed basis. First 2 endopodal segments each with distolateral corner produced into a spiniform process. Third endopodal segment with 8 setae. Third leg (Fig.8-k) similar to 2nd except that terminal and outer spines of exopod relatively small. Fourth leg (Fig.9-a) similar to 3rd except that basis with a small outer seta at distolateral corner of posterior surface, exopod with relatively small terminal and outer spines, and endopod with 7 setae on 3rd segment.

Fifth leg (Fig.9-b) with relatively narrow basipod; coxa without setae; basis with a small outer seta at distolateral corner of posterior surface. Second exopodal segment with well-developed inner spine. First 2 endopodal segments each with distolateral corner produced into a spiniform process; 3rd segment with 6 setae.

Additional description of male. Similar in habitus to female except that urosome 5-segmented. Right antennule as in female. Left antennule (Fig.12-c) geniculated with knee joint between 18th and 19th segments; first 16 segments similar to those of female except that first 2 segments fused; segments 19-21 fused to form 1st postgeniculate segment and segments 22-23 fused to form 2nd postgeniculate segment.

Fifth pair of legs asymmetrical (Fig.9-f), each with 3-segmented exopod and endopod. Left and right exopods form a pair of tongs with 2 distal segments of each turned clockwise or counterclockwise bringing their posterior surfaces to face mediad and both terminal spines turned mediad. Coxa relatively small. Basis with a small outer seta at distolateral corner of posterior surface and a large plumose inner lobe, which in right leg (Fig.9-h) is larger than that of left leg (Fig.9-i). First endopodal segment without a seta; 2nd with an inner seta only in right endopod; 3rd with 6 setae. In left exopod, 1st and 2nd segments each with an outer spine; 3rd with an outer, an inner, and a terminal spine. Right exopod with an outer spine on each segment in addition to an elongate, plumose medial projection on 2nd segment and a terminal spine on 3rd.

Remarks. Sars's (1905) original description of the genus *Mesorhabdus* was based on *M. annectens* Sars 1905, the only species originally included in the genus. Subsequently, Sars (1925) found *M. annectens* to be a junior synonym of *Heterorhabdus brevicaudatus* Wolfenden 1905.

Mesorhabdus is distinguishable from *Disseta* by the following features: 1) Antenna without a seta on 1st exopodal segment, 2) mandible with elongated teeth on masticatory edge, 0 or 2 small setae on basis, and only 2 or 3 setae on 1st endopodal segment, 3) maxillule with reduced numbers of setae, 1 of setae on 1st inner lobe developed into a large spine, small unsegmented endopod, and greatly elongated exopod occupying distal end of appendage, 4) in maxilla, saberlike spines on 5th and 6th lobes greatly developed; endopod small, poorly segmented, with 6 setae, 5) coxa of maxilliped with 4-6 rather poorly developed setae, and 6) first leg without an outer seta on basis. Male's geniculated antennule and 5th pair of legs are basically the same as those of the *Disseta* male, but the male can be distinguished from that of *Disseta* by the medial projection of the 2nd exopod of the right 5th leg, which is relatively long and extends at right angle to the longitudinal axis of the appendage.

Nominal species of the genus *Mesorhabdus*

Heterorhabdus brevicaudatus Wolfenden 1905
Mesorhabdus annectens Sars 1905
    Synonym of *Heterorhabdus brevicaudatus* Wolfenden 1905 (see Sars 1925, p. 234)
Mesorhabdus gracilis Sars 1907
Mesorhabdus angustus Sars 1907
Mesorhabdus truncatus Scott 1909 =*Neorhabdus truncatus* (Scott 1909) (see Heptner 1972b, p. 58)
Mesorhabdus paragracilis, new species
Mesorhabdus poriphorus, new species

    Remarks. Only 3 of the 5 previously described species are properly referable to this genus as *M. annectens* Sars 1905 has been synonymized with *M. brevicaudatus* (Wolfenden 1905) and *M. truncatus* Scott 1909 transferred to the new genus *Neorhabdus* Heptner 1972. Two new species are described in this study bringing the total number of species in the genus to 5.

Key to species of *Mesorhabdus*

Females

1a. Genital somite of female with spinules along posterior margin (Fig.7-f); oral
    cone with high anterior labral lobe (Fig.7-e); coxa of Mxp with 1+2+3
    setae (Fig.8-g). . . . .*M. angustus*
1b. Genital somite of female without spinules along posterior margin; oral cone
    with low anterior labral lobe; coxa of Mxp with either 1+2+2 or 1+1+2 setae. 2

2a. Coxa of Mxp with 1+1+2 setae; endopod of Mx1 with 3 setae, exopod with 5+4
    setae (Fig.13-e); Mx2 with 5 setae on 1st lobe and 3 setae on 2nd lobe (Fig.13-f);
    dorsally, anal segment with 1 pore on each side; caudal ramus with a large
    pore close to posterior end (Fig.13-b) . . . .*M. gracilis*
2b. Coxa of Mxp with 1+2+2 setae; endopod of Mx1 with 4 setae, exopod with
    6+4 setae ; Mx2 normally with 6 setae on 1st lobe and 2 or 3 setae on 2nd
    lobe; dorsally, caudal ramus without a large pore close to posterior end. 3

3a. Dorsally, posterolateral corners of prosome pointing posteriad (Fig.10-b);
    dorsally, caudal ramus normally with a pore at middle; Mx2 with 3 setae
    on 2nd lobe (Fig.11-b); lateral margin of 1st exopodal segment of P1 not
    conspicuously bulging (Fig.11-e) . . . .*M. brevicaudatus*
3b. Dorsally, posterolateral corners of prosome diverging (Figs.15-c, 17-a);
    Mx2 with 2 setae on 2nd lobe; 1st exopodal segment of P1 conspicuously
    bulging (Fig.16-a) 4

4a. Dorsally, anal segment with 1 pore on each side; caudal ramus
    without pores (Fig.15-c) . . . .*M. paragracilis*, n. sp.
4b. Dorsally, anal segment with 2 pores on each side; caudal ramus with
    pores close to anterior end (Fig.17-a) . . . .*M. poriphorus*, n. sp.

Males

1a. In right P5, 2nd exopodal segment with medial projection at proximal end
    of segment (Fig.9-g). . . . .*M. angustus*
1b. In right P5, 2nd exopodal segment with medial projection at distal end of
    segment (Fig.12-d). 2

2a. Maxillule with 3 setae on endopod and 5+4 setae on exopod (Fig.13-e);
    dorsally, caudal ramus with a pore close to posterior end (Fig.14-g) . . . .*M. gracilis*

2b. Maxillule with 4 setae on endopod and 6+4 setae on exopod (Fig.11-a);
     dorsally, caudal ramus without pores on posterior half (Fig.11-k) 3

3a. Dorsally, caudal ramus without pores; anteriorly, coxa of 5th leg
     with a single pore (Fig.16-f) . . . . *M. paragracilis*, n. sp.
3b. Dorsally, caudal ramus with pores (Figs.11-k, 17-c); anteriorly, coxa
     of 5th leg with 2 or more pores (Figs.12-e, 17-d). 4

4a. Dorsally, caudal ramus with a single pore (Fig.11-k); Mx2 with 3 setae
     on 2nd lobe (Fig.11-b) . . . . *M. brevicaudatus*
4b. Dorsally, caudal ramus with 2 or 3 pores (Fig.17-c); Mx2 with 2 setae
     on 2nd lobe . . . . *M. poriphorus*, n. sp.

Mesorhabdus angustus Sars 1907
Figures 7-9

Mesorhabdus angustus Sars 1907, p. 19.—Sars 1925, p. 236, pl. 66, figs. 14-20.—Sewell
    1932, p. 308.—Rose 1933, p. 207, fig. 247.—Sewell 1947, p. 182, text fig. 47.—Wilson
    1950, p. 263.

    Description of female. Prosome length, 4.16-6.41 mm; body length, 6.00-8.75 mm. Body elongate (Fig.7-a), with prosome approximately 2.3 times as long as urosome. Laterally, forehead (Fig.7-e) produced ventrad into a conical rostrum bearing 2 filaments; cervical groove clearly visible halfway along dorsal margin of cephalosome; oral cone with a large anterior labral lobe pointing anteriad and armed with a crown of radiating long setules. Dorsally as well as ventrally, forehead more or less truncate (Fig.7-d). Laterally, posterior end of prosome with a cluster of large cuticular pores and produced posteriad into an obtuse angle (Fig.7-f). Dorsally, posterolateral corners of prosome obtusely angular, overlapping anterior 1/4 length of genital somite (Fig.7-c).

    Dorsally, genital somite (Fig.7-c) symmetrical with smoothly curved lateral margins, with posterior margin fringed with spinules, and approximately 1.7 times length of 2nd urosomal segment. Third urosomal segment a little shorter than 2nd and about as long as anal segment. No spines or spinules found along posterior margins of 2nd and 3rd urosomal somites. Left caudal ramus partially separated from anal segment with line of joint visible only on lateral side and extending beyond posterior end of right by 1/5 its length as measured along medial margin. Line of articulation clearly visible between anal segment and right caudal ramus. Dorsal appendicular setae of caudal rami small. Fourth marginal seta of left caudal ramus unarmed, nearly twice as long as body. Third marginal seta of left caudal ramus and 3rd and 4th marginal setae of right caudal ramus armed with numerous fine spinules in addition to normal long setules.

    Laterally, genital somite (Fig.7-f) about as long as deep, with genital prominence roughly halfway along ventral margin. Dorsally as well as ventrally, genital somite widest at middle with smoothly curved lateral margins (Fig.7-g). In some individuals, genital somite seemed to be inflated, with enlarged genital prominence (Fig.7-h) and high lateral swellings in anterior half (Fig.7-i). Ventrally, genital field with a cluster of 30-40 cuticular pores on each side and a few scattered pores on anterior side.

    Antennule extending beyond posterior end of caudal ramus by its last 7 segments (Fig.7-a). First segment with well-developed aesthetes (Fig.7-d). Aesthetes on segments 2-19 showing a gradual reduction in size from proximal to distal.

    In antenna (Fig.7-j), marginal setae of coxa and basis rather poorly developed. Endopod approximately 1.3 times length of basipod; marginal setae of 1st segment similar in size to those of basis; posterior appendicular setae of inner and outer lobes of 2nd segment equally short. Exopod about as long as basipod; 1st segment without setae, 2nd with a small marginal seta, 3rd-7th each with a well-developed marginal seta, and 8th with a moderately developed marginal seta and 3 well-developed terminal setae.

Mandibular blade elongate (Fig.8-a), about as long as basis; masticatory edge with a basal spine, 3 short, contiguous dorsal teeth, and 3 or 4 large ventral teeth. In left Md (Fig.8-b), 1st and 3rd dorsal teeth (counted from basal spine) of similar size, bicuspid and longer than 2nd, which is tricuspid. Only 3 large ventral teeth on left Md; they are equally long, separated from each other by a similar gap, ventralmost tooth is monocuspid and the other 2 are tricuspid. Masticatory edge of right Md (Fig.8-c) with 3 relatively small dorsal teeth and 4 large ventral teeth. First dorsal tooth longer than the other 2. First ventral tooth relatively small, bicuspid; 2nd longest, tricuspid, and 3rd and 4th of similar size and monocuspid; gap between 1st and 2nd wider than gaps between the others.

In maxillule (Fig.8-d), outer lobe with 3 short setae proximally and 6 long setae terminally. First inner lobe with a large clawlike spine and 11 rather delicate setae—1 posterior submarginal, 9 marginal, and 1 anterior submarginal. In some individuals, posterior submarginal seta is missing and setae close to spine are curved (Fig.8-e). Second and 3rd inner lobes with 1 and 2 setae, respectively. Basis with 2 setae. Endopod relatively small, with a short and 4 long setae. Exopod nearly twice as long as wide, with 5 small medial and 6 long lateral setae.

Maxilla relatively short, with 1st segment longer than rest of appendage (Fig.8-f). First lobe with 2 short proximal setae, 1 medium-sized posterior subterminal seta, and 3 long terminal setae in addition to a small toothlike process; 2nd with 1 short posterior subterminal and 2 long terminal setae; 3rd and 4th each with 1 short posterior subterminal and 1 long terminal seta. Saberlike spine on 5th lobe much thicker than that of 6th lobe; one of 2 setae on 5th and 6th lobes very small. Endopod with 6 well-developed setae.

In Maxilliped (Fig.8-g), 3 lobes of coxa with 1, 2, and 3 setae, respectively; seta of 1st lobe very small; only 1st seta of 2nd lobe armed with setules and shorter than 2nd; anterior seta of 3rd lobe unarmed and relatively short; 2 posterior setae of similar length, furnished with setules, and about twice length of anterior seta. Basis elongate, nearly 1.5 times length of coxa, with middle group of 3 setae on distal half of segment. Endopod with first 2 segments of similar length, 3rd segment only 2/3 length of 2nd and a little longer than 4th, and 5th approximately 1/2 length of 4th. One of 3 terminal setae of 5th segment poorly developed.

Basis of P1 (Fig.8-h) with distomedial corner of posterior surface produced distad into a large, pointed lobe, from which inner marginal seta originates, and hooklike process on posterior surface (Fig.8-i) relatively well developed. Exopod with 1st outer spine extending beyond base of 2nd, which is far short of reaching base of 3rd, and 3rd barely reaching base of 4th. Endopod extending beyond distal end of 2nd exopodal segment and each segment produced distolaterally into a large spiniform process.

Second leg (Fig.8-j) with a large curved outer spine on 1st exopodal segment. Outer spine of 2nd exopodal segment also large, only a little shorter than that of 1st segment, and reaching close to base of following outer spine on 3rd segment. Three outer spines of 3rd exopodal segment all small, of which 3rd the largest. Terminal spine of 3rd exopodal segment approximately 3/5 length of segment. Second and 3rd exopodal segments with large cuticular pores close to lateral margins. Endopod relatively small, reaching distal end of 2nd exopodal segment; only first 2 segments with distolateral corners produced into a spiniform process.

Third leg (Fig.8-k) similar to P2, but first 2 outer spines of exopod smaller than those of P2, terminal spine of 3rd exopodal segment approximately 1/2 length of segment, and cuticular pores of 2nd and 3rd exopodal segments are scattered widely, although they are more numerous along lateral sides of segments.

Fourth leg (Fig.9-a) similar to 3rd except that basis with a short outer seta, first 2 outer spines of exopod much smaller that those of P3, terminal spine of 3rd exopodal segment shorter than 1/2 length of segment. Anteriorly, coxa with 3 large cuticular pores close to proximomedial corner.

In 5th leg (Fig.9-b), combined lengths of first 2 exopodal segments about equal to length of 3rd exopodal segment when measured along medial margins; of 4 outer spines of exopod, 2nd largest and

3rd smallest; terminal spine of 3rd exopodal segment only slightly shorter than 1/2 length of segment. Inner spiniform seta of 2nd exopodal segment approximately 1.5 times length of terminal spine of 3rd exopodal segment. Inner setae of 1st and 2nd endopodal segments fringed with long setules over proximal half and with short spinules over distal half; setae of 3rd endopodal segment armed with long setules for whole length. Anteriorly, 2nd exopodal segment with 1 cuticular pore close to outer spine; 3rd with 6 scattered cuticular pores.

Description of male. Prosome length, 4.75-6.08 mm; body length, 6.75-8.50 mm. Similar in habitus to female except that urosome 5-segmented (Figs.9-c, d), first 3 urosomal somites armed with fine spinules along posterior margins, and all urosomal somites and caudal rami with large cuticular pores. Genital somite shorter than 2nd urosomal somite, which is about equal in length to 3rd. Fourth urosomal somite about as long as anal segment and slightly shorter than 3rd. Left caudal ramus completely fused with anal segment, their combined lengths a little longer than combined lengths of 2 preceding urosomal somites. Dorsally, anal segment with 1 cuticular pore on each side and caudal ramus with 2-4 cuticular pores in anterior half.

Right antennule extending posterior end of caudal ramus by last 6 segments. In left, geniculated antennule (Fig.9-e), 17th segment with a large spiniform process distally along anterior margin in addition to a small seta and an aesthete; 19th segment with 2 spiniform processes along anterior margin and an aesthete at anterior distal end of segment; 20th segment with a small anterior seta marking distal end of segment.

In 5th pair of legs (Figs.9-f, g), inner lobe of right basis elongate, pointing directly distad (Fig.9-h); inner lobe of left basis short, with distal end produced laterad into a small lobe (Fig.9-i). Exopod of right leg with relatively small outer spines; measured along medial margins, terminal spine of 3rd segment roughly equal in length to segment itself; medial projection of 2nd segment (Fig.9-g) arising from proximal end of medial margin of segment, spiniform, fringed with setules along proximal margin. Left exopod with outer spines of 2nd and 3rd segments much larger than outer spine of 1st; inner spine of 3rd segment nearly twice as long as outer spine of same segment; terminal spine pointing mediad and almost twice as long as segment itself.

Distribution. Mesorhabdus angustus is a circumglobal bathypelagic species. It has been recorded from the eastern North Atlantic (between 31°21'N and 46°31'N) by Sars (1907, 1925) and from the Laccadive Sea, the northern Arabian Sea, and the Gulf of Oman by Sewell (1932, 1947). Wilson (1950) reported the species from 49°06'N, 153°06'E in the Sea of Okhotsk and from 10°05'N, 122°18'E between Panay and Negros Islands in the Philippines. In the present study, the species was rather rare and represented by only 40 specimens including 27 females and 13 males found in 16 samples—3 from the Atlantic and 13 from the Pacific. The 3 Atlantic samples were taken between 18°58'S and 26°33'N. Of the 13 Pacific samples, 7 were taken along the west coast of America between 01°25'N and 38°23'N, 3 in the central Pacific between 00°02'S and 30°40'N, 2 in the southwestern Pacific (at 09°25'S, 159°06'E and 14°14'S, 150°54'E, respectively), and 1 in the South China Sea (at 17°48'N, 119°55'E). The sample yielding the largest number of specimens (6 females and 4 males) was taken at 00°02'S, 165°44'W in equatorial waters of the central Pacific.

Remarks. Sars's (1907) brief original description was based on a single female 5.80 mm long obtained from 36°17'N, 28°53'W in the North Atlantic. Sars (1925) redescribed the female with figures and recorded the species from 2 additional localities in the North Atlantic (31°21'N, 19°09'W; 46°31'N, 5°13'W). Sewell (1947) redescribed the species from females (7.25 mm long) and males (6.72 mm long) collected from the northern part of the Arabian Sea and the Gulf of Oman. Sars's (1907) original female specimen (5.80 mm long) was much smaller than the specimens found by Sewell (1947) from the Arabian Sea (7.25 mm long in the female, 6.72 mm long in the male) and those found in present study (6.00-8.75 mm long in the female, 6.75-8.50 mm long in the male). The smallest specimen found in the present study was 6.00 mm long and was taken at 26°33'N, 21°27'W in the eastern Atlantic, not too far from the type locality. It is similar in size to Sars's original specimen and in the anatomical details agrees well with the large specimens found elsewhere in the present study. The largest specimens were found in waters off the west coast of North America. The specimens from the Atlantic and the central and western Pacific were mostly small.

The female of *M. angustus* can readily be distinguished from those of its congeners by its large body size, long antennule, and prominent spinules along the posterior margin of the genital somite. The male can be distinguished by its 5th pair of legs, in which the medial projection of the 2nd exopodal segment of the right leg originates from the proximal end of the segment.

Mesorhabdus brevicaudatus (Wolfenden 1905)
Figures 10-12

Heterorhabdus brevicaudatus Wolfenden 1905, p. 12, pl. 4, figs. 1, 2.
Mesorhabdus brevicaudatus; Sars 1925, p. 234, pl. 65.—Rose 1933, p. 206, fig. 246.—Tanaka
    1964, p. 30, fig, 189.—Grice and Hulsemann 1965, p. 247.
Mesorhabdus annectens Sars 1905, p. 9.
Mesorhabdus gracilis, not of Sars; Park 1970, p. 477.

Description of female. Prosome length, 2.64-3.00 mm; body length, 3.76-4.24 mm. Body somewhat robust (Fig.10-a). Prosome approximately 2.1 times length of urosome. Laterally, forehead (Fig.10-d) produced ventrad into a large rostrum bearing 2 filaments; cervical groove not apparent; anterior labral lobe low, with a crown of radiating setules. Dorsally, forehead with bilobed midanterior tubercular process (Fig.10-c). Laterally, posterior end of prosome broadly rounded with a few scattered cuticular pores (Fig.10-f). Dorsally, posterolateral corner of prosome (Fig.10-b) bluntly angular, closely embracing anterior 1/5 length of genital somite.

None of urosomal somites armed with spinules along posterior margin. Dorsally, genital somite (Fig.10-b) only slightly longer than wide, with smoothly curved lateral margins, and widest 1/3 its length from anterior end. Laterally, genital somite (Fig.10-f) only slightly longer than deep, with its entire ventral margin bulging out to form genital prominence. Dorsal margin of genital somite more or less smoothly curved and, in some individuals, with a low hump midway. Ventrally, genital field close to posterior end of somite (Fig.10-e), with its anterior half encircled by a belt of cuticular pores.

Second urosomal somite approximately 3/5 length of genital somite, only a little longer than 3rd. Anal segment about as long as 3rd urosomal somite; its joint with right caudal ramus is clearly visible in dorsal view but no trace is found for its joint with left ramus. Combined lengths of anal segment and left caudal ramus a little longer than combined lengths of 2 preceding urosomal somites. Dorsally, anal segment with 2 cuticular pores on each side; each caudal ramus normally with a cuticular pore halfway along lateral margin (Fig.10-b). Fourth marginal seta of left caudal ramus nearly as long as body. Third marginal seta of left ramus and 3rd and 4th marginal setae of right ramus armed with small but conspicuous spines in addition to normal setules.

Antennule extending beyond posterior end of caudal ramus by its last 2 segments (Fig.10-a). A long anterior seta on segments 3, 7, 14, 18, 21, and 24 (Fig.10-g). Most of anterior setae more or less modified into the form of a lancet. Posterior setae on segments 23 and 24 plumose.

In antenna (Fig.10-h), basipod about equal in length to 1st endopodal segment; coxa with a large tubercular outgrowth medially. Posterior appendicular seta of inner lobe of 2nd endopodal segment about 3 times length of posterior appendicular seta of outer lobe. Exopod about as long as 1st endopodal segment; inner marginal seta of 2nd segment poorly developed; inner marginal seta of 8th segment very small.

Mandible (Fig.10-i) with mandibular blade as long as mandibular palp. Masticatory edge with only 2 small dorsal teeth next to basal spine; 1st tricuspid, 2nd spiniform, both armed proximally with spinules. Left masticatory edge (Fig.10-j) with 3 large ventral teeth of similar length; first 2 multicuspid; 3rd or the ventralmost tooth monocuspid and separated from 2nd by a wide gap. Right masticatory edge (Fig.10-k) with 4 large ventral teeth; first 2 multicuspid and of similar length; 3rd monocuspid and shorter than 2nd; 4th or the ventralmost tooth also monocuspid, similar in height to

1st and separated from 3rd by a wide gap. Mandibular palp without setae on basis, with 2 setae on 1st endopodal segment. Exopod far short of reaching distal end of endopod.

Maxillule (Fig.11-a) with 13 setae on 1st inner lobe—2 anterior submarginal, 2 posterior submarginal, and 9 marginal setae; one of marginal setae developed into a large spine. Second inner lobe with 2 setae, 3rd with a single seta, basis with 2 setae, endopod with 4 setae. Exopod with 6 long lateral setae in addition to 4 small medial setae. Outer lobe with 5 long terminal setae.

In maxilla (Fig.11-b), 1st segment about as long as rest of appendage; 1st lobe with 2 small proximal, 1 small posterior subterminal, and 3 long terminal setae; 2nd to 4th lobes each with 1 small posterior subterminal and 2 relatively long terminal setae.

Coxa of Mxp (Fig.11-c) with 1+2+2 setae, all rather poorly developed. Basis only a little longer than coxa, with a patch of short spinules midway on anterior surface and a row of relatively short, fine setules along proximal half of inner margin. First of 3 middle setae small, attached at middle of segment; 2nd moderately developed, with its proximal half provided with long setules and distal half with short spinules; long setules of proximal portion of seta straight but those of distal portion curved distad (Fig.11-d); 3rd longest, armed with normal setules. Long seta at distal end of basis and 2 distal setae of 1st endopodal segment armed with same types of setules as in 2nd middle seta of basis. Endopod elongate; combined lengths of first 3 segments as long as basis.

First leg (Fig.11-e) with well-developed outer spines on exopod; 1st spine approximately 1.5 times as long as 2nd, overreaching distal end of 2nd exopodal segment; 2nd spine reaching base of 3rd spine; 3rd spine extending beyond base of 4th. Only 1st and 2nd outer spines each preceded by a spiniform process. Terminal spine of exopod as long as combined lengths of 2 distal segments. Endopod extending far beyond distal end of 2nd exopodal segment; each endopodal segment with its distolateral corner produced into a spiniform process. Hooklike process of basis well developed (Fig.11-f).

In 2nd leg (Fig.11-g), 1st and 2nd exopodal segments with outer spines of equal length, which are approximately 1.5 times as long as those of 3rd exopodal segment. Terminal spine of 3rd exopodal segment as long as segment itself. Endopod reaching distal end of 2nd exopodal segment; its 2 proximal segments each produced at distolateral corner into a spiniform process. Anteriorly, 1st exopodal segment with a pore close to base of outer spine; 2nd with 2 pores, 1 at base of outer spine and another some distance from it; 3rd with a pore at base of each outer spine and 2 between 2nd outer spine and last inner seta.

In 3rd leg (Fig.11-h), outer spines of first 2 exopodal segments of similar length, only slightly longer than outer spines of 3rd segment. Terminal spine of 3rd exopodal segment approximately 4/5 length of segment itself. Endopod reaching distal end of 2nd exopodal segment. All 3 endopodal segments each produced at distolateral corner into a spiniform process. Anteriorly, 1st exopodal segment with a pore close to base of outer spine; 2nd with 3 pores, one of which is at base of outer spine; 3rd with a pore at base of each of first 2 outer spines, 2 at base of 3rd outer spine, a large pore close to 4th inner seta and a very small pore close to base of 5th inner seta.

Fourth leg (Fig.11-i) with all exopodal outer spines similarly small; terminal spine of 3rd exopodal segment only 2/3 length of segment itself. Endopod extending far beyond distal end of 2nd exopodal segment, with first 2 segments each produced at distolateral corner into a spiniform process. Pore pattern of exopod similar to that of 3rd leg exopod except that 3rd segment with a pore close to base of each of 2nd, 3rd, and 4th inner setae and 1 additional pore close to base of 2nd outer spine.

In 5th leg (Fig.11-j), basipod as long as combined lengths of first 2 exopodal segments; 3rd exopodal segment only a little shorter than combined lengths of 2 preceding segments; endopod reaching distal end of 2nd exopodal segment. In exopod, 1st outer spine thinner and shorter than 2nd; 3rd and 4th of similar size and approximately 1/2 length of 2nd; terminal spine about as long as 3rd segment; inner spine of 2nd exopodal segment slender, a little shorter than terminal spine of 3rd segment. First 2 endopodal segments each with distolateral corner produced into a spiniform process and a well-developed inner seta, which is armed with long setules only for proximal half and its distal half serrated with short spinules. Anteriorly, coxa with 4 pores; first 2 exopodal segments each with a pore close to base of outer spine; 3rd segment with 7 pores; 1 at base of 1st outer spine, 2 at base of 2nd outer spine, 1 close to base of each of 2nd and 3rd inner setae, 2 between 1st outer spine and 4th inner seta.

Description of male. Prosome length, 2.52-2.76 mm; body length, 3.60-3.88 mm. Similar in habitus to female except that urosome (Fig.11-k) 5-segmented and first 3 urosomal somites armed with small spinules along posterior margin. First, 2nd, and 4th urosomal somites similar in length and a little shorter than 3rd. Anal segment as long as 4th urosomal somite. Combined lengths of anal segment and left caudal ramus a little longer than combined lengths of 2 preceding urosomal somites. Dorsally, 2nd urosomal segment with 2 pores. Pore pattern of anal segment and caudal rami as in female. Laterally, anal segment with 5 or 6 pores (Figs.12-a, b); caudal ramus with a row of 4 pores along ventral margin. Fourth marginal seta of left caudal ramus as long as body. Third marginal seta of left ramus and 3rd and 4th marginal setae of right ramus armed with short spines in addition to normal setules as in female.

Right antennule as in female. Left antennule (Fig.12-c) with first 2 segments fused; setation of 1st to 18th segments as in female except that distal setae on 10th to 13th segments small and middle seta on 18th segment missing. Postgeniculate segment composed of fused segments 19-21 with an aesthete marking distal end of segment 19 and with an aesthete and a long seta at distal end of segment 21. Segments 22 and 23 fused; otherwise, segments 22-25 as in female. All other cephalosomal appendages and 1st to 4th legs as in female.

In 5th pair of legs (Figs.12-d, e), coxa with 4 or 5 large pores on anterior surface. Inner lobe of right basis only a little higher than that of left basis. In right exopod (Figs.12-f-h), 2 distal segments of similar length; turned counterclockwise bringing their outer spines toward medial side of leg. Medial projection of 2nd segment nearly as long as segment itself, only slightly curved distad, arising from distal end of segment, and extending mediad at right angle to long axis of segment, its distal margin plumose. Outer spines of right exopod moderately developed and sharply pointed; terminal spine of right exopod as long as combined lengths of 2 distal segments and tapering like a fine needle. In left exopod (Figs.12-i, j), outer spine of 2nd segment in form of a lancet, originating halfway on lateral margin of segment and reaching distal end of segment. Third segment tapering distally to merge into terminal spine without a trace of joint between them, with a relatively short outer spine and long inner spine, each terminated with a blunt distal end; terminal spine also terminated with obtuse distal end.

Distribution. Wolfenden's (1905) original specimen was from off Valencia, Ireland, in the northeastern Atlantic. Subsequently, Sars (1925) recorded the species from many localities between 26°37'N and 46°38'N and between 05°13'W and 41°08'W in the northeastern Atlantic. Grice and Hulsemann (1965) also recorded the species from the northeastern Atlantic. Upon reexamination of the specimens, Park's (1970) record of *M. gracilis* from the Caribbean Sea is found to be referable to *M. brevicaudatus.* In the Pacific *M. brevicaudatus* is known only from the Izu region of Japan (Tanaka 1964).

In the present study, *M. brevicaudatus* was found to have a circumglobal distribution. A total of 61 specimens including 52 females and 9 males were found at 26 localities—3 localities in the Atlantic between 32°26'S and 01°10'N, 22 in the Pacific from the west coasts of the Americas to the East and South China Seas and the Malay Archipelago between 25°17'S (the central South Pacific) and 38°13'N (off California), and 1 in the eastern Indian Ocean (02°28'S, 86°23'E). The species was so rare that only 1 or 2 individuals were found at most of the localities. However, one sample taken from 1720-0 m at 09°25'S, 159°06'E in the western tropical Pacific yielded 18 specimens, 13 females and 5 males.

Remarks. Mesorhabdus brevicaudatus is very similar to *M. gracilis* and the two new species described below, but can be distinguished from them by the relatively small body size and by the shape of the posterolateral corners of the prosome, which in dorsal view are closely applied to the lateral margins of the genital somite as shown by Wolfenden (1905, pl. 4, fig. 1). Characters also useful for species identification are the cuticular pore pattern of the anal segment and caudal rami, the setation of Mx1, Mx2, and Mxp, and the shape of the medial projection of the 2nd exopodal segment of the male right P5.

Sars's (1925) descriptions and figures for *M. brevicaudatus* generally agree with specimens examined in this study, but his figures show the posterolateral corners of the prosome (pl. 65, fig. 1) widely diverging from the lateral margins of the genital somite and the maxilliped coxa (pl. 65, fig. 8) with 1+1+2 setae, instead of 1+2+2. However, according to the body size and details of the other appendages including Mx1 and Mx2, Sars's specimens seem to be referable to *M. brevicaudatus*. Tanaka's (1964) descriptions and figures are insufficient to confirm the identity of his specimens with those referred to *M. brevicaudatus* in the present study, but they are believed to be conspecific mainly because of the similarity in body size. Tanaka's female specimens had a prosome length of 2.63 mm (assuming prosome=2.63 mm and urosome=1.24 mm given by Tanaka are correct and body length=3.38 mm is a miscalculation) and a female specimen found in the East China Sea in the present study measured 2.68 mm in prosome length. No other species similar to *M. brevicaudatus* have been found in the northwestern Pacific.

Mesorhabdus brevicaudatus differs from *M. angustus* in the following aspects: 1) Antennule relatively short, 2) anterior labral lobe inconspicuous, 3) genital somite in female not armed with spinules along posterior margin, 4) in antenna, coxa with a large tubercular outgrowth; inner marginal seta of 8th exopodal segment poorly developed, 5) in mandible, basis without setae; 1st endopodal segment with 2 setae, 6) maxillule with 13, 2, and 1 seta, respectively on 1st, 2nd, and 3rd inner lobes; endopod with 4 setae; exopod with 4 small medial setae; outer lobe with 5 large setae, 7) in maxilla, 2nd to 4th lobes each with 3 setae, 8) coxa of Mxp with 1+2+2 poorly developed setae, 9) basis of P1 without lobular process at distomedial corner of anterior surface, 10) in legs 2-4, exopod with a relatively small number of cuticular pores on anterior surface, 11) fifth leg of female with a long exopodal terminal spine, 12) geniculated antennule of male without a seta attributable to segment 20; segment 17 without a large spiniform process at anterodistal corner, and 13) in male 5th pair of legs, inner lobe of right basis similar to that of left basis; medial projection of 2nd exopodal segment of right leg originating from distal end of segment; exopodal terminal spine of left leg terminating with a blunt distal end.

Mesorhabdus gracilis Sars 1907
Figures 13 and 14

Mesorhabdus gracilis Sars 1907, p. 18.—Sars 1925, p. 235, pl. 66, figs. 1-13.

Description of female. Prosome length, 3.41-4.00 mm; body length, 4.83-5.58 mm. Prosome approximately 2.2 times as long as urosome. Similar in habitus to *M. brevicaudatus* but posterolateral corners of prosome in dorsal view widely diverging with their medial margins separated from lateral margins of genital somite by wide angles (Fig.13-b). Dorsally, genital somite with smoothly curved lateral margins, typically with a length:width ratio of 100:80 and widest about 2/5 its length from anterior end. Laterally, genital somite (Figs.13-a, c) longer than deep, with a hump on dorsal margin about 2/3 its length from anterior end. Ventrally, genital operculum (Fig.13-d) longer and wide, close to posterior end of somite, and its anterior 1/3 surrounded by a band of cuticular pores. Second urosomal somite approximately 1/2 length of genital somite and a little longer than 3rd. Combined lengths of anal segment and left caudal ramus a little longer than genital somite. Anal segment about as long as 3rd urosomal somite, with 2 pores on each side. Caudal ramus with a large pore on dorsal surface next to 2nd marginal seta.

Antennule extending beyond posterior end of caudal ramus by its last 2 segments. In anatomical details, A1, A2, and Md as in *M. brevicaudatus*. Maxillule, Mx2, and Mxp differing from those of *M. brevicaudatus* in the following features: 1) Mx1 (Fig.13-e) with only 3 setae on endopod and 5 long lateral setae on exopod, 2) first lobe of Mx2 (Fig.13-f) normally with only 1 proximal seta in addition to 1 short subterminal and 3 long terminal setae; posterior subterminal setae of 2nd to 4th lobes relatively shorter, and 3) coxa of Mxp (Fig.13-g) with 1+1+2 setae, all poorly developed; basis with a relatively long band of spinules on proximal half of anterior surface.

Swimming legs similar to those of *M. brevicaudatus* except for the details described below. In exopod of P1 (Fig.14-a), outer margin of 1st segment highly bulging close to proximal end; outer spines well developed, with 2nd and 3rd each overreaching base of following outer spine. In exopod of P2 (Fig.14-b), all outer spines well developed, 3rd only slightly shorter than the others, which are all similar in length; anteriorly, 2nd and 3rd segments with a pore at base of each outer spine in addition to a pore on distal half of 3rd segment. Exopod of P3 (Fig.14-c) with well-developed outer spines, with a single pore at base of outer spine of 2nd segment, a pore next to each of 3rd to 5th inner setae of 3rd segment, and terminal spine of 3rd segment approximately 5/6 length of segment. In exopod of P4 (Fig.14-d), outer spines with conspicuously serrated margins and all more or less similar in size; terminal spine of 3rd segment about 3/4 length of segment; anteriorly, 2nd segment with 2 pores at base of outer spine. In 5th leg (Fig.14-e),1st, 2nd, and 4th outer spines of exopod similar in length and twice as long as 3rd, and 2nd to 4th with conspicuously serrated margins; terminal spine of 3rd segment only slightly longer than segment itself; inner spine of 2nd exopodal segment about as long as endopod; anteriorly, coxa with a large pore close to medial margin; 1st exopodal segment without pores and 3rd segment with 6 pores.

Description of male. Prosome length, 3.41 mm; body length, 4.91 mm. Similar in habitus to male of *M. brevicaudatus* except for the following details: 1) Posterolateral corners of prosome in dorsal view (Fig.14-g) widely divergent and genital somite with strongly bulging lateral margins, 2) second and 3rd urosomal somites of similar length and only a little shorter than genital somite, 3) fourth urosomal somite and anal segment of similar length, 4) dorsally, genital somite with a pore on each side; anal segment with 2 pores on each side, and caudal ramus with a large pore between 1st and 2nd marginal setae, and 5) laterally, genital somite and anal segment each with 2 pores (Fig.14-f).

Fifth pair of legs (Fig.14-h) similar to that of *M. brevicaudatus,* but can be distinguished from it by the following features: 1) Anteriorly, coxa with only 1 large pore close to medial margin, 2) medial projection of 2nd exopodal segment of right P5 (Figs.14-j, k) tapering into a rather sharp point, distal margin nearly straight, and 3) left exopod (Fig.14-i) with all outer and terminal spines tapering into sharp points, and inner spine of 3rd segment reaching about 2/3 length of terminal spine.

Distribution. This species has been recorded only from the North Atlantic. Sars's (1907) original specimens were obtained at 2 stations (34°02'N, 12°21'W; 31°41'N, 42°40'W) in the eastern and central temperate North Atlantic, respectively. Sars (1925) reported the species at 9 stations, including 2 stations previously reported on (Sars 1907), in the temperate eastern and central North Atlantic (between 27°36'N-43°04'N, 10°44'W-42°40'W). Park's (1970) record of this species from the Caribbean Sea is found to be referable to *M. brevicaudatus.*

The species seems to be quite rare. In the present study, only 11 specimens including 10 females and 1 male referable to this species were found at 8 stations—3 in the South Atlantic between 32°29'S and 10°20'S, 4 in the East Pacific along the west coast of America between 05°44'N and 31°58'N, and 1 at 35°00'S, 60°00'E in the southwestern Indian Ocean. Including the previous records, the species is now known to occur in the whole Atlantic, the eastern Pacific, and the southwestern Indian Ocean.

Remarks. The original description by Sars (1907) was based on female specimens 4.60 mm long obtained in the temperate North Atlantic. Sars (1925) redescribed the species with figures from both female and male specimens collected in the type locality. The specimens found in the present study are identified mainly on the basis of the body size and the setation of Mx1, which invariably had only 3 large setae on the endopod and 5 large lateral setae on the exopod as shown by Sars (1925, pl. 66, fig. 6). In the other members of the genus, however, Mx1 has 4 or 5 large setae on the endopod and 6 large lateral setae on the exopod. The other appendages described and figured by Sars (1925) seem to provide no specific characters useful in verifying the identity between the species here referred to *M. gracilis* and that originally described by Sars.

Mesorhabdus paragracilis, new species
Figures 15 and 16

Type material. Holotype female (USNM 286991), allotype male (USNM 286992) and 1 paratype female (USNM 286993) found in IKMT sample from 2000-0 m, SIO MV 73-I, station 34, 15°35'N, 98°10'W in the eastern tropical Pacific, April 7-8, 1973. Two paratype females (USNM 286994) found in IKMT sample from 4000-0 mwo, SIO MV 65-I, station 53, 22°35'N, 110°13'W in the eastern tropical Pacific, June 29, 1965.

Description of female. Prosome length, 3.17-3.75 mm; body length, 4.41-5.25 mm. Body robust (Fig.15-a). Prosome approximately 2.2 times as long as urosome. Laterally, forehead (Fig.15-d) produced ventrad into a conical rostrum bearing 2 filaments; cephalosome with a cervical groove halfway along dorsal margin; posterior margin of prosome broadly rounded (Fig.15-b). Dorsally, posterolateral corners of prosome (Fig.15-c) widely divergent with their medial margins separated from lateral margins of genital somite by a wide angle. Dorsally, genital somite about as long as wide, widest about 2/5 length of somite from anterior end, and 1.8 times the length of 2nd urosomal somite. Third urosomal somite about as long as anal segment and only a little shorter than 2nd. Combined lengths of anal segment and left caudal ramus approximately 1.2 times length of genital somite. Dorsally, anal segment with 2 pores on each side; caudal rami without pores. Laterally, genital somite (Figs.15-b, e) about as long as deep, with a hump halfway along dorsal margin; anal segment with 3 pores; caudal ramus with a row of 4 pores along ventral margin. Ventrally, genital operculum (Fig.15-f) wider than long, close to posterior end of somite, and its anterior half surrounded by a band of pores.

When applied close to body, antennule extending beyond posterior end of caudal ramus by its last 2 segments (Fig.15-a), its setation as in *M. brevicaudatus*. Antenna and mandible also as in *M. brevicaudatus*. Maxillule as in *M. brevicaudatus* with 4 setae on endopod and 6 large lateral and 4 small medial setae on exopod. Maxilla (Fig.15-g) similar to that of *M. brevicaudatus* but 2nd lobe normally with only 2 setae. Maxilliped (Fig.15-h) also similar to that of *M. brevicaudatus* but in 3rd coxal lobe, 1st seta approximately 2.5 times length of 2nd, whereas in *M. brevicaudatus* 1st seta is less than twice length of 2nd.

First leg (Fig,16-a) similar to that of *M. gracilis* with well-developed outer spines on exopod, but terminal spine of 3rd exopodal segment longer than combined lengths of 2 distal exopodal segments. Second leg (Fig.16-b) also similar to that of *M. gracilis* with inner margin of basis bulging out as a large protuberance, with large outer spines of exopod, and with terminal spine of 3rd exopodal segment as long as segment itself, but 1st exopodal segment normally with a pore close to base of its outer spine and 2nd normally with 3 pores at base of its outer spine. Third leg similar to that of *M. gracilis* except that in 3rd exopodal segment (Fig.16-c), pore at base of 1st outer spine and pore next to 3rd inner seta are absent. Fourth leg also similar to that of *M. gracilis* except that 1st exopodal segment (Fig.16-d) normally with a pore close to its outer spine and 3rd exopodal segment with 2 pores next to each of 2nd and 3rd inner setae. Fifth leg (Fig.16-e) differing from that of *M. gracilis* in having 2 large pores close to inner margin of coxa, pore next to 2nd inner seta of 3rd exopodal segment absent, terminal spine of 3rd exopodal segment about as long as segment itself, and inner spine of 2nd exopodal segment longer than terminal spine of 3rd exopodal segment.

Description of male. Prosome length, 3.08-3.87 mm; body length, 4.36-5.50 mm. Similar in habitus to *M. gracilis* with posterolateral corners of prosome in dorsal view widely diverging, but can readily be distinguished from it by the absence of pores on dorsal surface of caudal ramus as in female. Geniculated left antennule similar in morphological details to that of *M. brevicaudatus*. All other cephalosomal appendages and 1st to 4th swimming legs as in female.

In 5th pair of legs (Fig.16-f), each coxa with a large pore close to medial margin on anterior surface. Basal inner lobes of left and right legs similar in shape. Endopod reaching about distal end of 2nd exopodal segment; 2nd and 3rd endopodal segments each with a large pore on anterior surface. In right exopod (Figs.16-g, h), 1st segment with a large pore close to outer spine on anterior surface; 2nd segment relatively deep, with medial projection arising along distal margin of segment, tapering into a bluntly pointed end. Distal margin of medial projection straight and fringed with setules and is therefore distinct from curved distal margin of segment. Terminal spine of 3rd segment tapering distally like a fine needle and as long as combined lengths of 2nd and 3rd segments. First segment of left exopod (Figs.16-f, i) also with a large pore close to outer spine on anterior surface; 2nd and 3rd segments of similar length, each with a relatively large outer spine; 3rd segment with terminal spine about as long as segment itself and tapering distally into a sharp point; inner spine short of reaching middle of terminal spine.

Etymology. The specific name, formed with the Greek prefix *para*—near—refers to the similarity of this new species to *M. gracilis.*

Distribution. A total of 19 specimens including 13 females and 6 males were found in 10 samples, of which 2 were taken in the southeastern Atlantic between 30°08'S and 32°29'S, 6 along the ·west coast of America between 02°52'N and 31°58'N, 1 at 09°25'S, 159°06'E off New Guinea, and 1 at 35°00'S, 60°00'E in the southwestern Indian Ocean. This rare species apparently has a circumglobal distribution.

Remarks. This species is very close in size and morphological details to *M. gracilis* but can be distinguished from it by seemingly minor but consistent differences. The most important differences are: 1) Caudal ramus without pores on dorsal surface, 2) maxillule with 4 setae on endopod and 6 large lateral setae on exopod, 3) maxilla with only 2 setae on 2nd lobe, and 4) coxa of Mxp with 1+2+2 relatively well-developed setae.

Mesorhabdus poriphorus, new species
Figure 17

Type material. Holotype female (USNM 286995) and 2 paratype females (USNM 286997) found in IKMT sample from 1720-0 m, SIO Crosspac, station 3, 09°25'S, 159°06'E off New Guinea in the southwestern Pacific, March 3, 1975. Allotype male (USNM 286996) found in IKMT sample from 3000-0 mwo, SIO Aries I, station 3, 00°55'N, 112°50'W in the eastern tropical Pacific, November 22, 1970.

Description of female. Prosome length, 3.66-4.00 mm; body length, 5.25-5.75 mm. Prosome approximately 2.2 times as long as urosome. Similar in habitus to *M. gracilis* and *M. paragracilis,* new species, but posterolateral corners of prosome in dorsal view only moderately divergent (Fig.17-a). Dorsally, genital somite typically with a length:width ratio of 100:90, conspicuously bulging out 1/3 its length from anterior end. Laterally and ventrally, genital somite similar to that of *M. paragracilis* described above. Second urosomal somite approximately 1/2 length of genital somite. Third urosomal somite about as long as anal segment and only a little shorter than 2nd. Combined lengths of anal segment and left caudal ramus longer than combined lengths of 2nd and 3rd urosomal somites. Dorsally, anal segment with 2 pores on each side in addition to 1 on each lateral margin; caudal ramus with 2 or 3 pores close to anterior end. Laterally, anal segment with 7 pores.

Antennule extending beyond posterior end of caudal ramus by its last 2 segments. Antennule, A2, and Md agree in anatomical details with those of *M. brevicaudatus* and *M. paragracilis* described above. Maxillule as in *M. paragracilis* with 4 setae on endopod and 6 large lateral and 4 small medial setae on exopod. Maxilla also as in *M. paragracilis* with 2nd lobe bearing only 2 setae, but 1 of 2 proximal setae on 1st lobe normally very small or missing. Coxa of Mxp with 1+2+2 relatively well developed setae as in *M. paragracilis.* All 5 pairs of swimming legs show no discernible differences from those of *M. paragracilis.*

Description of male. Prosome length, 3.36-3.83 mm; body length, 5.41-5.50 mm. Similar in habitus as well as anatomical details of appendages to *M. paragracilis* but can be distinguished from it by the following features: 1) Dorsally, anal segment with 2 pores on each side and a pore on lateral margin, 2) each caudal ramus (Fig.17-c) with 2 or 3 pores close to anterior end as in female, and 3) laterally, anal segment (Fig.17-b) with 7 pores and caudal ramus with 2 or 3 pores close to dorsal margin and a row of 4 pores on ventral margin.

Anteriorly, 5th leg (Fig.17-d) with 2 or 3 large pores on coxa close to medial margin, 2-4 pores on basis, and 2 or 3 pores on each of 2 distal endopodal segments and 1st exopodal segment. Posteriorly, 5th leg (Fig.17-e) with 5 or 6 pores on coxa, 4 pores on basis, 1st exopodal and 2 distal endopodal segments each with 1-3 pores. Medial projection of 2nd exopodal segment of right P5 with an obtuse terminal end showing ridges and grooves; its distal margin somewhat wavy and separated from distal margin of segment by a notch (Fig.17-g). Terminal spine of right exopod much longer than combined lengths of 2nd and 3rd segments. In left exopod (Fig.17-f), 3rd segment gradually tapering into a terminal spine with an obtuse ending. Inner spine of 3rd segment extending far beyond 1/2 distance to distal end of terminal spine.

Etymology. The specific name *poriphorus*, from Greek *poros* meaning pore and suffix *phor* meaning bear, alludes to the many cuticular pores found in the species.

Distribution. The species is represented by 8 specimens including 6 females and 2 males found in 6 samples—1 in the South Atlantic (18°44'S, 10°15'W), 3 in the eastern Pacific between 16°09'S and 22°35'N, 1 in the southwestern Pacific (09°25'S, 159°06'E), and 1 in the South China Sea (17°44'N, 119°55'E).

Remarks. The species is very close in morphology to *M. paragracilis* described above but can easily be distinguished from it mainly by the pore patterns of the anal segment and caudal rami in both the female and male and the pore patterns of the 5th pair of legs in the male. In the present study, *M. paragracilis* and *M. poriphorus* occurred together at 2 stations, one off the southern end of Baja California and another east of New Guinea, where *M. poriphorus* (female PL=3.83 mm at both stations) was considerably larger than *M. paragracilis* (female PL=3.41-3.50 mm, off Baja California; female PL=3.17-3.25 mm, east of New Guinea). The females of these two species can also be distinguished from each other by the minor but consistent differences in shape of the posterolateral corners of the prosome and the genital somite in dorsal view and the males by the shape of the medial projection and the relative length of the terminal spine of the right P5 exopod and the shape and relative length of the terminal and inner spines of the 3rd exopodal segment of the left P5.

GENUS *HETEROSTYLITES* SARS 1920

Heterostylites Sars 1920, p. 11.

   Type by original designation: *Heterostylites longicornis* (Giesbrecht 1889)

   Diagnosis. None of marginal setae of caudal rami armed with conspicuous spines in addition to normal setules. First segment of A1 (Fig.18-h) with 4 lobes bearing 1, 3, 2, 2 setae/aesthetes, respectively. In exopod of A2 (Fig.18-j), first 2 exopodal segments without setae. In mandible (Fig.19-a), masticatory edge with a basal spine and 3 or 4 long spiniform teeth, without a group of short, contiguous teeth next to basal spine; basis with a long seta; 1st endopodal segment with 2 setae; 2nd endopodal segment with only 8 terminal setae, without appendicular setae. Maxillule (Fig.19-d) with 3rd inner lobe missing, exopod greatly elongated, making up distal half of appendage. In maxilla (Fig.19-e), 5th and 6th lobes each with a large saberlike spine; 1 of 3 setae on 4th lobe transformed into a long spine. Coxa of Mxp (Fig.19-f) with 1 middle and 3 distal marginal setae—1 anterior and 2 posterior; first posterior distal seta greatly elongated. Basis of P1 (Fig.19-g) without outer seta. Third endopodal segment of P2 (Fig.19-h) with 7 setae. Anteriorly, 2nd exopodal segment of female 5th leg (Fig.19-k) with distal margin extended distad into a serrated lappet. Left, geniculated A1 (Fig.20-c) of male with 2 setae belonging to 20th segment. Both right and left endopods of male 5th pair of legs (Fig.20-d) with inner marginal seta on 2nd segment. Second exopodal segment of right P5 (Fig.20-f) much wider than long, with short medial projection normally divided distally into 2 processes—a spiny and a plumose process (Fig.20-i).

   Additional description of female. Body elongate. First pedigerous somite separated from cephalosome; 4th and 5th pedigerous somites fused. Forehead produced ventrad into a rostrum bearing 2 filaments. Anal segment partially fused with caudal rami; its joint with right caudal ramus is indicated by a short line, but no trace is visible for its joint with left caudal ramus (Fig.18-c). Left caudal ramus longer than right, its 4th marginal seta naked and about as long as body. All other marginal setae of caudal rami plumose with normal setules and without short spines as found in *Disseta* and *Mesorhabdus*. Dorsal appendicular seta of left ramus longer than that of right ramus.

   Antennule (Figs.18-h, i) longer than body. In 1st segment, 1st lobe with 1 small seta; 2nd with 1 aesthete and 1 small and 1 large setae; 3rd with 1 small and 1 large setae; 4th with 1 aesthete and 1 large seta. Second to 19th segments each with 3 setae/aesthetes—1 middle and 2 distal. Twentieth and 21st segments each with 2 distal setae/aesthetes; 22nd and 23rd segments each with 3 distal setae—2 anterior and 1 posterior; 24th segment with 1 anterior and 1 posterior setae; 25th segment with 6 terminal setae.

   In antenna (Fig.18-j), coxa and basis partially fused, with 1 and 2 setae, respectively. Endopod and exopod of similar length. First endopodal segment with 2 inner marginal setae; 2nd with 8 marginal and 1 posterior appendicular setae on inner lobe and 6 marginal and 1 posterior appendicular setae on outer lobe. Exopod 8-segmented; first 2 segments without setae; 3rd to 7th segments each with 1 well-developed inner marginal seta; 8th segment with 1 inner marginal and 3 terminal setae.

   Mandibular blade about as long as palp (Fig.19-a). Basis with 1 long seta. Endopod with 2 inner marginal setae on 1st segment and 8 terminal setae on 2nd. Exopod originating 2/3 length of basis from proximal end and short of reaching distal end of endopod. Left masticatory edge (Fig.19-b) with 1 serrated basal spine and 3 spiniform teeth of similar length. First 2 teeth, counted from basal spine, bicuspid and close to each other; 3rd, the ventralmost tooth monocuspid, separated from 2nd by a wide gap. Right masticatory edge (Fig.19-c) with 1 serrated basal spine and 4 spiniform teeth. First 2 tricuspid and 3rd and 4th monocuspid. First 3 close to each other, 4th separated from 3rd by a relatively wide gap.

   Maxillule (Fig.19-d) with well-developed 1st inner lobe and exopod and rest of the appendages rather poorly formed. First inner lobe with 12 setae—2 anterior submarginal, 9 marginal and 1 posterior submarginal setae. Second inner lobe with a single seta, 3rd inner lobe missing. Basis with a single seta. Endopod small, unsegmented, with 5 setae. Exopod making up distal half of appendage, with

5 long setae in addition to 2 very small setae. Outer lobe with 5 long setae, the proximalmost of which is very small.

In maxilla (Fig.19-e), 1st segment about as long as rest of appendage and 5th lobe greatly elongated; 1st lobe with 1 short and 3 long setae; 2nd and 3rd each with 2 setae of equal length; 4th with 1 long spiniform seta and 2 relatively short setae of similar length; 5th and 6th lobes each with a large saberlike spine and 2 relatively short setae. Endopod with 7 slender setae. Only 3 long setae of 1st lobe, all setae of 2nd and 3rd lobes, and 1 of 2 short setae of 4th lobe armed with long setules.

Maxilliped elongate (Fig.19-f). Coxa with 2 lobes—1 middle and 1 distal; middle lobe with a single seta; distal lobe with 1 anterior and 2 posterior setae; 1st posterior seta of distal lobe much longer than the others. Basis approximately 1.5 times length of coxa, with a band of spinules extending longitudinally on anterior surface. Three middle marginal setae increasing in size in order from proximal to distal. Endopod 5-segmented and about as long as basis. First 4 endopodal segments each with 3 marginal setae; 5th segment with 3 terminal setae plus 1 small outer seta.

First leg (Fig.19-g) with long coxa bearing 1 inner marginal seta. Basis relatively short, with a curved inner seta at distomedial corner of anterior surface and a hooklike process close to lateral margin on posterior surface. All outer spines of exopod and marginal setae of both rami well developed. Anteriorly, all endopodal segments without any morphological features that could be referred to von Vaupel-Klein's organ. Second to 4th legs (Figs.19-h-j) each with relatively long coxa and relatively short basis; outer spines of exopods relatively short, while setae of both rami are well developed. Fifth leg (Fig.19-k) with relatively short coxa bearing no setae. Basis with a small outer seta. First exopodal segment without inner seta. Anteriorly, 2nd exopodal segment with distal margin extending distad into a serrated lappet.

Additional description of male. Similar in habitus to female except that urosome 5-segmented. In left, geniculated antennule (Figs.20-b, c), first 2 segments fused; 19th to 21st segments fused to form 1st postgeniculate segment; 22nd and 23rd segments fused. Setation of segments 1-16 and segments 20-25 as in female; segment 17 with 1 of 2 distal setae missing; segment 18 with middle seta missing; segment 19 with only 1 seta. All other cephalosomal appendages and 1st to 4th legs as in female.

Fifth pair of legs strongly asymmetrical (Figs.20-d-f); each leg with short coxa and relatively large basis bearing plumose inner lobe, which on right basis normally much elongated. Endopods in both legs with 1 inner marginal seta on 2nd segment. Two distal segments of exopods turned clockwise or counterclockwise bringing their posterior sides toward mediad. Second exopodal segment of right P5 greatly enlarged, produced into a small toothlike process at distomedial corner, with a small but highly characteristic medial projection at distomedial corner (Fig.20-j).

Remarks. Heterostylites differs from *Disseta* and *Mesorhabdus* in the following features: 1) First antennular segment with 1+3+2+2 setae/aesthetes, instead of 1+3+3+3 setae/aesthetes, 2) first 2 exopodal segments of A2 without setae, 3) masticatory edge of Md without a group of short, contiguous teeth next to basal spine and 2nd endopodal segment with only 8 terminal setae, 4) maxillule with 3rd inner lobe missing; exopod greatly elongated, making up distal half of appendage, 5) in maxilla, 1 of 3 setae on 4th lobe transformed into a long spine, 6) coxa of Mxp with 4 setae—3 on distalmost lobe and 1 seta on penultimate lobe; 1st posterior seta of distalmost lobe much longer than the others, 7) in basis of P1, outer seta missing as in *Mesorhabdus*, 8) in P2, 3rd endopodal segment with 7 setae, instead of 8, 8) anteriorly, distal margin of 2nd exopodal segment of female 5th leg extending distad into a serrated lappet, 9) in male 5th pair of legs, 2nd endopodal segment in both legs with an inner marginal seta, and 10) in right 5th leg of male, 2nd exopodal segment greatly enlarged, with a small but highly specialized medial projection.

The females of *Heterostylites* can readily be distinguished from those of the other calanoid copepods by a unique feature of P5—the serrated distal lappet of the 2nd exopodal segment. The male can be distinguished from those of the other heterorhabdid genera by the greatly enlarged 2nd exopodal segment of the right P5.

List of species of the genus *Heterostylites*

Heterostylites longicornis (Giesbrecht 1889)
Heterostylites major (Dahl 1894)
Heterostylites nigrotinctus (Brady 1918)
Heterostylites submajor, new species
Heterostylites longioperculis, new species
Heterostylites echinatus, new species

Remarks. Only 2 species have been previously referred to this genus (Tanaka 1964). In the present study, *Heterorhabdus nigrotinctus* Brady 1918 is transferred to *Heterostylites* and these 3 previously described species are redescribed. With 3 new species found in this study, a total of 6 species are now known in the genus.

Alloiorhabdus medius Wolfenden 1911 originally described from the Indian Ocean sector of the Antarctic has been regarded (Farran 1929, Vervoort 1951) as synonymous with *Heterostylites major.* The descriptions and figures given by Wolfenden (1911) are insufficient to allow specific identification, and *A. medius* should be regarded as a *nomen dubium.*

Key to species of *Heterostylites*

Females

1a. Dorsally, genital somite widest posteriorly (Fig.22-c)....................*H. nigrotinctus*
1b. Dorsally, genital somite widest at middle....................................2

2a. Ventrally, genital operculum much wider than 1/2 width of somite
  (Fig.24-f)...*H. longicornis*
2b. Ventrally, genital operculum about as wide as 1/2 width of somite
  (Fig.18-g) 3

3a. Laterally, ventral margin of genital somite extending straight beyond
  posterior end of genital operculum (Fig.18-f).................................4
3b. Laterally, straight ventral margin of genital somite not extending beyond
  posterior end of genital operculum (Fig.26-e) 5

4a. Laterally, 2nd urosomal somite with swollen ventral margin (Fig.18-b);
  setae of mouthparts conspicuously plumose..............................*H. major*
4b. Laterally, 2nd urosomal somite with straight ventral margin (Fig.21-b);
  setae of mouthparts not conspicuously plumose.....................*H. submajor,* n. sp.

5a. Mxp basis without spinules on distal part (Fig.26-f)..............*H. longioperculis,* n. sp.
5b. Mxp basis with spinules on distal part (Fig.28-g)....................*H. echinatus,* n. sp.

Males

1a. Basal lobe of right P5 long (Fig.23-b); basal lobe of left P5 relatively
  long and rounded...2
1b. Basal lobe of right P5 relatively short (Fig.20-e); basal lobe of left P5
  low and wide...3

2a. Basal lobe of left P5 relatively small (Fig.25-c)........................*H. longicornis*
2b. Basal lobe of left P5 relatively large (Fig.23-b) ....*H. nigrotinctus*

3a. In 2nd exopodal segment of right P5, medial projection with a long serrated
  process (Figs.20-h, 21-h)..4
3b. In 2nd exopodal segment of right P5, medial projection with a short serrated
  process (Figs.27-g, 29-f) 5

4a.   Setae of mouthparts highly plumose . . . .*H. major*

4b.   Setae of mouthparts not highly plumose. . . . .*H. submajor*, n. sp.

5a.   Mxp basis without spinules on distal part (Fig.26-f) . . . .*H. longioperculis*, n. sp.

5b.   Mxp basis with spinules on distal part (Fig.28-g). . . . .*H. echinatus*, n. sp.

Heterostylites major (Dahl 1894)
Figures 18-20

Heterochaeta major Dahl 1894, p. 79.

Heterorhabdus major; Giesbrecht and Schmeil 1898, p. 117.

Heterostylites major; Sars 1925, p. 239, pl. 67, figs. 17, 18.

Description of female. Prosome length, 2.68-3.41 mm; body length, 4.00-5.00 mm. Body slender (Fig.18-a). Prosome approximately 1.9 times as long as urosome. Dorsally, forehead (Fig.18-d) with a low midanterior tubercular process, which continues ventrad onto rostrum. Two rostral filaments relatively short (Fig.18-e). Cephalosome with a cervical groove halfway along dorsal margin and separated from 1st pedigerous somite by a clear line of articulation. Fourth and 5th pedigerous somites fused without a visible trace of articulation. Laterally, prosome with a broadly rounded posterior margin (Fig.18-f). Dorsally, posterolateral corners of prosome (Fig.18-c) bluntly angular, overlapping only about 1/10 length of genital somite. Dorsally, genital somite widest at middle, typically with a length:width ratio of 100:93, and approximately 1.5 times as long as 2nd urosomal somite. Third urosomal somite about 4/5 length of 2nd and approximately 1.4 times length of anal segment. Combined lengths of anal segment and left caudal ramus about equal to combined lengths of 2nd and 3rd urosomal somites. Dorsally, anal segment with a pore on each side and each caudal ramus with 2 pores anterior to 1st marginal seta. Dorsal appendicular seta of left ramus more than twice length of that of right ramus.

Laterally, genital somite (Fig.18-f) with slightly undulate dorsal and ventral margins, which are roughly parallel to each other; genital operculum produced beyond ventral margin of somite and short of reaching posterior end of somite. Ventrally, genital operculum (Fig.18-g) more or less ovoid with a shallow constriction at middle on each side and approximately 1/2 as wide as somite itself; anterior and posterior halves of genital operculum more or less symmetrical; a band of pores on each side of genital field. Laterally, 2nd urosomal somite with a hollow dorsal margin and a conspicuously bulging ventral margin (Fig.18-b).

Antennule extending beyond posterior end of caudal ramus by its last 7 segments. In anatomical details, all cephalosomal appendages typical of the genus except that long setae of antennal exopod, mandibular exopod, and maxillular outer lobe and exopod are conspicuously feathery with densely packed long setules.

In 1st leg (Fig.19-g), basipod about as long as exopod. In exopod, outer spine of 1st segment extending beyond base of following outer spine. Outer spine of 2nd segment similar in size to following outer spine on 3rd segment and barely reaching base of following outer spine. First outer spine of 3rd segment reaching base of following outer spine. Terminal spine about as long as combined lengths of 2nd and 3rd exopodal segments. Endopod reaching beyond distal end of 2nd exopodal segment; 1st segment with a rounded distolateral corner; 2nd segment with distolateral corner produced into a small spiniform process.

In 2nd leg (Fig.19-h), basipod approximately 2/3 length of exopod. All exopodal outer spines of similar size. Combined lengths of 1st and 2nd exopodal segments only a little shorter than 3rd exopodal segment. Terminal spine of 3rd exopodal segment about 2/3 length of segment. Endopod extending beyond distal end of 2nd exopodal segment; first 2 endopodal segments each produced into a short spiniform process at distolateral corner. Anteriorly, first 2 exopodal segments each with a pore close to outer spine; 3rd segment normally with 6 pores.

Third leg (Fig.19-i) with basipod shorter than 2/3 length of exopod; combined lengths of 1st and 2nd exopodal segments equal to about 4/5 length of 3rd exopodal segment; terminal spine of 3rd exopodal segment approximately 1/2 length of segment. Endopod extending beyond distal end of 2nd exopodal segment by about 1/4 its length. First 2 endopodal segments each with distolateral corner produced into a spiniform process. Anteriorly, 1st exopodal segment with a pore close to outer spine, 2nd with 2 pores close to outer spine, 3rd normally with 9 pores.

In 4th leg (Fig.19-j), basipod only slightly shorter than 2/3 length of exopod; combined lengths of 1st and 2nd exopodal segments approximately 4/5 length of 3rd exopodal segment; terminal spine of 3rd exopodal segment slightly shorter than 1/2 length of segment itself; endopod extending beyond distal end of 2nd exopodal segment by 1/3 its length. Anteriorly, 1st exopodal segment with 1 pore close to outer spine; 2nd with 2 pores close to outer spine; 3rd normally with 8 pores.

Fifth leg (Fig.19-k) with relatively short basipod, approximately 1/2 length of exopod. Combined lengths of 1st and 2nd exopodal segments excluding distal serrated lappet about as long as 3rd segment. Endopod reaching about distal end of serrate lappet of 2nd exopodal segment. Inner spine of 2nd exopodal segment about as long as 3rd exopodal segment. Terminal spine of 3rd exopodal segment approximately 2/3 length of segment itself. Inner marginal setae of first 2 endopodal segments bearing long setules on proximal half and short spinules on distal half. Anteriorly, 1st and 2nd exopodal segments each with a pore close to outer spine; 3rd with 3 pores.

Description of male. Prosome length, 2.84-3.58 mm; body length, 4.12-5.16 mm. Prosome about twice as long as urosome. First, 2nd, and 4th urosomal somites of similar length and a little shorter than 3rd (Fig.20-a). Anal segment shorter than 4th urosomal segment. Combined lengths of anal segment and left caudal ramus a little longer than combined lengths of 3rd and 4th urosomal somites. Dorsally, anal segment with a pore on each side and caudal ramus with 2 pores as in female. Dorsal appendicular seta of left ramus longer than that of right ramus. Fourth marginal seta of left ramus about as long as body and naked; all other principal marginal setae plumose with normal setules.

Antennule extending beyond posterior end of caudal ramus by its last 7 segments as in female. Seventeenth and 18th segments each with anterodistal corner produced into a spiniform process (Fig.20-c); 19th segment with 2 spiniform processes along anterior margin.

In 5th pair of legs (Figs.20-d-f), coxa shorter than basis. Right basal inner lobe longer than segment and short of reaching 1/2 length of 2nd endopodal segment. Left basal inner lobe relatively low. In right exopod, 3rd segment about as long as combined lengths of 1st and 2nd segments. Terminal spine of 3rd segment approximately 2/3 length of segment itself. Greatly expanded mediad, 2nd segment (Figs.20-g-j) about 1.5 times as wide as long, with distomedial corner produced into a small toothlike process armed with spinules; its medial projection arising from distomedial corner of segment, relatively large, and divided into 2 processes—a long proximal process armed with spinules and a short distal process fringed with setules. Anteriorly, 2nd segment (Fig.20-j) with a band of short setules enclosed by a circular ridge close to distomedial corner. In left exopod, combined lengths of 1st and 2nd segments about as long as 3rd segment including terminal spine.

Distribution. This species has a very wide distribution. It is represented in the study by 141 specimens including 103 females and 38 males from 39 samples—7 in the Atlantic, 103 in the Pacific and 1 in the Indian Ocean. The species occurred between 32°29'S (off the western coast of South Africa) and 63°24'N (east of Iceland) in the Atlantic, between 46°57'S (off Chile) and 52°20'N (in the Gulf of Alaska) in the eastern Pacific, between 9°25'S and 35°01'N in the central and western Pacific, and at 35°00'S, 60°00'E in the Indian Ocean. Although the species occurs widely in all three major oceans, the number of specimens found per sample was relatively low—only one or 2 specimens were found in most samples. The largest number (17 specimens) was found in a sample from off the coast of southern Chile.

Heterostylites major has been recorded from the Antarctic by Farran (1929) and Vervoort (1951, 1957) but was not found in the Antarctic in the present study. The record of Vervoort (1951) is referable to *H. nigrotinctus* (Brady 1918), a closely related Antarctic species.

Remarks. Dahl's (1894) original description of *H. major* was based on a female specimen from the Atlantic but the exact location was not given. It was distinguished from *H. longicornis* by its large body—*H. major* over 5 mm long, *H. longicornis* 3 mm long. Sars (1925) redescribed the species from female specimens 4.70 mm long obtained in the northeastern Atlantic between 31°41'N-46°31'N, 5°13'W-42°40'W, and indicated that it is distinguishable from *H. longicornis* by its large body size and relatively long antennule.

Subsequently the species has been recorded in the Atlantic as far north as Baffin Bay and the sea south of Iceland (Jespersen 1940) and also in the Gulf of Guinea (Vervoort 1965) and the Gulf of Mexico (Park 1970). Outside the Atlantic, it has been recorded from the northwestern Pacific by Brodsky (1950), off Japan by Tanaka (1964), the Malay Archipelago by Wilson (1950), the Indian Ocean (07°37'N, 84°19'E) by Sewell (1932), the Ross Sea of the Antarctic by Farran (1929), and the Atlantic and Indian sectors of the Antarctic by Vervoort (1951, 1957). Of these records, only Vervoort's (1951) record from the Atlantic sector of the Antarctic based on 2 male specimens is accompanied by descriptions and figures allowing a positive identification of the species. According to the figure of the 5th pair of legs, his male is definitely referable to *H. nigrotinctus* (Brady 1918) redescribed below. The descriptions and figures given by the other authors, however, are mostly too general to allow a positive identification of the species reported on.

In the present study, 2 species fitting the descriptions of *H. major* by Dahl (1894) and Sars (1925) and clearly distinguishable from *H. longicornis* were found. One of these is here referred to *H. major* because it occurs widely and is more common in the Atlantic, and the other is described below as a new species (*H. submajor*). The most reliable characters by which these 2 species can be distinguished from all other species of the genus are: 1) Ventrally, genital operculum of female about 1/2 as wide as genital somite, 2) laterally, genital operculum of female far short of reaching posterior end of genital somite, and 3) in male right P5 exopod, spiny process of medial projection of 2nd segment elongated.

Heterostylites major can be distinguished from *H. submajor* described below by the following features: 1) Genital somite of female with only moderately swollen lateral margins, while it has markedly swollen lateral margins in *H. submajor*, 2) second urosomal somite of female has a conspicuously bulging ventral margin but it has a straight ventral margin in *H. submajor*, 3) in 2nd exopodal segment of male right P5, spiny process of medial projection relatively long, armed with relatively large spines and plumose process pointed, while the spiny process of medial projection is relatively short, armed with small spines and the plumose process is small and not pointed in *H. submajor*, 4) basal inner lobe of male left P5 relatively small, with a small distolateral extension, and 5) long setae of A2, Md, and Mx1 conspicuously feathery.

Heterostylites submajor, new species
Figure 21

Type material. Holotype female (USNM 286998) and 4 paratype females (USNM 287000) found in IKMT sample from 2100-0 m, SIO Scan, station 117-10, 05°39'N, 83°04'W in the eastern tropical Pacific, December 31, 1969. Allotype male (USNM 286999) and 3 paratype males (USNM 287001) found in IKMT sample from 2000-0 m, SIO MV 73-I, station 53, 13°24'N, 92°04'W in the eastern tropical Pacific, April 14, 1973.

Description of female. Prosome length, 3.17-3.58 mm; body length, 4.75-5.50 mm. Similar in habitus to *H. major* but body (Fig.21-a) somewhat bulkier than that species. Prosome approximately 1.9 times as long as urosome. Two rostral filaments relatively short. Dorsally, forehead with a relatively low midanterior tubercular process, which continues onto rostrum. A distinct cervical groove halfway along dorsal margin of cephalosome. Laterally, posterior margin of prosome broadly rounded. Dorsally, posterolateral corners of prosome (Fig.21-c) angular, covering about 1/10 length of genital somite. Dorsally, genital somite about as wide as long, widest at middle with strongly bulging lateral margins, and approximately 1.4 times as long as 2nd urosomal somite. Third urosomal somite about 7/10 length of 2nd and about 1.4 times length of anal segment. Combined lengths of anal segment and left caudal ramus distinctly longer than combined lengths of 2nd and 3rd urosomal somites. Dorsally, anal segment with a single pore on each side and each caudal ramus also with a single pore next to 1st marginal seta. Dorsal appendicular seta of left caudal ramus about twice length of that of right ramus.

Laterally, genital somite (Fig.21-b) as in *H. major*, with genital operculum far short of reaching posterior end of somite. However, 2nd urosomal somite with a straight ventral margin, without a conspicuous swelling as found in *H. major*. Ventrally, genital operculum (Fig.21-d) widest in anterior half and posterior half somewhat tapering posteriorly. Cuticular pores on right side of genital field are fewer than left side.

Antennule extending beyond posterior end of caudal ramus by at least its last 8 segments, clearly longer than in *H. major*. In anatomical details, all cephalosomal appendages and swimming legs showed no detectable differences from those of *H. major* except that long setae of A2, Md, and Mx1 are not so conspicuously feathery.

Description of male. Prosome length, 3.20-3.50 mm; body length, 4.56-5.08 mm. Similar in habitus to *H. major* except that body is larger and somewhat bulkier. Antennule extending beyond posterior end of caudal ramus by at least its last 8 segments as in female. All cephalosomal appendages and 1st to 4th swimming legs as in female. Fifth pair of legs (Figs.21-e-g) very similar to that of *H. major* but can be distinguished from it by the following features: 1) Basal inner lobe of right leg relatively long, extending beyond 1/2 length of 2nd endopodal segment; basal inner lobe of left leg relatively well developed, extending distolaterally into a large process, 2) in 2nd exopodal segment (Figs.21-h, i) of right P5, spiny process of medial projection relatively short and armed with relatively small spinules, plumose process very low and inconspicuous; band of short setules on anterior surface of segment relatively long and enclosed by a triangular ridge (Figs.21-j, k), and 3) none of setae on A2, Md, and Mx1 are conspicuously feathery.

Etymology. The species name alludes to the close similarity of the new species to *H. major.*

Distribution. This species is represented in the study by 26 specimens including 21 females and 5 males from 8 samples—1 in the South Atlantic at 32°27'S, 08°48'E, 5 off the west coast of America between 02°52'N and 38°13'N, 1 in the East China Sea at 25°35'N, 124°30'E, and 1 in the southwestern Indian Ocean at 35°00'S, 60°00'E. The species is quite rare but appears to have a worldwide distribution.

Remarks. This species closely resembles *H. major* but can be distinguished from it in the female by the genital somite, which has strongly bulging lateral margins, the genital operculum in ventral view, of which the posterior half is relatively narrow and tapering posteriorly, and the 2nd urosomal somite, which has a straight ventral margin. Furthermore, it has a larger body and longer antennules than *H. major*. The largest individuals were found in a sample from off Panama in the eastern tropical Pacific and the specimens from off Guatemala in the eastern tropical Pacific, the East China Sea, and the southwestern Indian Ocean were the smallest. *Heterostylites submajor* and *H. major* have occurred together in 4 samples, where the former was substantially larger than the latter. A sample from the South Atlantic contained a female of *H. submajor* and 5 females of *H. major* whose body lengths were 5.00 mm and 4.16-4.48 mm, respectively. A sample from off Guatemala in the eastern tropical Pacific contained 7 females of *H. submajor* and 2 females of *H. major* with body lengths of 5.00-5.33 mm and 4.50-4.58 mm, respectively. A sample from off Acapulco, Mexico, yielded 2 females of *H. submajor*, measuring 4.83 and 4.91 mm long, respectively, and 1 female of *H. major*, measuring 4.44 mm long. *Heterostylites submajor* and *H. major* were each represented by a single female specimen in a sample from the southwestern Indian Ocean and their body lengths were 4.75 mm and 4.08 mm, respectively.

Characters by which the male of this species can be distinguished from that of *H. major* are quite subtle, and the positive identification of these males therefore requires careful examination of anatomical details. However, if they occur together, they are easily distinguished from each other by the differences in body size, length of A1, and appearance of the long setae on A2, Md, and Mx1. In the present study, the males of *H. submajor* and *H. major* occurred together in a sample from off Guatemala in the eastern tropical Pacific, and their body lengths were 5.00-5.08 mm in *H. submajor* and 4.66 mm in *H. major*. The long setae of A2, Md, and Mx1 in the latter were strikingly more plumose with densely packed long setules than in the former.

Heterostylites nigrotinctus Brady 1918
Figures 22 and 23

Heterorhabdus nigrotinctus Brady 1918, p. 27, pl.6, figs. 1-8.
Heterostylites major, not of Dahl; Vervoort 1951, p. 140, figs. 78, 79.

Description of female. Prosome length, 3.04-3.12 mm; body length, 4.44-4.64 mm. Similar in habitus to *H. major* but body relatively thick (Fig.22-a). Prosome about twice as long as urosome. Dorsally, forehead (Fig.22-d) with a low midanterior tubercular process, which continues onto rostrum as in *H. major*. Laterally, prosome with broadly rounded posterior margin (Fig.22-b). Dorsally, posterolateral corner of prosome (Fig.22-c) produced into a short, bluntly angular process covering only about 1/10 length of genital somite. Dorsally, genital somite gradually widening toward posterior end; widest part close to posterior end is about 1.7 times as wide as narrowest, anterior end. Laterally, posteroventral corner of genital somite produced into a large, rounded lobe (Fig.22-b); genital operculum far short of reaching posterior end of genital somite. Ventrally, genital operculum (Fig.22-e) only slightly longer than wide and nearly as wide as anterior end of somite. Second urosomal somite approximately 4/5 length of genital somite; dorsally, widest at anterior end, with a tubercular swelling on each side; laterally, its ventral wall with a tubercular outgrowth at posterior end (Fig.22-b). Third urosomal somite about 7/10 length of 2nd and about 1.7 times as long as anal segment. Combined lengths of anal segment and left caudal ramus a little shorter than combined lengths of 2nd and 3rd urosomal somites. Dorsally, anal segment with a pore on each side; left caudal ramus normally with 3 pores and right with 4 pores. Dorsal appendicular seta of left caudal ramus approximately 3 times as long as that of right ramus.

Antennule extending beyond posterior end of caudal ramus by its last 7 segments. All cephalosomal appendages and 1st to 4th swimming legs similar in anatomical details to those of *H. major*. In 5th leg (Fig.23-a), basipod, endopod, combined lengths of 1st and 2nd exopodal segments, and 3rd exopodal segment all similar in length. In 2nd exopodal segment, outer spine reaching almost 2/3 way to base of following outer spine on 3rd exopodal segment; inner spine about as long as 3rd exopodal segment. Terminal spine of 3rd exopodal segment approximately 2/3 length of segment itself. Inner setae of 1st and 2nd endopodal segments similar in length and armed with only small spinules on distal half.

Description of male. Prosome length, 3.00 mm; body length, 4.36 mm. Similar in habitus to *H. major* male except that body is relatively thick. Left, geniculate A1 as in *H. major* male. All the other cephalosomal appendages and 1st to 4th swimming legs as in female. In right P5 (Fig.23-b), basal inner lobe about as long as basipod. Combined lengths of 1st and 2nd exopodal segments measured along lateral margins approximately 3/4 length of 3rd exopodal segment. Second exopodal segment (Fig.23-e) about as long as wide; posteriorly, its distomedial corner not produced into a small spiniform process as in *H. major* but with a short row of spinules; its medial projection very small, divided into a small spiny process, which is triangular and armed with a longitudinal row of spinules, and a low plumose ridge. Anteriorly, 2nd exopodal segment (Fig.23-d) with a circular band of short setules enclosed by a roughly circular ridge. In 3rd exopodal segment (Fig.23-c), outer spine about 2/3 length of segment from proximal end; terminal spine about as long as 1/3 length of segment itself. In left P5 (Fig.23-b), basal inner lobe wide and long, extending beyond distal end of 1st endopodal segment. Second exopodal segment partially fused with 3rd, with outer spine close to proximal end. Combined lengths of 1st and 2nd exopodal segments about as long as 3rd segment including terminal spine.

Distribution. The type locality of this species is 64°34'S, 127°08'E in the Indian Ocean sector of the Antarctic. Vervoort's (1951) record, as *H. major*, was based on 2 males collected from 900-0 m at 66°30'S, 11°00'W and 100-0 m at 66°58'S, 16°03'W, respectively, in the Atlantic sector of the Antarctic. Three females and one male representing this species in the present study were found in the following localities: 1 female from 3221-0 mwo at 56°37'S, 24°15'W in the southern Atlantic; 1 female and 1 male from 2000-0 mwo at 60°42'S, 59°48'W in the Drake Passage; 1 female from 3248-0 mwo at 54°22'S, 150°54'E south of Tasmania. This species is therefore believed to be endemic to the Southern Ocean.

Remarks. This species was originally described from a single male specimen 5 mm long obtained at 64°34'S, 127°08'E in the Indian Ocean sector of the Antarctic. The male described by Vervoort (1951) as *H. major* from 2 specimens collected from the Atlantic sector of the Antarctic undoubtedly belongs to this species according to the close similarity in the 5th pair of legs. The female is described here for the first time and can readily be distinguished from those of the congeners by its highly characteristic genital somite. The male can be distinguished from those of the congeners by its 5th pair of legs, in which the basal inner lobe and 3rd exopodal segment of the right leg is relatively long, the medial projection of the 2nd exopodal segment of the right leg is very short, and the basal inner lobe of the left leg is large, extending beyond the distal end of the 1st endopodal segment.

Heterostylites longicornis (Giesbrecht 1889)
Figures 24 and 25

Heterochaeta longicornis Giesbrecht 1889, p. 812.—Giesbrecht 1892, p. 373, pl. 20, figs. 14, 21, 25, 26; pl. 39, fig. 44.
Heterorhabdus longicornis; Giesbrecht and Schmeil 1898, p. 116.
Heterostylites longicornis; Sars 1925, p. 238-239, pl. 67, figs. 1-16.

Description of female. Prosome length, 1.72-2.36 mm; body length, 2.60-3.44 mm. Body rather robust (Fig.24-a). Prosome nearly twice as long as urosome. Laterally, cephalosome with a distinct cervical groove halfway along dorsal margin. Posterior margin of prosome broadly rounded (Fig.24-d). Dorsally, forehead (Fig.24-b) somewhat rectangular, with a low midanterior tubercular process, which continues onto rostrum as in *H. major.* Dorsally, posterolateral corners of prosome rounded (Fig.24-c), overlapping about 1/7 length of genital somite. Dorsally, genital somite about as long as wide, with lateral margins only slightly arched, and 1.6 times as long as 2nd urosomal somite. Third urosomal somite about 2/3 length of 2nd and about 1.4 times as long as anal segment. Combined lengths of anal segment and left caudal ramus a little longer than combined lengths of 2nd and 3rd urosomal somites. Dorsally, anal segment with a pore on each side; each caudal ramus with 2 pores. Fourth marginal seta of left caudal ramus about as long as body. Dorsal appendicular seta of left ramus approximately 3 times as long as that of right ramus.

Laterally, genital somite (Figs.24-d, e) excluding arthrodial membrane, which often extends beyond posterior margin of somite, slightly longer than deep and much deeper than 2nd urosomal somite, with posteroventral corner produced into a large lobe; dorsal margin of somite more or less wavy. Genital operculum far short of reaching posterior end of somite. Ventrally, genital operculum (Fig.24-f) more or less ovoid with a relatively deep constriction on each side, a little longer than wide, and approximately 3/4 as wide as somite. Genital field with about 20 pores on each side.

When applied against body, antennule extending beyond posterior end of caudal ramus by its last 7 segments. All cephalosomal appendages and 1st to 4th swimming legs similar in anatomical details to those of *H. major.* Long setae of A2, Md, and Mx1 are conspicuously feathery with long, densely packed setules. In P5 (Fig.25-a), basipod, combined lengths of 1st and 2nd exopodal segments excluding distal serrated lappet, and 3rd exopodal segment all of similar length and slightly longer than endopod. Inner spine of 2nd exopodal segment about as long as 3rd exopodal segment. Terminal spine of 3rd exopodal segment approximately 2/3 length of segment itself. Anteriorly, 1st and 2nd exopodal segments each with a pore; 3rd exopodal segment with 3 pores. Only 2nd endopodal segment with spiniform process at distolateral corner. Inner setae of 1st and 2nd endopodal segments with proximal half naked and distal half serrated with short spinules.

45

Description of male. Prosome length, 1.80-2.32 mm; body length, 2.56-3.40 mm. Similar in habitus to *H. major.* Prosome a little more than twice length of urosome. Left, geniculate antennule similar in anatomical details to that of *H. major.* All the other cephalosomal appendages and 1st to 4th swimming legs as in female. Long setae of A2, Md, Mx1 also conspicuously feathery as in female.

In 5th pair of legs (Fig.25-b), basal inner lobe of right leg elongate, about as long as basipod, extending straight distad far beyond 1/2 length of 2nd endopodal segment. Basal inner lobe of left leg relatively narrow, with distolateral end produced into a rounded process. Second exopodal segment of right leg (Figs.25-d, e) greatly expanded mediad and much wider than long, with distomedial corner pointed and armed with a few small spines and a small patch of short setules. Some of the small spines and the patch of short setules seem to represent the greatly reduced medial projection of the segment. Anteriorly, the segment (Fig.25-d) with a band of short setules enclosed by a circular ridge close to distomedial corner and a lobular outgrowth close to proximal end. Measured along lateral margins, combined lengths of 1st and 2nd exopodal segments of right leg about as long as 3rd segment. Outer spine of 3rd segment about 2/3 length of segment from proximal end; terminal spine approximately 1/3 length of segment itself. In exopod of left P5 (Fig.25-f), combined lengths of 1st and 2nd exopodal segments shorter than 3rd segment including terminal spine.

Distribution. This species is represented in the study by 173 specimens including 124 females and 49 males from 41 samples—4 in the Atlantic between 33°47'S and 01°25'N, 20 in the eastern Pacific between 35°11'S and 40°35'N, 8 in the central Pacific between 41°27'S and 31°13'N, 4 in the western Pacific including the Malay Archipelago and South China Sea between 09°25'S and 35°01'N, and 5 in the Indian Ocean between 35°00'S and 02°28'S. These findings confirm *H. longicornis* as a circumglobal species.

Since its original discovery by Giesbrecht (1889) in the eastern tropical Pacific, *H. longicornis* has been recorded throughout the world's oceans—the North Atlantic (Wolfenden 1902, 1904; Sars 1925; Wilson 1950), the Bay of Biscay (Farran 1926), the tropical South Atlantic (Wolfenden 1911), the Gulf of Guinea (Vervoort 1965), the Caribbean Sea (Park 1970), the eastern tropical Pacific (Giesbrecht 1892; Wilson 1950), off San Diego (Esterly 1905), the central North Pacific (Park 1968), the Malay Archipelago (Scott 1909; Wilson 1950), the northwestern Pacific (Brodsky 1950; Tanaka 1964), and the Indian Ocean (Sewell 1932; Vervoort 1957). However, none of these authors has provided sufficient descriptions or adequate figures with which to accurately identify species reported.

Remarks. This species was originally described briefly from female specimens 3.00 mm long obtained in the eastern tropical Pacific. It has since been redescribed by Giesbrecht (1892 as *Heterochaeta longicornis*), Wolfenden (1902 as *Heterochaeta zetesios*, and 1904 as *Heterorhabdus longicornis*), Esterly (1905 as *Heterorhabdus longicornis*), Sars (1925), and Tanaka (1964). However, none of these authors noted any anatomical differences between this species and *H. Major.* The distinction between the two species was mainly based on the body size—the body of *H. longicornis* was known to be equal to or less than 3.5 mm, while that of *H. major* was recorded as equal to or more than 4.7 mm.

In the present study, two species similar in body size to *H. longicornis* as described by Giesbrecht (1889, 1892) were found. The species occurring in all three major oceans is referred here to *H. longicornis* and the other, occurring only in the Pacific, is described below as a new species (*H. longioperculis*).

The female of *H. longicornis* can readily be distinguished from the congeners by the genital operculum in ventral view, which is much wider than 1/2 width of the somite itself, and by the long setae of A2, Md, and Mx1, which are more conspicuously feathery than in *H. major.* Laterally, the genital operculum is short of reaching the posterior end of the somite as in *H. major.* The male of *H. longicornis* can be distinguished from those of the congeners by the right P5, of which the basal inner lobe is long and straight and the 2nd exopodal segment has a lobular outgrowth on the anterior surface and a greatly reduced medial projection at the distomedial corner. Long setae of A2, Md, and Mx1 are conspicuously feathery as in the female.

Heterostylites longioperculis, new species
Figures 26 and 27

Type material. Holotype female (USNM 287002), 38 paratype females (USNM 287004), allotype male (USNM 287003), and 4 paratype males (USNM 287005) found in IKMT sample from 3000-0 mwo, SIO Aries I, station E, 09°52'N, 113°50'W in the eastern tropical Pacific, November 19, 1970.

Description of female. Prosome length, 1.88-2.40 mm; body length, 2.76-3.56 mm. Body slender (Fig.26-a). Prosome about twice as long as urosome. Dorsally, forehead with a low midanterior tubercular process as in *H. longicornis,* which continues onto rostrum. Laterally, posterior margin of prosome broadly rounded (Fig.26-b). Dorsally, posterolateral corners of prosome bluntly angular (Fig.26-c), overlapping about 1/10 length of genital somite. Genital somite only slightly longer than wide, with broad lateral swellings midway. Laterally, dorsal and ventral margins of genital somite somewhat wavy (Fig.26-e), but more or less parallel to each other; posteroventral corner of genital somite normally rounded; genital operculum reaching posterior end of ventral margin of somite excluding arthrodial membrane, which often protrudes beyond posterior end of somite. Ventrally, genital operculum (Fig.26-d) elliptical, with shallow lateral constrictions, about 1.5 times as long as wide, and about 1/2 as wide as somite itself. Genital field with a long band of pores on each side.

Laterally, 2nd urosomal somite (Fig.26-b) with a tubercular outgrowth on dorsal margin close to anterior end and straight ventral margin; about 2/3 length of genital somite and about 1.3 times as long as 3rd urosomal somite. Anal segment approximately 2/3 length of 3rd urosomal somite. Combined lengths of anal segment and left caudal ramus about as long as combined lengths of 2nd and 3rd urosomal somites. Dorsally, anal segment with a pore on each side; caudal ramus also with a pore, close to 1st marginal seta. Dorsal appendicular seta of left caudal ramus about twice as long as that of right ramus.

Antennule extending beyond posterior end of caudal ramus by its last 7 segments. In anatomical details, all cephalosomal appendages similar to those of *H. major* except for some minor features of Mxp as follows: Middle seta of coxa (Fig.26-f) relatively short and anterior distal seta about as long as 2nd posterior distal seta, which is approximately 2/5 length of 1st posterior distal seta; band of spinules on basis extending up to base of 3rd seta.

Swimming legs similar to those of *H. major* except for the features described below. In P1 exopod (Fig.26-g), 1st segment as long as 3rd; terminal spine longer than combined lengths of 2nd and 3rd segments. In P2 exopod (Fig.26-h), combined lengths of 1st and 2nd segments about 6/7 length of 3rd; terminal spine about 3/5 length of 3rd segment. In P3 exopod (Fig.27-a), combined lengths of 1st and 2nd segments approximately 4/5 length of 3rd; terminal spine slightly longer than 1/2 length of 3rd segment; anteriorly, 2nd segment with 1 pore close to outer spine. In P4 exopod (Fig.27-b), combined lengths of 1st and 2nd segments approximately 4/5 length of 3rd; terminal spine slightly shorter than 1/2 length of 3rd segment; anteriorly, 2nd segment with 1 pore close to outer spine. In P5 (Fig.27-c), terminal spine of exopod approximately 3/4 length of 3rd segment; inner marginal setae of 1st and 2nd endopodal segments without long setules; 2nd exopodal segment without pores on anterior surface.

Description of male. Prosome length, 1.96-2.40 mm; body length, 2.92-3.44 mm. Prosome approximately 7/10 length of body and urosome slightly shorter than 1/2 length of prosome. Similar in habitus to female except that urosome is 5-segmented. Left A1 geniculated and similar in anatomical details to that of *H. major* male. All the other cephalosomal appendages and 1st to 4th swimming legs similar in anatomical details to those of female.

Measured along medial margins, basal inner lobe of right P5 (Fig.27-d) approximately 1.2 times length of segment and overreaching middle of 2nd endopodal segment. Second exopodal segment about 1.2 times as wide as long; its distomedial corner rounded with a very small pointed process (Fig.27-f), from which a band of spinules extends proximolaterad on posterior surface (Fig.27-h). Posteriorly, the segment with a platelike medial projection, which is divided into a short proximal spinulose process and a short distal plumose process (Fig.27-g). Anteriorly, the segment with a band of short setules enclosed by a circular ridge at distomedial corner (Fig.27-f). Measured along lateral margins, combined lengths of 1st and 2nd exopodal segments about as long as 3rd segment (Fig.27-e). Terminal spine approximately 2/5 length of 3rd segment. In left P5, basal inner lobe low and wide (Fig.27-d), with distolateral end produced into a short process; 3rd exopodal segment including terminal spine longer than combined lengths of 1st and 2nd exopodal segments (Fig.27-e).

Etymology. The specific name, *longioperculis*, alludes to the elongate genital operculum of the female.

Distribution. This species is represented in the study by 258 specimens including 154 females and 81 males from 45 samples all taken in the Pacific including the Malay Archipelago. Thirty-six of these samples were taken in the eastern Pacific between 39°26'S and 37°40'N, 6 in the central Pacific between 09°25'S and 15°03'N, and 3 in the Malay Archipelago between 05°27'S and 06°08'N. The species was particularly common in the eastern tropical Pacific, where as many as 40 specimens were found in a single sample (from 3000-0 mwo at 09°52'N, 113°50'W).

Remarks. The female of this species can readily be distinguished from the congeners described above by the genital operculum, which in lateral view reaches the posterior end of the genital somite, and in ventral view is elongated and about 1/2 as wide as the genital somite. The male is characterized by a relatively short basal inner lobe in the right P5 and a wide platelike medial projection of the 2nd exopodal segment of the right P5. This species, however, is very similar to *H. echinatus* described below and can be distinguished from it mainly by body size and the armature of the Mxp basis.

Heterostylites echinatus, new species
Figures 28 and 29

Type material. Holotype female (USNM 287006), 11 paratype females (USNM 287008), allotype male (USNM 287007), and 8 paratype males (USNM 287009) found in IKMT sample from 1300-0 mwo, SIO Krill II, station 16, 13°33'N, 98°36'W in the eastern tropical Pacific, June 10, 1974.

Description of female. Prosome length, 2.28-2.64 mm; body length, 3.32-3.92 mm. Body slender (Fig.28-a). Prosome almost 7/10 length of body and 1.9 times as long as urosome. Dorsally, forehead with a low midanterior tubercular process as in *H. major*. Posterior margin of prosome in lateral view broadly rounded (Fig.28-e). Dorsally, posterolateral corners of prosome (Fig.28-c) bluntly angular, overlapping 1/10 length of genital somite. Genital somite only slightly longer than wide and widest about 2/3 its length from anterior end. Laterally, genital somite (Fig.28-d) with a hump halfway along dorsal margin; genital operculum reaching posterior end of somite excluding arthrodial membrane, which often extends beyond posterior margin of somite (Fig.28-e). Ventrally, genital operculum (Fig.28-f) elongate, with lateral constrictions and about 1/2 as wide as somite itself. Genital field with a band of about 20 pores on each side. Second urosomal somite approximately 7/10 length of genital somite and laterally (Fig.28-b), somewhat sigmoid with a dorsal hump close to anterior end and a ventral swelling in posterior half. Third urosomal somite approximately 7/10 length of 2nd urosomal somite. Combined lengths of anal segment and left caudal ramus only slightly longer than combined lengths of 2nd and 3rd urosomal somites. Caudal ramus with a pore on dorsal side. Dorsal appendicular seta of left caudal ramus more than twice as long as that of right ramus.

Antennule extending beyond posterior end of caudal ramus by its last 7 segments. All cephalosomal appendages similar in anatomical details to those of *H. major* and *H. longioperculis* described above except that band of spinules on anterior side of Mxp basis (Fig.28-g) extending for nearly the whole length of the segment.

In P1 exopod (Fig.28-h), 1st and 3rd segments of equal length; 1st outer spine overreaching base of 2nd, which reaches base of 3rd; terminal spine approximately 7/10 length of exopod. In P2 exopod (Fig.28-i), outer spines relatively well developed; anteriorly, 1st and 2nd segments without pores, 3rd segment normally with 4 pores; 3rd segment approximately 5/9 length of exopod when measured along medial margins; terminal spine of 3rd segment about 7/10 length of segment. In P3 exopod (Fig.29-a), 3rd segment slightly shorter than 3/5 length of exopod; terminal spine of 3rd segment approximately 3/5 length of segment; anteriorly, 1st segment without pores, 2nd with 1 pore close to outer spine, 3rd normally with 7 pores. In P4 exopod (Fig.29-b), 3rd segment about 5/9 length of exopod; terminal spine of 3rd segment approximately 3/5 length of segment; 1st segment without pores, 2nd with 1 pore close to outer spine, 3rd normally with 5 pores. Basipod, endopod, combined lengths of 1st and 2nd exopodal segments, and 3rd exopodal segment of P5 all similar in length (Fig.29-c). Inner spine of 2nd exopodal segment about as long as terminal spine of 3rd expodal segment and almost 4/5 length of 3rd exopodal segment. No pores found on anterior surface of exopod. Inner seta of 1st endopodal segment devoid of long setules; inner seta of 2nd endopodal segment armed with long setules on proximal half and small spinules on distal half.

Description of male. Prosome length, 2.28-2.68 mm; body length, 3.28-3.88 mm. Similar in habitus to *H. longioperculis* male described above. Left, geniculate antennule similar in anatomical details to that of *H. major.* All other cephalosomal appendages, including the armature of Mxp basis, and 1st to 4th swimming legs as in female. In right 5th leg (Fig.29-d), inner lobe of basis extending straightly distad, about 1.2 times as long as basis, overreaching 1/2 length of 2nd endopodal segment. Basal inner lobe of left P5 wide (Fig.29-d), with distolateral end produced into an angular process. Second exopodal segment of right leg approximately 1.5 times as wide as long, its distomedial corner pointed and armed with a band of spinules extending proximolaterad on posterior surface (Fig.29-g). Posteriorly, the segment (Fig.29-f) with a platelike medial projection at distomedial corner, which is divided into a proximal, spinulose lobe and a distal plumose edge. Anteriorly, the segment (Fig.29-h) with a band of short setules extending proximad from distomedial corner. Measured along lateral margins, combined lengths of 1st and 2nd exopodal segments of right P5 (Fig.29-e) a little shorter than 3rd exopodal segment; terminal spine of 3rd exopodal segment approximately 1/2 length of segment. In left P5, combined lengths of 1st and 2nd exopodal segments approximately 2/3 length of 3rd exopodal segment including terminal spine.

Etymology. The specific name *echinatus*, meaning spiny, refers to the long band of spinules on the basis of Mxp.

Distribution. A total of 255 specimens including 147 females and 108 males were found in 27 samples all taken along the west coast of the Americas between 27°05'S and 22°41'N. The species was particularly common between the southern end of Baja California and the Gulf of Panama. A MOCNESS sample taken from 592-393 m at 13°00'N, 102°00'W in the eastern tropical Pacific contained 78 specimens. According to these findings, this species is believed to be endemic to the eastern tropical Pacific.

Remarks. This species is very similar to *H. longioperculis* described above, but can be distinguished from it by the differences described below. *H. echinatus* is larger than *H. longioperculis* and their size differences are particularly pronounced when they occur together. A sample from 05°38'N, 83°03'W off Ecuador, for example, contained both species—*H. echinatus* females measuring 2.44-2.52 mm in prosome length and 3.60-3.72 mm in body length (2 specimens) and *H. longioperculis* females measuring 1.92-2.20 mm in prosome length and 2.96-3.24 mm in body length (10 specimens). In Mxp basis, the band of spinules on the anterior surface extends up to the base of the 3rd seta in *H. longioperculis,* but it extends far beyond the base of the 3rd seta and close to the distal end of the segment in *H. echinatus* . The females of these 2 species can also be distinguished from each other by the shape of the 2nd urosomal somite, which in lateral view has a straight ventral margin in *H. longioperculis,* but the same margin in *H. echinatus* has a swelling in the posterior half. The outer exopodal spines of P1 and P2 are better developed in *H. echinatus* than in *H. longioperculis.* The males of these 2 species show a minor but consistent difference in shape of the medial projection of the 2nd exopodal segment of the right P5, which is relatively large and divided into a large spiny proximal lobe and a plumose distal lobe in *H. longioperculis* . The 2 species are also different in distribution. *Heterostylites longioperculis* occurs throughout the Pacific while *H. echinatus* is restricted to coastal waters of the lower latitudes of the eastern Pacific.

GENUS *HEMIRHABDUS* WOLFENDEN 1911

Hemirhabdus Wolfenden 1911, p. 308.
Macrorhabdus Sars 1920, p. 11.

Type by page priority. *Hemirhabdus grimaldii.* (Richard 1893).

Diagnosis. Body massive. First segment of A1 (Fig.30-f) with 1+3+3+3 setae/aesthetes. First and 2nd exopodal segments of A2 (Fig.30-h) each with a small inner marginal seta. Basis and 1st endopodal segment of Md with 1 and 2 setae, respectively (Fig.30-i); 2nd endopodal segment without appendicular setae; masticatory edge with a basal spine followed by 3 teeth in left Md and 4 teeth in right Md, which are separated by an extremely wide gap from last, ventralmost tooth. Posteriorly, ventralmost tooth with longitudinal ridges resembling a reinforcement rib (Figs.30-i, j). In Mx1 (Fig.31-a), 3rd inner lobe missing; endopod 1-segmented but relatively well developed; exopod elongate, occupying at distal end of appendage. In Mx2 (Fig.31-b), 2nd lobe missing, 5th and 6th each with a large, coarsely serrated saberlike spine in addition to 2 small setae; endopod with 2 long setae. Coxa of Mxp (Fig.31-c) with a middle seta and 2 or 3 distal setae. Third endopodal segment of P2 (Fig.31-f) with 7 setae. Third exopodal segment of P3 broad (Fig.31-g), with small terminal spine. In male 5th pair of legs (Fig.32-d), 2nd endopodal segment in both legs with an inner marginal seta. Second exopodal segment of right P5 with toothlike or fingerlike medial projection close to proximal end (Fig.32-e).

Additional description of female. Rostrum (Fig.30-d) consisting of 2 relatively short filaments borne on a large base, which is not visible dorsally. First pedigerous somite fully separated from cephalosome. Fourth and 5th pedigerous somites completely fused. Laterally, cephalosome with a cervical groove halfway along dorsal margin. Posterolateral corners of prosome rounded in dorsal and lateral view (Figs.30-b, c). Dorsally, left caudal ramus fused with anal segment and longer than right caudal ramus (Fig.30-c). Numerous cuticular pores on whole urosome. None of marginal setae of caudal rami armed with spines in addition to normal setules. Fourth marginal seta of left caudal ramus about 3 times as long as body.

Antennule longer than body. Endopod of A2 (Fig.30-h) much thicker than exopod. In masticatory edge of Md (Figs.30-i, j), 1st tooth, counted from basal seta, small, spiniform; 2nd and 3rd elongate and tricuspid; 4th in right Md short, triangular; last, ventralmost tooth sickle shaped, separated from preceding tooth by a gap about 1/2 as wide as masticatory edge itself. First inner lobe of Mx1 with 9 setae (Fig.31-a), of which distal 7 similar in shape and size and proximal 2 of unequal size; 2nd inner lobe and basis each with a single seta; outer lobe with 5 setae of equal length; endopod normally with 6 setae; exopod nearly 1/2 length of appendage when measured along outer margins, with 5 long lateral setae in addition to 3 small medial setae. In Mx2 (Fig.31-b), 1st lobe with 4 plumose setae of about equal length; 2nd missing; 3rd and 4th with 2 and 3 relatively small unarmed setae, respectively; saberlike spine on 5th coarsely serrated with relatively large teeth for proximal 2/3 its length and very finely serrated for distal 1/3; saberlike spine on 6th longer but thinner than that of 5th and coarsely serrated with relatively small teeth for whole length. Maxilliped slender (Fig.31-c); basis approximately 1.5 times length of coxa, with all 3 middle setae arising in distal half; 1st to 4th endopodal segments each with 3 inner marginal setae; 5th with 3 terminal setae and 1 small outer marginal seta; basis and some endopodal segments with large cuticular pores on anterior surface.

All swimming legs (Figs.31-d-i) relatively wide, with large cuticular pores on anterior surface. In each of first 4 legs, coxa relatively large, with a well-developed inner marginal seta; basis relatively small, with inner marginal seta only in P1. Posteriorly, basis of P1 (Fig.31-e) with a hooklike process close to lateral margin. Anteriorly, 1st endopodal segment (Fig.31-d) with a cuticular pore and 2nd with 2 cuticular pores and without any morphological features referable to von Vaupel-Klein's organ. Outer and terminal spines of exopods rather poorly developed in 3rd to 5th legs. Third endopodal segment with 7 setae in P2 and P4, 8 setae in P3. Inner spine of 2nd exopodal segment of P5 relatively small.

Additional description of male. Similar in habitus to female except that urosome 5-segmented. Dorsally, lines representing joints between anal segment and both caudal rami clearly visible (Fig.32-a). Left, geniculate antennule (Fig.32-b) with first 2 segments only partially fused; 18th segment with a low angular process on anterior margin; 19th and 20th segments fused to form 1st postgeniculate segment; 21st to 25th segments all separate.

In 5th pair of legs (Figs.32-d, e), bases with relatively low inner lobes; 2nd endopodal segment in each leg with an inner marginal seta; terminal spines of both exopods rather poorly developed; that of left exopod relatively thick, with transverse ridges (Fig.32-g); that of right exopod relatively thin, tapering into a sharp point. Second exopodal segment of right P5 with a rounded tubercular outgrowth proximal to medial projection (Fig.32-f).

Remarks. Hemirhabdus can readily be distinguished from *Disseta, Mesorhabdus,* and *Heterostylites* by the following features: Body massive; masticatory edge of Md with ventralmost tooth separated from the other teeth by an extremely wide gap; in Mx2, 2nd lobe missing, 3rd and 4th lobes with only small and unarmed setae, and endopod with only 2 setae; in left A1 of male, first 2 segments only partially fused and last 5 (21st to 25th) segments all separate. *Hemirhabdus* shares the following features with *Heterostylites:* Mx1 with 3rd inner lobe missing; masticatory edge of Md without a group of small, contiguous teeth next to basal spine; coxa of Mxp with only 2 distal lobes bearing 1 seta on 1st and 2 or 3 setae on 2nd; 3rd endopodal segment of P2 with 7 setae.

Nominal species of the genus *Hemirhabdus*

Heterochaeta grimaldii Richard 1893 = *Hemirhabdus grimaldii* (Richard 1893)
Hemirhabdus falciformis Wolfenden 1911, now *Neorhabdus falciformis* (Wolfenden 1911)
    (see Heptner 1972b, p. 59)
Hemirhabdus amplus, new species

    Remarks. In the original description of the genus *Hemirhabdus,* Wolfenden (1911) included the following 2 species in the genus: *Heterochaeta grimaldii* Richard 1893 and *Hemirhabdus falciformis,* new species. Sars (1925) added *Heterorhabdus latus* Sars 1905 to *Hemirhabdus.* However, Heptner (1972b) recognized the difference of the last 2 species from the first and erected a new genus, *Neorhabdus,* for them by designating *Heterorhabdus latus* Sars 1905 as type species. In the present study, a new species was found in addition to *H. grimaldii,* the only species previously known in the genus.

Key to species of *Hemirhabdus*

Females and Males

1a. Coxa of Mxp normally with 1+3 setae (Fig.31-c); genital somite of female
     in dorsal view with smoothly bulging lateral margins (Fig.30-c) and in
     ventral view with genital operculum at middle of somite (Fig.30-e); 2nd
     exopodal segment of male right P5 with somewhat triangular medial
     projection (Fig.32-h). . . . . *H. grimaldii*
1b. Coxa of Mxp normally with 1+2 setae (Fig.33-e); genital somite of female
     in dorsal view with nearly straight lateral margins (Fig.33-c) and in ventral
     view with the genital operculum close to anterior end of somite (Fig.33-d);
     2nd exopodal segment of male right P5 with elongate medial projection
     (Fig.33-i). . . . . *H. amplus,* n. sp.

Hemirhabdus grimaldii (Richard 1893)
Figures 30-32

Heterochaeta grimaldii Richard 1893, p. 151.
Heterorhabdus grimaldii; Giesbrecht and Schmeil 1898, p. 117.—Wolfenden 1905, p. 10,
    pl. 4, figs. 3-5.—Farran 1908, p. 66.
Hemirhabdus grimaldii; Wolfenden 1911, p. 309, fig. 56.—Sars 1925, p. 230, pl. 63.—Farran
    1926, p. 284.—Sewell 1932, pl. 304, text figs. 99, 100.—Sewell 1947, p. 181.—Wilson
    1950, p. 238.—Tanaka 1964, p. 26, fig. 187.—Heptner 1972b, p. 58.

Description of female. Prosome length, 6.08-7.82 mm; body length, 8.41-11.00 mm. Body
massive (Fig.30-a). Prosome approximately 7/10 length of body and about 2.3 times as long as urosome. Urosome relatively short, approximately 3/10 length of body. Dorsally, posterolateral corner of
prosome (Fig.30-c) rounded, covering about 1/7 length of genital somite. Dorsally, genital somite
with smoothly curved lateral margins and about as long as wide. Laterally, genital somite (Fig.30-b)
deeper than long, with smoothly curved dorsal and ventral margins. Ventrally, genital operculum
(Fig.30-e) wider than long, about 2/5 as wide as somite and located at center of somite.

Second urosomal somite approximately 3/5 length of genital somite and about 1.3 times as
long as 3rd urosomal somite. Anal segment only slightly shorter the 3rd urosomal somite. Combined
lengths of anal segment and left caudal ramus about as long as genital somite. Dorsally, left caudal
ramus (Fig.30-c) slightly wider than right and extending beyond posterior end of right by about 1/10
its length as measured along medial margins. Dorsal appendicular setae of both caudal rami similar
in size. Large cuticular pores found over whole surface of genital somite and clustered on anterior and
lateral sides of genital field (Fig.30-e). In 2nd and 3rd urosomal somites, cuticular pores found mainly
on dorsal and lateral sides; in anal segment they are found on lateral and ventral sides. Each caudal
ramus with 3 pores on dorsal surface.

Antennule extending beyond posterior end of caudal ramus by last 3 segments when applied
close to body. Antenna (Fig.30-h) with stocky basipod and endopod, each with relatively small inner
marginal setae; 1st endopodal segment a little longer than basipod; both inner and outer lobes of 2nd
endopodal segments with well-developed marginal setae and poorly developed posterior appendicular setae. Exopod a little longer than 1st endopodal segment; inner marginal setae of first 2 exopodal
segments very small. In Md (Fig.30-i), mandibular palp shorter than mandibular blade; exopod reaching close to distal end of endopod. In Mxp (Fig.31-c), coxa with 1 middle and 3 distal setae; 1st distal seta about as long as middle seta, nearly twice as long as 2nd distal seta, and approximately 1.3
times as long as 3rd. Basis about 1.2 times as long as coxa; its 2nd and 3rd setae of about equal length
and longer than its 1st seta; its anterior surface with a band of spinules extending down to base of 2nd
seta. Second endopodal segment with 3 pores and 3rd with 1 pore on anterior surface.

First leg (Figs.31-d, e) with large basipod, approximately 1.6 times as long as endopod and
only a little shorter than exopod. In exopod, each outer spine preceded by a small spiniform process;
1st outer spine overreaching base of 2nd; 2nd to 4th outer spines of similar size; terminal spine about
as long as combined lengths of 2nd and 3rd segments. Distolateral corner of 1st endopodal segment
rounded, that of 2nd produced into a spiniform process. Sixteen pores are found on anterior surface
of P1—2 each on coxa, basis, 1st exopodal segment, and 2nd endopodal segment; 1 each on 2nd exopodal segment and 1st endopodal segment; 6 on 3rd exopodal segment.

In P2 (Fig.31-f), basipod approximately 1.5 times length of endopod and about 7/10 length of
exopod; 3rd exopodal segment about as long as combined lengths of 1st and 2nd; terminal spine of
3rd exopodal segment about 2/5 length of segment. In exopod, first 3 outer spines of similar size; last
2 of similar size and much smaller than first 3. Distolateral corners of 1st and 2nd endopodal segments each produced into a spiniform process. Anterior surface with 1 pore each on basis and 3rd endopodal segment, 2 pores each on 1st and 2nd exopodal segments, and 3 pores on 3rd.

Third leg (Fig.31-g) similar to P2, but basipod is only 3/5 length of exopod; 3rd exopodal segment longer than combined lengths of 1st and 2nd; terminal spine of 3rd exopodal segment less than

1/5 length of segment. In exopod, first 2 outer spines of similar size and larger than 3rd; last 2 equally smallest. Anteriorly, 3rd endopodal segment without pores; 3rd exopodal segment with 9 pores.

Fourth leg (Fig.31-h) similar to P3 except that basipod about 1.3 times as long as endopod, 3rd exopodal segment relatively narrow and much longer than combined lengths of 1st and 2nd; terminal spine of 3rd exopodal segment approximately 3/10 length of segment. Anteriorly, 2nd exopodal segment with 3 pores, 3rd with 11 pores.

Fifth leg (Fig.31-i) with basipod as long as endopod; combined lengths of 1st and 2nd exopodal segments about as long as 3rd; inner spine of 2nd exopodal segment a little shorter than 1/2 length of 3rd exopodal segment; terminal spine of 3rd exopodal segment approximately 1/3 length of segment itself. First 2 endopodal segments each with distolateral corner produced into a spiniform process and their inner setae of similar size and armed with long setules for proximal half and short spinules for distal half. Anteriorly, exopod with 2 pores on 1st segment, 1 on 2nd, and 4 on 3rd.

Description of male. Prosome length, 6.16-7.32 mm; body length, 8.33-10.33 mm. Prosome approximately 7/10 length of body and about 2.3 times as long as urosome. Urosome about 3/10 length of body. Body massive as in female. First 3 urosomal somites of similar length (Fig.32-a), each about 1.3 times as long as 4th urosomal somite and about 1.4 times as long as anal segment. Combined lengths of anal segment and left caudal ramus about as long as combined lengths of 3rd and 4th urosomal somites. Left caudal ramus extending beyond posterior end of right by 1/7 its length as measured along medial margins. Dorsally, 1st to 4th urosomal somites and caudal rami with pores. Anal segment with pores only on lateral sides.

In left, geniculated A1 (Fig.32-b), distal 3 lobes of 1st segment each with 1 aesthete and 2 setae; each of 2nd to 17th segments also with 1 aesthete and 2 setae; 18th with 1 aesthete and 1 seta; 19th and 20th fused, with 1 aesthete and 1 seta, respectively; 21st with 1 aesthete and 1 seta; 22nd and 23rd each with 1 aesthete and 1 seta anterodistally and 1 seta posterodistally; 24th with 1 anterodistal and 1 posterodistal seta; 25th (Fig.32-c) with 1 seta on anterior margin, 1 aesthete and 1 seta at anterodistal end, and 3 setae at posterodistal end. Two of the 3 posterodistal setae attached to small piece that is clearly separated from the segment by an articulation and therefore could be regarded as a separate segment.

Right A1 and all other cephalosomal appendages and 1st to 4th swimming legs as in female. In 5th pair of legs (Figs.32-d, e), inner lobe of right basis wide, extending in a distomedial direction a little beyond distal end of 1st endopodal segment, and much larger than that of left basis. Endopod a little short of reaching distal end of 2nd exopodal segment in right leg but far short of reaching it in left leg. Combined lengths of first 2 exopodal segments about as long as 3rd exopodal segment including terminal spine in right leg but much longer than it in left leg. In left exopod, terminal spine of 3rd segment fused with segment, digitiform with blunt distal end tipped with a very small spine and under a high magnification, revealing many fine lines resembling annuli (Fig.32-g). Anteriorly, terminal spine, as measured from lateral indentation presumably marking distal end of segment, approximately 2/5 length of segment. Second exopodal segment of right leg with a rounded tubercular outgrowth at proximal end of medial margin followed by a well-developed medial projection, which is more or less triangular but varies in shape to a certain extent depending on the angle of view (Figs.32-f, h-j).

Distribution. Since the original description from the North Atlantic (48°N, 21°W), this species has been recorded from the North Atlantic by Farran (1908, 1926) and Sars (1925), the tropical Atlantic by Wolfenden (1905), the eastern tropical Pacific and Philippine Islands by Wilson (1950), the Izu region of Japan by Tanaka (1964), the Indian Ocean by Sewell (1932), and the Arabian Sea by Sewell (1947).

In the present study, the species is represented by 110 specimens including 63 females and 47 males from 51 samples—16 taken in the Atlantic between 18°58'S and 53°15'N, 29 in the central and eastern Pacific between 46°57'S and 40°35'N, 1 at 40°05'S, 160°55'E in the Tasman Sea, and 5 in the Indian Ocean between 35°00'S and 02°28'S. The species was not found in samples taken in the northwestern Pacific and the Malay Archipelago.

Remarks. The body size showed a wide range of variation, 8.41-11.00 mm in the female and 8.33-10.33 mm in the male. However, no morphological variation associated with the body sizes was detected. The largest specimens were found off California in the eastern Pacific and the smallest off northwestern Africa in the eastern Atlantic, as in some other species. This species closely resembles *H. amplus* described below but can be distinguished from it by careful comparisons of the body size, the shape of the genital somite, the setation of Mxp, and details of the male 5th pair of legs as pointed out in the following description of that species.

Hemirhabdus amplus, new species
Figure 33

Type material. Holotype female (USNM 287010), 2 paratype females (USNM 287012), and allotype male (USNM 287011) found in IKMT sample from 5600-0 mwo, SIO MV 69-VI, station 7, 31°12'N, 119°17'W off San Diego in the eastern Pacific, December 17-18, 1969.

Description of female. Prosome length, 7.50-8.16 mm; body length, 10.50-11.83 mm. Similar in habitus to *H. grimaldii.* Prosome approximately 7/10 length of body and 2.3 times as long as urosome. Urosome approximately 3/10 length of body. Genital somite in dorsal view (Fig.33-c) about as long as wide, with more or less straight lateral margins; in lateral view (Figs.33-a, b), it is about as deep as long, with moderately bulging dorsal margin and more or less rounded ventral margin, which shows minor variation in outline depending on whether the somite is inflated or not. Ventrally, genital operculum (Fig.33-d) wider than long, approximately 1/3 as wide as somite, and located close to anterior end of somite. Second urosomal somite about 3/5 length of genital somite and about 1.3 times as long as 3rd urosomal somite. Anal segment approximately 4/5 length of 3rd urosomal somite. Combined lengths of anal segment and left caudal ramus about as long as genital somite. Dorsal appendicular seta of left caudal ramus longer than that of left ramus. Urosome with a pore pattern similar to that of *H. grimaldii* except that only 2 pores were normally found on dorsal surface of 3rd urosomal somite.

All cephalosomal appendages and swimming legs similar to those of *H. grimaldii* except for the following features of Mxp (Fig.33-e): Coxa normally with 1 middle seta and 2 distal setae of similar size and a little shorter than middle seta; anteriorly, coxa and basis each with 3 large pores, 2nd endopodal segment with a single pore.

Description of male. Prosome length, 7.50-7.66; body length, 10.66-11.16 mm. Similar in habitus to *H. grimaldii.* Prosome approximately 7/10 length of body and about 2.2 times as long as urosome. Urosome approximately 3/10 length of body. Left, geniculate A1 as in *H. grimaldii* male and all other cephalosomal appendages and 1st to 4th swimming legs as in female.

Fifth pair of legs (Figs.33-f, g) similar to that of *H. grimaldii* male except for the following features: 1) In left exopod, terminal spine (Fig.33-h) of 3rd segment tapering into a sharp point and approximately 1/3 length of segment itself; in right exopod, medial projection (Figs.33-i, j) of 2nd segment elongate, digitiform, and much narrower than in *H. grimaldii* male; and 2) posteriorly, left basis without pores; 1st exopodal segment with 2 pores close to its outer spine; 3rd endopodal segment with a pore (Fig.33-g).

Etymology. The species name *amplus*, from the Latin meaning large, refers to the large body size.

Distribution. This species was infrequently encountered. Only 14 specimens, 11 females and 3 males, were found in the present study. They were taken in 7 samples collected exclusively in the Indo-Pacific—4 in the eastern Pacific between 17°46'S and 31°12'N, 1 at 24°39'S, 155°01'W in the central South Pacific, 1 at 14°14'S, 150°54'E in the Coral Sea, and 1 at 20°43'S, 57°51'E in the western Indian Ocean.

Remarks. The female of this species differs from that of *H. grimaldii* as described above mainly in details of the genital somite and Mxp in addition to body size. The genital somite of this species is characteristic in having rather straight lateral margins, a depth about equal to the length, and a genital operculum close to its anterior end. The maxilliped of this species differs from that of *H.*

grimaldii in that the coxa of Mxp normally has only 2 distal setae and 3 large pores on the anterior surface and the endopod normally has only 1 pore anteriorly.

The male can be distinguished from that of *H. grimaldii* by the terminal spine of the left exopod, which is relatively short and tapers to a sharp spine, and by the fingerlike medial projection of the 2nd segment of the right exopod.

Hemirhabdus amplus is larger than *H. grimaldii*. They co-occurred in 5 samples and *H. amplus* was always the larger. In a sample from the eastern tropical Pacific, for example, *H. amplus* females were 11.33-11.50 mm long, while an *H. grimaldii* female was 10.33 mm long; in a sample from off the east coast of Madagascar, an *H. amplus* female was 11.33 mm long, while an *H. grimaldii* female was 8.83 mm long.

GENUS *NEORHABDUS* HEPTNER 1972

Neorhabdus Heptner 1972b, p. 58.

Type by original designation: *Neorhabdus latus* (Sars 1905)

Diagnosis. Body massive. In 1st segment of A1 (Fig.34-h), 1st lobe with 1 seta, 2nd with 2 setae and 1 aesthete, 3rd with 2 setae and 1 or more aesthetes, and 4th with 2 setae and 2 or more aesthetes. First and 2nd exopodal segments of A2 without setae (Fig.34-i); inner and outer lobes of 2nd endopodal segment without appendicular setae. In Md (Figs.34-j, k), basis without setae, 1st endopodal segment with 2 setae, 2nd endopodal segment with 7 terminal setae and without appendicular seta; masticatory edge with a basal spine followed by 3 teeth in left Md and 4 teeth in right Md, which are separated from last, ventralmost tooth by a wide gap. Second inner lobe of Mx1 (Fig.35-a) with or without a seta, 3rd inner lobe without setae; endopod 1-segmented and small; exopod elongate, occupying at distal end of appendage. In Mx2 (Fig.35-b), all setae on 1st to 4th lobes modified into spines, either unarmed or armed with short spinules only; 5th and 6th lobes of equal length, each with a large saberlike, serrated spine in addition to a small spine; endopod vestigial, with 2 setae of equal length. Coxa of Mxp (Fig.35-c) with 1 middle seta and 2 or 3 distal setae. Third endopodal segment of P2 with 7 setae (Fig.35-f). In male 5th pair of legs (Fig.36-f), 2nd endopodal segment with a seta in both legs. Second exopodal segment of right P5 with a digitiform medial projection (Fig.36-g).

Additional description of female. Rostrum consisting of 2 relatively slender filaments on a large base, which is not visible in dorsal view (Fig.34-e). First pedigerous somite fully separate from cephalosome. Fourth and 5th pedigerous somites completely fused. Laterally, cephalosome with a smoothly curved dorsal margin, without cervical groove (Fig.34-a); posterior margin of prosome rounded (Fig.34-b). Dorsally, forehead slightly produced with a rounded or wide angular tip (Fig.34-e); posterolateral corners of prosome (Fig.34-c) bluntly angular, covering about 1/10 length of genital somite. Laterally, genital somite with large genital prominence (Fig.34-b). Articulation between anal segment and left caudal ramus fully or partially formed (Figs.34-c; 40-c). Right caudal ramus a little shorter than or nearly equal to left ramus. None of caudal setae armed with spines in addition to normal setules. Fourth marginal seta of left caudal ramus unarmed, thicker than other marginal setae and about twice as long as body.

Antennules in some species with many small aesthetes. Endopod and exopod of A2 (Fig.34-i) relatively slender. Last exopodal segment with inner marginal seta reduced in some species (Fig.34-i). Palp of Md shorter than mandibular blade (Fig.34-j). In Mx1 (Fig.35-a), 1st inner lobe relatively large as compared with other parts of appendage, with 6 or 7 setae; outer lobe with 3-5 setae of similar length; basis with 2 or 3 setae; endopod with 5 or 6 setae; exopod with 3 or 4 setae. In Mx2 (Fig.35-b), 1st lobe with 3 moderately developed spines—2 serrated posterior spines and 1 terminal spine completely covered with spinules; 2nd lobe with a large terminal spine completely covered with spinules and a small, unarmed posterior spine; 3rd and 4th each with a large terminal spine completely covered with spinules, 1 anterior spine of varying size normally armed with spinules for whole surface, and a small terminal spine. Basis of Mxp (Fig.35-c) with a relatively short band of spinules.

All swimming legs relatively wide. Each of first 4 legs with relatively large coxa and small basis. Basis of P1 (Figs.35-d, e) with inner marginal seta at distomedial corner of anterior surface and a hooklike process close to lateral margin of posterior surface. Each outer spine of P1 exopod preceded by a spiniform process. Anteriorly, none of 3 endopodal segments with any visible features referable to von Vaupel-Klein's organ. Third endopodal segment with 7 setae in P2 and P4, with 8 setae in P3. Fifth leg (Fig.36-a) without inner marginal seta on coxa or basis, with a small outer seta on basis. Second exopodal segment with a relatively short inner spine. First 2 endopodal segments each with a relatively short and thick inner marginal seta.

Additional description of male. Similar in habitus to female except that urosome 5-segmented. Dorsally, lines marking joints between anal segment and caudal rami clearly visible (Fig.36-b). Antennules (Fig.36-d) with many small aesthetes in some species. In left, geniculate A1, first 2 segments completely fused; 19th-21st segments also completely fused to form 1st postgeniculate segment.

In 5th pair of legs (Fig.36-f), right basal inner lobe much larger than left basal inner lobe. Second endopodal segment in both legs with an inner marginal seta. Right exopod with an outer spine on 1st and 2nd segments and 2 outer spines on 3rd; medial projection of 2nd segment fingerlike (Fig.36-g). Left exopod with an outer spine on first 2 segments and an outer and an inner spine on 3rd. Terminal spine of 3rd exopodal segment relatively small and articulated with segment in both legs.

Remarks. The species of this genus had been referred to *Hemirhabdus* until Heptner (1972b) recognized their differences from the latter and placed them in a new genus, *Neorhabdus*. However, *Neorhabdus* is most closely related to *Hemirhabdus* and they share the following common features: 1) Dorsally, forehead without midanterior tubercular process extending from rostral base, 2) caudal rami wide; left ramus only a little longer than right, 3) on masticatory edge, basal spine is followed by 3 teeth in left Md and 4 in right, which are separated from last, ventralmost tooth by an extremely wide gap, 4) ventralmost tooth with longitudinal ridges resembling a reinforcement rib, 5) in Mx1, 1st inner lobe elongate, with 9 or fewer setae; 3rd inner lobe without setae, 6) in Mx2, large spines of 5th and 6th lobes well developed, saberlike and conspicuously serrated with large teeth; endopod vestigial, with only 2 setae, 7) coxa of Mxp with 1 middle and 2 or 3 distal setae, and 8) in male 5th pair of legs, each endopod with an inner seta on 2nd segment; 2nd segment of right exopod with digitiform medial projection.

The differences by which the genera can readily be distinguished from each other are found in Mx2, which in *Neorhabdus* has all setae on the first 4 lobes modified into spines and the 5th and 6th lobes of similar length. In Mx2 of *Hemirhabdus*, however, the 2nd lobe is missing, none of the setae on the first 3 remaining lobes modified into spines, and the 5th lobe is much larger and longer than the 6th. Additional differences between the genera are found in the following structures: 1) Cephalosome in lateral view with a cervical groove in *Hemirhabdus* but without it in *Neorhabdus*, 2) antenna with a small inner marginal seta on 1st and 2nd exopodal segments and a posterior appendicular seta on inner and outer lobes of 2nd endopodal segment in *Hemirhabdus*; they are absent in *Neorhabdus*, 3) mandible with a seta on basis and 8 terminal setae on 2nd endopodal segment in *Hemirhabdus*, but no seta on basis and 7 terminal setae on 2nd endopodal segment in *Neorhabdus*, 4) maxillule with 9 setae on 1st inner lobe, 1 seta on basis, and 5 long lateral and 3 small medial setae on exopod in *Hemirhabdus*, but 6 or 7 setae on 1st inner lobe, 2 or 3 setae on basis, and only 3 or 4 long lateral setae on exopod in *Neorhabdus*, 5) in P1, each outer exopodal spine is attached in a socket formed by spiniform processes of segment in *Neorhabdus*, but no such sockets in P1 of *Hemirhabdus*, 6) exopodal terminal spine of P3 is much smaller than that of P4 in *Hemirhabdus*, but the same spine of P3 is slightly larger than that of P4 in *Neorhabdus*, and 7) in geniculate A1 of male, 19-20th segments fused in *Hemirhabdus*, but 19th-21st segments fused in *Neorhabdus*.

List of species of the genus *Neorhabdus*

Neorhabdus latus (Sars 1905)
Neorhabdus falciformis (Wolfenden 1911)
Neorhabdus brevicornis, new species
Neorhabdus capitaneus, new species
Neorhabdus subcapitaneus, new species

Remarks. In his original description of the genus *Neorhabdus*, Heptner (1972b) included the following 3 species in the genus: *Heterorhabdus latus* Sars 1905, *Mesorhabdus truncatus* Scott 1909, and *Hemirhabdus falciformis* Wolfenden 1911. In the present study, 5 species referable to this genus were found including *N. latus* (Sars 1905), *N. falciformis* (Wolfenden 1911), and 3 new species. As suggested by Sewell (1932), *Mesorhabdus truncatus* Scott 1909 is believed to be synonymous with *Neorhabdus latus* (Sars 1905).

Key to species of *Neorhabdus*

Females and Males

1a. In 3rd lobe of Mx2, anterior spine much smaller than 1st terminal spine
(Fig.35-b); Mx1 with 2nd inner lobe missing (Fig.35-a) 2

1b. In 3rd lobe of Mx2, anterior spine about as long as 1st terminal spine
(Fig.38-j); Mx1 with 2nd inner lobe bearing a seta (Fig.38-i) 3

2a. Second distal seta of Mxp coxa about twice as long as 1st (Fig.35-c);
in 3rd lobe of Mx2, 1st terminal spine more than twice length of anterior
spine (Fig.35-b) . . . .*N. latus*

2b. Second distal seta of Mxp coxa approximately 1.5 times length of 1st
(Fig.37-h); in 3rd lobe of Mx2, 1st terminal spine approximately 1.5
times length of anterior spine (Fig.37-g) . . . .*N. falciformis*

3a. Last exopodal segment of A2 with inner marginal seta reduced;
2nd-7th segments of A1 each with more than 20 small aesthetes . . . . *N. capitaneus*, n. sp.

3b. Last exopodal segment of A2 with normally developed inner
marginal seta (Figs.38-f, 42-e); 2nd-7th segments of A1 each with
less than 10 aesthetes. 4

4a. A1 reaching about posterior end of prosome; Mxp coxa with 1+3
setae (Fig.38-k); Mx1 exopod with 3 setae (Fig.38-i) . . . . *N. brevicornis*, n. sp.

4b. A1 extending beyond posterior end of prosome by last 3 segments;
Mxp coxa with 1+2 setae (Fig.42-h); Mx1 exopod with 4 setae
(Fig.42-f) . . . .*N. subcapitaneus*, n. sp.

Neorhabdus latus (Sars 1905)
Figures 34-36

Heterorhabdus latus Sars 1905, p. 9.
Hemirhabdus latus; Sars 1925, p. 232, pl. 64.—Wilson 1950, p. 238.—Tanaka 1964, p. 28,
fig. 188.
Neorhabdus latus; Heptner 1972b, p. 58.
Mesorhabdus truncatus Scott 1909, p. 132, pl. 39, figs. 12-21.
Hemirhabdus truncatus; Sewell 1932, p. 306, text fig. 101.
Neorhabdus truncatus; Heptner 1972b, p. 58.

Description of female. Prosome length, 5.25-6.58 mm; body length, 6.75-8.91 mm. Prosome approximately 3/4 length of body and about 2.7 times as long as urosome. Body massive (Fig.34-a). Laterally, prosome deepest at joint between cephalosome and 1st pedigerous somite, typically with a length-depth ratio of 100:50. Cephalosome approximately 2/5 length of prosome, gradually tapering anteriorly into a narrowly rounded forehead (Fig.34-d), which is produced ventrad into a rostral base bearing 2 long filaments. Dorsally and ventrally, forehead (Figs.34-e, f) rounded or slightly angular.

Urosome approximately 3/10 length of body. Dorsally, genital somite (Fig.34-c) with only slightly curved lateral margins, typically with a length-width ratio of 100:87. Laterally, genital somite (Fig.34-b) as deep as long, with a low hump halfway along dorsal margin; genital prominence high and rounded, with similarly curved anterior and posterior slopes. Ventrally, genital operculum (Fig.34-g) typically with a width-length ratio of 100:67 and about 1/2 as wide as somite. Genital field with about 10 pores on each lateral side and 2 on anterior side. Second urosomal somite slightly shorter than 1/2 length of genital somite and approximately 1.2 times length of 3rd urosomal somite. Anal segment about as long as 3rd urosomal somite. Combined lengths of anal segment and left caudal ramus about as long as genital somite. Dorsally, both caudal rami separated from anal segment by clear lines. Left caudal ramus only slightly longer than right. Dorsal appendicular setae of caudal rami similar in length.

Antennule reaching about posterior end of caudal ramus. Four lobes of 1st segment (Fig.34-h) with 1+2+2+2 setae and 0+1+1+2 aesthetes. Second to 19th segments each with 2 setae—a middle and a distal. Number of aesthetes seems variable to a certain extent and normal number of aesthetes on each segment appears to be as follows: 1 on 2nd segment, 2 on 3rd, 1 on 4th, 2 on 5th, 4 on 6th, 2 on 7th, 3 on 8th, 2 on 9th, 1 on each of 10th-19th.

In antenna (Fig.34-i), basipod a little shorter than 1st endopodal segment. Measured along lateral margins, 2nd endopodal segment a little longer than 1/2 length of 1st, without posterior appendicular setae on either inner or outer lobe. Exopod slender, much shorter than endopod, with inner marginal seta on last segment reduced to a vestige.

In mandible (Fig.34-j), first tooth from basal spine very small, 2nd and 3rd of similar size and shape, tricuspid, and about as long as basal spine. Fourth tooth, present only on right Md (Fig.34-k), is low and more or less triangular. Ventralmost tooth sicklelike, only a little longer than 3rd tooth, initially diverging somewhat from and then curved toward dorsal teeth. Mandibular palp approximately 7/10 length of mandibular blade, with exopod far short of reaching distal end of endopod.

Maxillule (Fig.35-a) relatively small, with 6 setae on 1st inner lobe, no setae on 2nd or 3rd inner lobe, 3 or 4 setae on outer lobe, 2 or 3 setae on basis, 6 setae on endopod, and 4 setae on exopod.

In maxilla (Fig.35-b), 1st lobe with 2 posterior spines, which are serrated with short spinules, and 1 terminal spine, which is completely covered with short spinules. First posterior spine about 2/3 length of 2nd. Terminal spine intermediate in size. Second lobe with a small, unarmed posterior spine and a large terminal spine, which is completely covered with small spinules and about as long as the corresponding spine on 1st lobe. Third lobe with 1 large, completely armed terminal spine, a small serrated terminal spine, and a small serrated anterior spine. First terminal spine much thicker and more than twice as long as anterior spine. Fourth lobe with 3 spines similar to but smaller than those of 3rd lobe. Large serrate terminal spine of 5th lobe approximately 3.5 times length of small posterior spine. Large serrate terminal spine of 6th lobe thinner than but about as long as terminal spine of 5th lobe; anterior spine very small. Two endopodal setae of equal length, approximately 1/2 length of terminal spine of 6th lobe.

In Mxp (Fig.35-c), coxa about as long as basis, with 1 middle and 2 distal setae. First distal seta a little shorter than middle seta and approximately 1/2 length of 2nd distal seta. Three middle setae of basis gradually increasing in size in order from proximal to distal. Band of spinules on basis extending up to base of 2nd middle seta. Endopod approximately 1.3 times length of basis; 1st to 4th segments each with 3 marginal setae, 5th with 3 terminal setae, and 1 small lateral seta.

Coxa in 1st to 4th legs large, about as wide as long. In P1 (Figs.35-d, e), coxa about as long as endopod and almost twice as long as basis when measured along medial margins. Endopod a little shorter than combined lengths of first 2 exopodal segments. Each outer spine of exopod preceded by a spiniform process. Distolateral corners of 2nd and 3rd exopodal segments each produced into a spiniform process. First outer spine of exopod reaching base of following outer spine. Second outer spine short of reaching base of following outer spine. Third and 4th outer spines of similar size and slightly shorter than 2nd. Terminal spine of 3rd segment only slightly shorter than combined lengths of 2nd and 3rd segments. Distolateral corners of 1st and 2nd endopodal segments each produced into a spiniform process.

Basis of P2 (Fig.35-f) approximately 1/3 length of coxa. Endopod about as long as coxa and a little shorter than combined lengths of first 2 exopodal segments as in P1. Each outer spine of exopod preceded by a large spiniform process. Distolateral corners of all 3 exopodal segments each produced into spiniform process. First outer spine of exopod reaching about 2/5 way to base of following outer spine, which is only slightly larger than 1st outer spine. Third and 4th outer spines of similar size and about as large as 1st. Fifth outer spine approximately 2/3 length of 4th. Terminal spine of 3rd segment slightly shorter than 3/5 length of segment itself. Anteriorly, 1st and 2nd exopodal segments and 3rd endopodal segment each with a large pore.

Third leg (Fig.35-g) similar to P2 except for the following differences: Basis approximately 2/5 length of coxa; outer spines of exopod relatively small; first outer spine reaching only 1/4 way to base of 2nd, which is slightly smaller than 1st and similar in size to 3rd; three outer spines of 3rd segment diminishing in size in order from proximal to distal; terminal spine of 3rd segment approximately 2/5 length of segment itself; anteriorly, first 2 exopodal segments respectively with 1 and 2 pores close to outer spine; 3rd exopodal segment with about 7 pores; 3rd endopodal segment with 1 pore.

Fourth leg (Fig.35-h) similar to 3rd except for the following features: Basis with a small outer seta; first 2 outer spines of exopod similar in size; last 3 outer spines similar in size and smaller than first 2; terminal spine of 3rd exopodal segment approximately 2/5 length of segment itself.

In P5 (Fig.36-a), coxa about as wide as long, and approximately 1.4 times as long as basis. Endopod about 1.3 times as long as coxa or 3/4 length of basipod, and a little shorter than combined lengths of first 2 exopodal segments. Third exopodal segment about as long as combined lengths of first 2 exopodal segments. All 3 exopodal segments each with distolateral corner produced into spiniform process. Each of 4 outer spines of exopod preceded by a spiniform process. First outer spine of exopod reaching about 1/3 way to base of 2nd and a little larger than the following 3 outer spines, which are all of similar size. Terminal spine of 3rd exopodal segment approximately 3/5 length of segment itself and about as long as inner spine of 2nd exopodal segment. First 2 endopodal segments with distolateral corners produced into spiniform processes; their inner setae thick, about 1/2 as long as inner setae of 3rd endopodal segment, and fringed with normal setules along proximal 2/3 and serrated along distal 1/3. Anteriorly, 1st and 2nd exopodal and 3rd endopodal segments each with a pore; 3rd exopodal segment with 4 pores.

Description of male. Prosome length, 5.33-6.00 mm; body length, 7.16-8.08 mm. Similar in habitus to female except that forehead in dorsal view (Fig.36-c) with a low midanterior tubercular process, which extends ventrad into rostral base, and urosome 5-segmented. Prosome approximately 3/4 length of body and about 2.7 times length of urosome. Laterally, posterior margin of prosome broadly rounded as in female. Dorsally, posterolateral corners of prosome bluntly angular (Fig.36-b), covering anterior 1/3 of genital somite. First 3 urosomal somites of about equal length. Fourth urosomal somite about as long as anal segment and only slightly shorter than 3rd. Left caudal ramus slightly longer than right. Combined lengths of anal segment and left caudal ramus about as long as combined lengths of first 2 urosomal somites and only a little longer than combined lengths of 3rd and 4th urosomal somites. Fourth marginal seta of left caudal ramus about twice as long as body. Dorsal appendicular setae of caudal rami similar in size.

Antennule reaching about posterior end of caudal ramus. In left, geniculate A1 (Figs.36-d, e), first 2 segments fused, with 5 groups of setae/aesthetes. First group with 1 seta, 2nd with 2 setae and 1 aesthete, 3rd with 1 seta and 5 aesthetes, 4th and 5th each with 2 setae and 6 aesthetes. Fifth group belongs to 2nd segment. Segment 3 with 1 short middle seta, 1 long distal seta, and 8 aesthetes. Segments 4 to 8 each with 2 setae and 9 aesthetes; distal seta of 7th segment large. Segment 9 with 2 setae and 7 aesthetes. Segments 10 to 17 each with 1 middle seta, 1 distal seta, and 1 distal aesthete; 1 additional aesthete was found close to middle seta only in segment 12. Segment 18 with a small spiniform process in addition to 1 seta and 1 aesthete at anterodistal end. Segments 19 to 21 fused, with 1 aesthete on 19th, a small seta on 20th, and 1 long seta and 1 aesthete on 21st. Segments 22 and 23 separate, each with a posterodistal seta, an anterodistal seta, and an anterodistal aesthete. Segment 24 with a posterodistal and a anterodistal seta. Segment 25 with 5 setae and 2 aesthetes.

In 5th pair of legs (Fig.36-f), coxa and basis excluding inner lobe, of similar length. Inner lobe of right basis with wavy outer margin, extending straight distad, reaching about 1/3 way to distal end of 2nd endopodal segment. Inner lobe of left basis wide, extending distad, reaching halfway along medial margin of 1st endopodal segment. In both legs, endopod, basipod, and combined lengths of first 2 exopodal segments similar in length. Second endopodal segment with distolateral corner produced into a small toothlike process in right leg, while the same corner produced into a relatively large toothlike process in left leg. In both exopods, first 2 exopodal segments of similar length, and each only a little shorter than 3rd segment. In right exopod (Fig.36-g), medial projection of 2nd segment arising from about middle of segment, fingerlike, with short setules on distal half; terminal spine of 3rd segment approximately 1/3 length of segment and about as long as 1st outer spine of that segment; outer spines of first 2 segments and 2nd outer spine of 3rd segment all relatively small. All 3 outer spines of left exopod (Fig.36-h) similar in length, although 1st thicker than the others. Inner spine of 3rd segment small and slender; terminal spine approximately 2/3 length of segment itself. Third exopodal segment in each leg with 3 elevated pores along lateral margin.

Distribution. Neorhabdus latus has been recorded from between 27°43'N-34°02'N and between 12°21'W-19°09'W in the North Atlantic by Sars (1905, 1925), the eastern tropical Pacific by Wilson (1950), and the Izu region of Japan by Tanaka (1964). It has also been recorded (as *N. truncatus*) from the Malay Archipelago by Scott (1909) and from the Laccadive Sea of the Indian Ocean by Sewell (1932).

Neorhabdus latus is represented in the present study by 81 specimens including 78 females and 3 males found in 25 samples—3 taken in the Atlantic between 00°55'N and 39°22'N, 15 in the eastern and central Pacific between 20°20'S and 31°13'N, 5 in the western Pacific including the Malay Archipelago between 14°14'S and 25°37'N, and 2 in the central Indian Ocean at 17°07'S and 02°28'S, respectively. Based on these and the previous findings, *N. latus* can now be regarded as a circumglobal species.

Remarks. The brief original description of this species was based on a single female 6.65 mm long taken at 27°43'N, 18°28'W in the North Atlantic. The species was redescribed in detail by Sars (1925) from both female and male specimens taken at the type locality, and it is this redescription that for the first time made its identification possible. The specimens obtained in the present study seem to agree with the descriptions and figures given by Sars (1925), except that a small seta on the basis of Md figured by Sars was not found in the present study, and the 3rd exopodal segment of the male right P5 was figured by Sars as having 3 outer spines, instead of 2 as found in the present study.

Neorhabdus truncatus originally described by Scott (1909) and redescribed by Sewell (1932) is here considered a junior synonym of *N. latus* despite some differences as noted below. *Neorhabdus truncatus* (Scott 1909) was originally described from a single female 7 mm long from 2000-0 m at 03°58'S, 128°20'E in the Banda Sea. It is similar in body size, habitus, Md, Mxp, and P5 to *N. latus* found in the present study but differs from it and the 3 new species found in the study in morphological details of Mx2 and P1. According to Scott's (1909) figures, in *N. truncatus* each of the first 4 lobes of Mx2 has 2 spines, instead of 3+2+3+3 spines as found in all of the species observed in the present study, and the basis of P1 has a well-developed outer seta, which is absent in all species found in the present study. In all *Neorhabdus* species, however, the basis of P1 has a hooklike process on its posterior surface close to the lateral margin. It is therefore possible that Scott (1909) considered the hooklike process of the basis of P1 a broken outer seta. It is also possible, as suggested by Sewell (1932), that Scott (1909) overlooked the small spines of the Mx2. In the present study, a female specimen (8.16 mm long) of *N. latus* was found at the type locality of *N. truncatus*.

Sewell (1932) redescribed *N. truncatus* from female specimens 7.44 mm long taken in the Laccadive Sea of the Indian Ocean, which seem to be identical to *N. latus* as found in the present study, although there are some differences in details of A1 and Mx2. Sewell's description and figure of A1 show many pores on each segment, which seem to represent scars left by lost aesthetes. If this is true, Sewell's species has many more aesthetes on A1 than in *N. latus* as found in the present study. According to Sewell's figure of Mx2, the 3rd and 4th lobe of this appendage each have a short and 2 long spines, while these same lobes each have only 1 long and 2 very short spines in *N. latus* found in the present study.

Sewell (1947) recorded *N. truncatus* from the Arabian Sea based on a single female 9.66 mm long that was said to be identical in morphology to the specimens he had previously found in the Laccadive Sea except body size. This female is obviously too large to be referable to *N. latus* as described in this study. Sewell's specimen is similar in size to *N. subcapitaneus*, new species described below.

The specimens found in the present study showed a wide range of size variation. The female specimens ranged from 5.25-6.58 mm in prosome length and 6.75-8.91 mm in body length. However, in the absence of any detectable morphological differences, they are all regarded here as belonging to a single species.

Neorhabdus latus can be distinguished from the other species of the genus by the combination of the following features: In A2, inner marginal seta of last exopodal segment reduced to a vestige; maxillule without setae on 2nd and 3rd inner lobes; third lobe of Mx2 with a large and 2 very small spines.

Neorhabdus falciformis (Wolfenden 1911)
Figure 37

Hemirhabdus falciformis Wolfenden 1911, p. 310, text fig. 57.

Description of female. Prosome length, 4.66-5.25 mm; body length, 6.16-6.83 mm. Very similar in habitus (Fig.37-a) to *N. latus* as described above. Prosome approximately 3/4 length of body and about 3.1 times as long as urosome. Laterally, prosome typically with a length-depth ratio of 100:46. Cephalosome approximately 2/5 length of prosome. Urosome about 1/4 length of body. Dorsally, genital somite (Fig.37-d) about as long as wide, with only slightly curved lateral margins. Laterally, genital somite (Figs.37-b, c) with a high genital prominence, deeper than long, typically with a depth-length ratio of 100:85; its dorsal margin smoothly curved or only slightly wavy; genital prominence with posterior slope much steeper than anterior slope; posterior slope somewhat variable depending on the extent of telescoping by 2nd urosomal somite. Ventrally, genital operculum (Fig.37-e) typically with a width-length ratio of 100:67 and 1/2 as wide as somite itself as in *N. latus*. Genital field with about 5 pores on each side. Second urosomal somite about 1/2 length of genital somite and about 1.3 times length of 3rd urosomal somite. Anal segment about as long as 3rd urosomal somite as in *N. latus*. Combined lengths of anal segment and left caudal ramus only slightly longer than genital somite. Dorsally, caudal rami separate from anal segment by distinct lines; left ramus only slightly longer than right; their setation as in *N. latus*.

Antennule reaching about posterior end of caudal ramus, its setation as in *N. latus* except for the following details: Segments 6-12 each with 2 setae as in *N. latus*, but the number of aesthetes on these segments are 2, 1, 3, 1, 0, 0, 0, instead of 4, 2, 3, 2, 1, 1, 1, as found in *N. latus*. Antenna and Md as in *N. latus*. Maxillule (Fig.37-f) with 6 setae on 1st inner lobe, no setae on 2nd and 3rd inner lobes, 3 setae on outer lobe, 2 setae on basis, 6 setae on endopod, and 4 setae on exopod.

Maxilla (Fig.37-g) similar to that of *N. latus* but spines of first 4 lobes are elongate. First posterior spine of 1st lobe short, approximately 1/3 length of 2nd posterior spine. Second lobe with terminal spine slender, much longer than terminal spine of 1st lobe. First terminal spine of 3rd lobe about as long as but thicker than terminal spine of 2nd lobe; anterior spine completely covered with spinules, elongate, approximately 2/3 length of 1st terminal spine. First terminal spine of 4th lobe slender, about as long as anterior spine of 3rd lobe. Posterior spine of 5th lobe elongate, approximately 2/5 length of large, serrated terminal spine. Endopod with 2 relatively short, unequal setae. Maxilliped (Fig.37-h) as in *N. latus* except that setae on coxa are better developed. First to 5th legs similar to those of *N. latus*.

Description of male. Prosome length, 4.58 mm; body length, 6.08 mm. Similar in habitus to female except for urosomal segmentation. Prosome approximately 3/4 length of body and about 3.0 times length of urosome. Left, geniculate A1, with many aesthetes on segments 1-9, similar in setation to that of *N. latus* male. Other cephalosomal appendages and 1st to 4th legs as in female.

Fifth pair of legs similar to that of *N. latus* male but differs from it in the following details: 1) Inner lobe of right basis relatively short (Fig.37-i), with convex outer margin, 2) third segment of right exopod (Fig.37-j) with only 1 outer spine and 2 elevated pores, instead of 2 outer spines and 3 elevated pores as found in *N. latus,* and 3) third segment of left exopod (Fig.37-k) has 2 elevated pores, instead of 3 as found in *N. latus*, and its inner and terminal spines better developed.

Distribution. Neorhabdus falciformis seems to be a rare species. Only 11 specimens, 10 females and 1 male, were found in the present study. They occurred at 4 localities in the Pacific between 25°17'S and 15°00'N and between 159°06'E and 106°06'W and 2 localities in the Indian Ocean at

17°05'S, 67°05'E and 12°40'S, 64°01'E, respectively. According to these findings plus the type locality in the South Atlantic, this species has a circumglobal distribution along the low latitudes.

Remarks. Neorhabdus falciformis (Wolfenden 1911) was originally based on a single female 5.6 mm long (PL=4.2 mm) taken from 3000-0 m in the South Atlantic (about 23°S, 21°W) and has not been rediscovered. The species described above is identified with *N. falciformis* mainly because its urosome, A2, Mx1, masticatory edges of Md, and Mxp agree with those described and figured by Wolfenden (1911). However, the identification is provisional because all specimens found in the present study were taken in the Pacific and Indian Oceans and none were from the Atlantic, the type locality, and all of them were larger than Wolfenden's type specimen, although they are closer to Wolfenden's type specimen than to any other species of the genus. Maxilla was found to be very useful for species identification in *Neorhabdus,* but it was not described or figured by Wolfenden (1911).

Neorhabdus falciformis as described above is very close in morphological details to *N. latus* but can be distinguished from it by the following features: 1) Spines on the first 4 lobes of Mx2 relatively slender and long; the anterior spine of the 3rd lobe nearly 2/3 length of the 1st terminal spine and about as long as the 1st terminal spine of the 4th lobe, 2) both distal setae of the Mxp coxa well developed, 3) urosome relatively small, only about 1/4 length of the body in both the female and male, 4) in the female, the genital somite deeper than long, typically with a depth-length ratio of 100:85, 5) segments 6-12 of A1 with fewer aesthetes than in *N. latus,* and 6) in the male, the 3rd exopodal segment with only 2 elevated pores along the lateral margin in both 5th legs, instead of 3 as found in *N. latus,* and with 1 outer spine, instead of 2, in the right 5th leg.

Furthermore, *Neorhabdus latus* is considerably larger (6.75-8.91 mm long in female) than *N. falciformis* (6.16-6.83 mm long in female). The size difference is particularly pronounced when they occur together. For example, in a sample from the central North Pacific (15°00'N, 160°20'W), 4 females of *N. latus* were 7.91-8.25 mm long, while a female of *N. falciformis* was only 6.50 mm long. In a sample from the eastern tropical Pacific (14°58'S, 106°06'W), 1 female of *N. latus* was 8.00 mm long and 3 females of *N. falciformis* were 6.41-6.75 mm long.

Neorhabdus brevicornis, new species
Figure 38 and 39

Type material. Holotype female (USNM 287013) and 1 paratype female (USNM 287015) found in IKMT sample from 1100-0 m, SIO MV 72, station II-38, 13°51'S, 77°41'W in the eastern tropical Pacific, May 12-13, 1972. One paratype female (USNM 287016) found in IKMT sample from 2100-0 m, SIO Scan, station 117-10, 05°39'N, 83°04'W in the eastern tropical Pacific, December 31, 1969. Allotype male (USNM 287014) found in IKMT sample from 1300-0 mwo, SIO Krill II, station 16, 13°33'N, 98°36'W in the eastern tropical Pacific, June 10, 1974.

Description of female. Prosome length, 6.16-7.16 mm; body length, 8.33-9.83 mm. Prosome approximately 3/4 length of body and about 2.8 times as long as urosome. Laterally, prosome (Fig.38-a) typically with a length-depth ratio of 100:48. Cephalosome approximately 2/5 length of prosome; laterally, with a smoothly curved dorsal margin and rounded forehead produced ventrad into a rather small rostrum bearing 2 filaments. Posterior margin of prosome in lateral aspect broadly rounded (Fig.38-c). Dorsally, forehead broadly rounded; posterolateral corners of prosome more or less truncated (Fig.38-b).

Urosome slightly longer than 1/4 length of body. Dorsally, genital somite (Fig.38-b) only slightly wider than long, with smoothly curved lateral margins. Laterally, genital somite (Fig.38-c) a little deeper than long, with smoothly curved dorsal margin; genital prominence large with similar anterior and posterior slopes. Ventrally, genital operculum (Fig.38-d) typically with a width-length ratio of 100:63 and about 1/2 as wide as somite itself. Genital field with about 15 pores on each side and 2 pores on anterior side. Second urosomal somite approximately 3/5 length of genital somite and about 1.1 times length of 3rd urosomal somite. Anal segment approximately 4/5 length of 3rd urosomal somite. Combined lengths of anal segment and left caudal ramus about 1.2 times length of genital somite. Left caudal ramus distinctly longer than right, its 4th marginal seta about twice length of body. Dorsally, caudal rami each with 3 or 4 pores and with appendicular setae of similar size. Laterally, anal segment (Fig.38-c) with 2 pores, left caudal ramus with 3 pores.

Antennule reaching about posterior end of prosome. Four lobes of 1st segment (Fig.38-e) with 1+2+2+2 setae and 0+1+1+3 aesthetes. Segments 2 to 19 each with 2 setae (a middle and a distal) as in *N. latus* female. Number of aesthetes as follows: 2 on 2nd segment, 3 on 3rd, 2 on 4th, 3 on 5th, 2 on each of 6th to 9th, 1 on each of 10th to 19th. Aesthetes better developed than in *N. latus*.

Antenna, Md, Mx1, Mx2, and Mxp similar to those of *N. latus* except for the features described below. In A2 (Fig.38-f), inner lobe of 2nd endopodal segment with a small posterior appendicular seta and inner marginal seta of last exopodal segment well developed. Masticatory edge of Md (Figs.38-g, h) with well-developed basal spine and ventralmost tooth thick, initially pointing in a direction about 155° from masticatory edge and then curved dorsad, or toward the other teeth. In Mx1 (Fig.38-i), 1st inner lobe with 6 to 8 setae, 2nd inner lobe well developed with a seta, outer lobe with 4 or 5 setae, endopod with 5 or 6 setae, and exopod with 3 or 4 setae. In 3rd lobe of Mx2 (Fig.38-j), anterior spine completely covered with spinules and as long as 1st terminal spine; in 4th lobe, anterior spine also completely covered with spinules and only a little shorter than 1st terminal spine; 2 endopodal setae well developed, 1 of which is nearly as long as large, terminal spine of 6th lobe. Coxa of Mxp (Fig.38-k) with 1 small middle seta and 3 equally well developed distal setae; anteriorly, basis with 5 pores, 1st and 2nd endopodal segment each with 2 pores, 3rd and 4th each with 1 pore.

First to 5th legs similar to those of *N. latus* except for the features described below. In 1st leg (Figs.39-a, b), last 3 outer spines of similar size and approximately 1/2 as long as 1st. Anteriorly, coxa with 1 pore, basis with 3 pores, 1st endopodal segment with a group of about 11 pores close to distal margin, 2nd and 3rd each with a pore, 1st and 3rd exopodal segments with 2 and 1 pore, respectively. In P2 exopod (Fig.39-c), 2nd to 4th outer spines of similar size and slightly smaller than 1st; terminal spine slightly longer than 1/2 length of segment; anteriorly 1st and 3rd segment each with 3 pores. In P3 exopod (Fig.39-d), 2nd to 4th outer spines of similar size and only slightly smaller than 1st; anteriorly, 1st and 2nd segments each with 2 pores and 3rd with 13 pores. In P4 exopod (Fig.39-e), 1st outer spine slightly larger than 2nd; terminal spine approximately 1/3 length of segment; anteriorly, 1st and 2nd each with 2 pores, 3rd with 11 pores. In P5 (Fig.39-f), endopod approximately 1.5 times length of coxa and about 4/5 length of basipod; terminal spine of 3rd exopodal segment about 5/9 length of segment and about 3/4 length of inner spine of 2nd segment. Distolateral corner of 1st endopodal segment produced into a lobular process, that of 2nd pointed; distal 1/3 of inner seta in 1st and 2nd endopodal segments serrated with relatively large teeth; anteriorly, 3rd exopodal segment with 5 pores, endopodal segments without pores.

Description of male. Prosome length, 6.25-6.58 mm; body length, 8.50-9.08 mm. Similar in habitus to *N. latus*. Prosome approximately 3/4 length of body and about 2.8 times length of urosome. Antennule reaching about posterior end of prosome as in female. Right A1 as in female. In left, geniculated A1 (Fig.39-g), first 2 segments fused; setation of segments 1 to 9 as in right A1 and segments 10 to 25 similar in setation and morphological details to male of *N. latus*. Other cephalosomal appendages and 1st to 4th legs as in female.

Fifth pair of legs (Figs.39-h, i) similar to that of *N. latus* male except for the features described below. Distolateral corner of 1st endopodal segment produced into a lobular process, which in right leg is divided distally into 2 rounded tips. Distolateral corner of 2nd endopodal segment produced into a bifurcated spiniform process. In right exopod, first 3 outer spines relatively well developed, 4th outer spine small; medial projection of 2nd segment tapering into a point; 3rd segment with terminal spine slightly longer than 1/2 length of segment and 4 elevated pores along distal half of lateral margin. In left exopod, all 3 outer spines of similar size; 3rd segment with inner spine about as large as outer spine, 4 elevated pores along distal half of lateral margin, and terminal spine approximately 2/3 length of segment.

Etymology. The species name is from the Latin *brevi*, short, and *cornis*, horn, alluding to the relatively short antennule.

Distribution. This species is circumtropical. It is represented by a total of 27 specimens, 24 females and 3 males, found in 20 samples—4 taken in the Atlantic between 30°09'S and 26°33'N; 9 along the Pacific coast of the Americas between 21°26'S and 22°35'N; 2 in the eastern tropical Pacific at 17°46'S, 110°19'W and 14°58'S, 106°06'W, respectively; 2 in the central Pacific at 00°02'S, 165°44'W and 04°59'N, 164°14'W, respectively; 3 in the Indian Ocean between 17°05'S and 02°28'S.

Remarks. Neorhabdus brevicornis is similar in habitus to *N. latus* but can readily be distinguished from it by the differences described below. *Neorhabdus brevicornis* is larger than *N. latus.* Their size differences are highly pronounced when they occur together. For example, a female of *N. brevicornis* from off northwestern Africa (26°33'N, 21°27'W) is 9.08 mm long, while a female of *N. latus* from the same sample is only 7.58 mm long. A female of *N. brevicornis* from the eastern tropical Pacific (17°46'S, 110°19'W) is 8.83 mm long, while 8 females of *N. latus* from the same sample are 7.75-8.16 mm long. A female from the Indian Ocean (17°05'S, 67°05'E) is 9.00 mm long, while 13 females of *N. latus* are 7.25-7.41 mm long. The antennule reaches only to the posterior end of the prosome and has fewer aesthetes in *N. brevicornis,* while it reaches about the posterior end of the caudal ramus and has more aesthetes in *N. latus.* The inner marginal seta of the last exopodal segment of A2 is well developed in *N. brevicornis,* vestigial in *N. latus.* The maxillule has a seta on the 2nd inner lobe in *N. brevicornis,* no seta on the 2nd inner lobe in *N. latus.* In the 3rd lobe of Mx2, the anterior spine is about as long as the 1st terminal spine in *N. brevicornis,* it is very small in *N. latus.* The coxa of Mxp normally has 1+3 setae in *N. brevicornis,* it has 1+2 setae in *N. latus.*

Neorhabdus capitaneus, new species
Figures 40 and 41

Type material. Holotype female (USNM 287017), 1 paratype female (USNM 287018), 1 paratype female (USNM 287019), and 2 paratype females (USNM 287020) found in IKMT sample from 2000-0 m, SIO Southtow IV, MV 72-II, station 36, 13°47'S, 77°49'W in the eastern tropical Pacific, May 12, 1972.

Description of female. Prosome length, 8.50-9.83 mm; body length, 11.66-13.16 mm. Prosome approximately 3/4 length of body and about 2.8 times length of urosome. Laterally, prosome (Fig.40-a) typically with a length-depth ratio of 100:47. Cephalosome approximately 2/5 length of prosome, with a smoothly curved dorsal margin, tapering anteriorly into a narrowly rounded forehead, which is produced ventrad into a small rostrum bearing 2 filaments. Posterior margin of prosome rounded (Fig.40-b). Dorsally, forehead rounded and slightly produced anteriad along midsagittal line as in *N. latus.* Posterolateral corners of prosome somewhat truncated (Fig.40-c).

Urosome approximately 1/4 length of body. Dorsally, genital somite (Fig.40-c) with smoothly curved lateral margins, only slightly wider than long, typically with a length-width ratio of 96:100. Laterally, genital somite (Fig.40-b) deeper than long, typically with a length-depth ratio of 89:100, with a hump halfway along dorsal margin. Genital prominence high with similar anterior and posterior slopes. Ventrally, genital operculum (Fig.40-d) typically with a width-length ratio of 100:69 and about 4/9 as wide as somite itself. Genital field with about 6 pores on each side.

Second urosomal somite approximately 1/2 length of genital somite and about 1.2 times length of 3rd urosomal somite. Anal segment only slightly longer than 3rd urosomal somite. Combined lengths of anal segment and left caudal ramus about 1.1 times length of genital somite. Dorsally, right caudal ramus separate from anal segment by a clear line but left ramus fused with anal segment. Left caudal ramus only slightly longer than right, its dorsal appendicular seta a little longer than that of right ramus.

Antennule reaching about posterior end of anal segment. Four lobes of 1st segment with 1+2+2+2 setae and 0+1+20+16 aesthetes. Second to 19th segments each with 2 setae (1 middle and 1 distal). Normal number of aesthetes on each segment seems to be as follows: segments 2-4 each with 21, segments 5 and 6 each with 25, segment 7 with 20, segment 8 with 17, segment 9 with 8, segment 10 with 2, segments 11-19 each with 1 or 2.

Antenna as in *N. latus*, with inner marginal seta of last exopodal segment reduced to a vestige. In masticatory edge of Md (Figs.40-e, f), ventralmost tooth initially strongly diverging from and then curved toward the other teeth as in *N. brevicornis* described above. Mandibular palp as in *N. latus*. Maxillule (Fig.40-g) with 7 setae on 1st inner lobe; 2nd lobe well developed with 1 seta, outer lobe with 5 setae, basis with 2 or 3 setae, endopod with 6 setae, and exopod with 4 setae.

Maxilla (Fig.40-h) similar to that of *N. brevicornis* in that 2 large spines of 3rd lobe similar in size and endopodal setae well developed but differs from it in that spines on 2nd to 4th lobes slender and relatively long and large spines of 5th and 6th lobes with many more teeth. Coxa of Mxp (Fig.40-i) with 1 middle seta and 2 distal setae of similar length; anteriorly, no pores found on basis or endopodal segments.

First to 5th legs similar to those of *N. brevicornis* except for the details described below. In exopod of P1 (Fig.41-a), 1st outer spine extending well beyond base of following outer spine; spiniform process in front of each outer spine separated from outer spine by a relatively wide space; 1st endopodal segment with rounded distolateral corner; no pores found on anterior surface. In P2 exopod (Fig.41-b), outer spines relatively large; 1st, 3rd, and 4th of similar size and slightly smaller than 2nd; terminal spine of 3rd segment approximately 5/9 length of segment; anteriorly, 1st segment with 2 pores, 3rd with 1 pore. Anteriorly, P3 exopod (Fig.41-c) with 3 pores on 1st segment, 2 pores on 2nd, and 11 pores on 3rd; terminal spine of 3rd segment about 2/5 length of segment. In P4 exopod (Fig.41-d), 1st and 2nd segments each with 2 pores anteriorly, 3rd with 10 pores; terminal spine of 3rd segment approximately 2/5 length of segment. In P5 (Fig.41-e), endopod about 4/5 length of basipod; both 1st and 2nd endopodal segments with a pointed distolateral corner; inner setae of first 2 endopodal segments serrated distally with relatively fine teeth; terminal spine of 3rd exopodal segment about 5/9 length of segment and about 3/5 length of inner spine of 2nd segment; anteriorly, exopod with 3 pores on 3rd segment.

Description of male. Prosome length, 8.66 mm; body length, 12.00 mm. Similar in habitus to males of *N. latus* and *N. brevicornis*. Prosome almost 3/4 length of body and approximately 2.6 times length of urosome. Antennules reaching about posterior end of anal segment as in female. In morphological details including setation, right A1 similar to that of female. In left, geniculated A1, first 2 segments fused; segments 1 to 9 with many aesthetes as in female; segments 10 to 25 similar in morphological details including setation to males of *N. latus* and *N. brevicornis*. Antenna, Md, Mx1, Mx2, Mxp, and 1st to 4th legs as in female.

In 5th pair of legs (Figs.41-f, g), inner lobe of right basis with smoothly curved medial and lateral margins and pointing in a distomedial direction, reaching about 1/3 way to distal end of 2nd endopodal segment. Second endopodal segment in both legs with distolateral corner produced into a bifurcated spiniform process. Endopod a little shorter than basipod, which is about as long as combined lengths of first 2 exopodal segments. Third exopodal segment longer than 2nd and about as long as combined lengths of last 2 endopodal segments, with 2 outer spines in right leg and 1 outer and 1 inner spine in left leg in addition a terminal spine, and with 3 elevated pores along lateral margin. Terminal spine of 3rd exopodal segment approximately 2/5 length of segment in right leg and about 3/4 length of segment in left leg. Medial projection of 2nd exopodal segment of right leg fingerlike, tapering distally into a pointed process curved proximad.

Etymology. The specific name *capitaneus*, Latin meaning large, alludes to the large body size.

Distribution. The species is represented in the study only by 11 specimens including 10 females and 1 male found in 6 samples—1 taken at 47°26'N, 43°10'W in the North Atlantic, 4 in the eastern tropical Pacific between 13°47'S and 00°55'N and between the west coast of South America and 112°50'W, and 1 at 25°25'S, 38°14'E in Mozambique Channel of the Indian Ocean. These findings seem to indicate this is a rare species with a circumglobal distribution.

Remarks. This species is the largest of the *Neorhabdus* species found in the present study and is unique in having numerous aesthetes on the first 9 antennular segments. It is similar to *N. brevi-*

cornis and different from *N. latus* and *N. falciformis* in having a 2nd inner lobe on Mx1, 2 similarly long spines on the 3rd lobe of Mx2, and well-developed endopodal setae of Mx2. However, it can readily be distinguished from *N. brevicornis* by its long antennule with numerous aesthetes, a vestigial inner marginal seta of the last exopodal segment of A2, slender and relatively long spines of the maxillary lobes, and only 2 distal setae of the coxa of Mxp. The female of this species is also distinct in having a deep genital somite with a high genital prominence.

Neorhabdus subcapitaneus, new species
Figure 42

? *Hemirhabdus truncatus*, not of Scott; Sewell 1947, p. 181.

Type material. Holotype female (USNM 287021) found in IKMT sample from 3000-0 mwo, SIO Piquero V, station 7, 17°46'S, 110°19'W in the eastern tropical Pacific, March 29, 1969. One paratype female (USNM 287022) found in IKMT sample from 3500-0 mwo, SIO Piquero V, station 5, 09°00'S, 107°50'W in the eastern tropical Pacific, March 24, 1969.

Description of female. Prosome length, 7.33-8.50 mm; body length, 9.50-10.83 mm. Similar in habitus to *N. capitaneus* described above. Prosome approximately 3/4 length of body and about 3 times length of urosome. Laterally, prosome (Fig.42-a) typically with a length-depth ratio of 100:48. Cephalosome approximately 2/5 length of prosome, with a smoothly curved dorsal margin, terminated anteriorly with a narrowly rounded forehead produced ventrad into a small rostrum bearing 2 slender filaments. Posterior margin of prosome rounded. Dorsally, posterolateral corners of prosome (Fig.42-c) produced into obtuse angles covering about 1/5 length of genital somite.

Urosome approximately 1/4 length of body. Dorsally, genital somite (Fig.42-c) with almost straight lateral margins, wider than long, typically with a length-width ratio of 90:100. Laterally, genital somite (Fig.42-b) deeper than long, typically with a length-depth ratio of 82:100, with a smoothly curved dorsal margin or with a barely noticeable dorsal hump. Genital prominence high with similar anterior and posterior slopes. Ventrally, genital operculum (Fig.42-d) typically with a width-length ratio of 100:51 and about 4/9 as wide as somite itself. Genital field with about 7 pores on each side.

Second urosomal somite approximately 5/9 length of genital somite and about 1.2 times length of 3rd urosomal somite. Anal segment approximately 1.1 times length of 3rd urosomal somite. Combined lengths of anal segment and left caudal ramus about 1.2 times length of genital somite. Dorsally, right caudal ramus separate from anal segment, but left caudal ramus only partially separate from it. Left caudal ramus only slightly longer than right, its dorsal appendicular seta a little longer than that of right.

Antennule reaching about posterior end of 2nd urosomal somite. Four lobes of 1st segment with 1+2+2+2 setae and 0+1+4+6 aesthetes. Second to 19th segments each with 2 setae. Number of aesthetes found on each of first 12 segments is as follows: 2 on 2nd segment, 5 on 3rd, 3 on 4th, 5 on each of 5th and 6th, 4 on 7th, 6 on 8th, 5 on 9th, 3 on 10th, 4 on 11th, 3 on 12th. Aesthetes on distal part of A1 were not well preserved in all specimens examined.

Antenna (Fig.42-e) similar to that of *N. capitaneus* but inner marginal seta of last exopodal segment well developed. Mandible with ventralmost tooth strongly diverging from the other teeth as in *N. capitaneus*. Maxillule (Fig.42-f) with a well-developed 2nd inner lobe and similar in setation to that of *N. capitaneus*. Spines of maxillary lobes elongate; anterior spine of 3rd lobe nearly as long as 1st terminal spine as in *N. capitaneus*, but endopodal setae less well developed, less than 1/3 length of large spine on 6th lobe. Maxilliped (Fig.42-h) also similar to that of *N. capitaneus,* but 2 distal setae of coxa not similar in size and basis with 5 pores on anterior surface. All legs similar to those of *N. capitaneus* .

Etymology. The specific name, formed with the Latin prefix *sub*—under—refers to the apparent relationship of this new species with *N. capitaneus.*

Distribution. This species is represented only by 3 female specimens found in 3 samples—1 taken at 39°04'N, 68°05'W in the northwestern Atlantic and 2 at 17°46'S, 110°19'W and 09°00'S, 107°50'W, respectively, in the eastern tropical Pacific.

Remarks. The species is very close to *N. capitaneus* described above but can be distinguished from it by the following differences: 1) It is considerably smaller (9.50-10.83 mm long in female) than *N. capitaneus* (11.66-13.16 mm long in female), 2) antennule reaching about posterior end of 2nd urosomal somite, instead of posterior end of anal segment as in *N. capitaneus,* with no more than 6 aesthetes per segment; in *N. capitaneus,* as many as 25 aesthetes were found on a single segment, 3) inner marginal seta on last exopodal segment of A2 well developed, while this seta is vestigial in *N. capitaneus,* 4) endopodal setae of Mx2 relatively short, and 5) coxa of Mxp with 2 distal setae of strongly different size.

Sewell's (1947) record of *Neorhabdus truncatus* (Scott 1909) from the Arabian Sea was probably based on this species. As noted above, Sewell's specimen (9.66 mm long) is closer in body size to this species than to Scott's original specimen (7 mm long).

GENUS *PARAHETERORHABDUS* BRODSKY 1950

Paraheterorhabdus Brodsky 1950, p. 347.

Type by original designation: *Paraheterorhabdus robustus* (Farran 1908)

Diagnosis. Dorsally, forehead (Fig.43-e) with a midanterior tubercular process that continues into rostral base. Third marginal seta of left caudal ramus and 3rd and 4th marginal setae of right caudal ramus armed with spines in addition to normal setules. In 1st segment of A1 (Fig.43-h), 1st lobe with 1 seta, 2nd to 4th each with 2 setae and 1 aesthete. Left, geniculated A1 (Fig.45-c) of male with segments 22 and 23 fused. In A2 (Fig.43-i), first 2 exopodal segments without setae, last exopodal segment with a normally developed inner marginal seta, inner and outer lobes of 2nd endopodal segment each with a posterior appendicular seta in addition to 8 and 6 marginal setae, respectively. In Md (Figs.44-a, b), basis with a single seta, 1st endopodal segment with 3 setae, 2nd with 1 anterior appendicular seta in addition to 8 terminal setae, masticatory edge with a basal spine followed by 2 or 3 dorsal teeth, ventralmost tooth with a longitudinal rib along ventral margin, much larger than and widely separated from the other teeth, ventralmost tooth of left Md either similar to or much larger than that of right Md. Maxillule (Fig.44-c) with 2nd inner lobe bearing a single seta. In Mx2 (Fig.44-d), 1st lobe of small papillary form bearing 2 unequal setae; 2nd represented by a single seta arising directly from segment; 3rd represented by 1 seta and 1 spine, both arising directly from segment, distal half of spine covered with spinules; 4th elongate, with 1 seta and 2 subequal spines covered distally with spinules; 5th elongate, with 1 small seta and 1 long, saberlike spine, serrated along medial margin except for distal part; 6th similar to 5th but short, with 2 small setae and a long, saberlike spine, which is thinner than the same spine on 5th; endopod with 7 slender setae. Coxa of Mxp (Fig.44-e) with a long middle spine and 3 distal spines—1 thick anterior and 2 relatively thin posterior spines. Third exopodal segment of P3 (Fig.44-i) broad, with small terminal spine.

Additional description of female. Rostrum with 2 slender filaments (Fig.43-d). First pedigerous somite fully separated from cephalosome; 4th and 5th pedigerous somites completely fused. Laterally, cephalosome (Fig.43-a) with cervical groove halfway along dorsal margin; posterior margin of prosome rounded. Dorsally, midanterior tubercular process of forehead (Fig.43-e) relatively low; posterolateral corners of prosome (Fig.43-c) produced into obtusely angular processes. Genital somite (Fig.43-b) with large genital prominence. Left caudal ramus fused with or separated from anal segment. Fourth marginal seta of left ramus unarmed, much longer than the others, which are typically plumose.

Antennule short of reaching posterior end of caudal ramus. In A2 (Fig.43-i), endopod much longer than exopod. First 2 exopodal segments without setae; last exopodal segment with a well-developed inner marginal seta. Palp of Md (Fig.44-a) much shorter than mandibular blade. Exopod reaching to or beyond distal end of endopod. In maxillule (Fig.44-c), exopod nearly as long as or only a little shorter than rest of appendage; 1st inner lobe with 11 or 12 setae, outer lobe with 5 setae, endopod with 5 setae, exopod with 5 long setae plus a very small seta. In Mx2 (Fig.44-d), proximal seta of 1st lobe more than twice length of distal seta; seta on 2nd lobe more or less similar in length to distal seta of 1st lobe; spine of 3rd lobe longer than seta; proximal spine of 4th lobe shorter than distal spine and longer than seta; 6th lobe about 1/2 or less than 1/2 length of 5th. Endopod small, with 7 setae of similar length. In Mxp (Fig.44-e), 3 middle setae of basis on distal half of segment, band of spinules on anterior surface of basis reaching close to 3rd middle seta of segment.

First leg (Fig.44-f) with slender, long exopodal outer spines; anteriorly, only 3rd endopodal segment armed with short spinules; basis (Fig.44-g) with well-developed hooklike process on posterior surface close to lateral margin. Third endopodal segments of P2, P3, and P4 (Figs.44-h-j) with 7, 8, and 7 setae, respectively. Terminal spines of 3rd exopodal segments in P2-P4 relatively small.

Additional description of male. Similar in habitus to female except that urosome 5-segmented. Antennule similar in setation to that of female. Left, geniculated A1 (Fig.45-c) with first 2 segments fused, 1st postgeniculate segment formed by the fusion of segments 19-21, segments 22 and 23 fused.

In 5th pair of legs (Fig.45-e), 2nd endopodal segment in both legs with a seta. Only right basal lobe well developed. Right exopod (Fig.45-d) with a small outer spine on each segment in addition to a small terminal spine on 3rd segment; medial projection of 2nd segment more or less fingerlike or toothlike. Left exopod (Fig.45-1) also with a small outer spine on each segment in addition to an inner spine and a terminal spine on 3rd segment; distolateral corner of 2nd segment produced into a large lobe bearing outer spine.

Remarks. The taxon *Paraheterorhabdus* was erected by Brodsky (1950) as a subgenus with *Heterorhabdus robustus* Farran 1908 as type species. Brodsky originally characterized *Para-heterorhabdus* as having a stocky body, massive legs, and short distal spines on the exopods of the male P5. Not all species subsequently referred by Heptner (1972b) to *Paraheterorhabdus* share such characters. *Paraheterorhabdus vipera* (Giesbrecht 1889) and *P. medianus* (Park 1970), for example, have an elongate body and slender legs, as in *Heterorhabdus spinifrons* (Claus 1863), the type species of the genus *Heterorhabdus*. However, *Paraheterorhabdus* is a valid taxon clearly distinguishable from all the other genera of the family Heterorhabdidae in morphological details mainly of Mx2 and Mxp. In Mx2, 1st segment elongate, about as long as the rest of the appendage; first 3 lobes with 2, 1, and 2 relatively small setae/spines, respectively; 5th highly elongated, with a small seta and a long spine; 6th lobe with 2 small setae and a long spine; long spines of 5th and 6th lobes saberlike, slender, gradually tapering into sharp points, finely serrated along medial margins except for distal sections. Coxa of Mxp with well-developed middle and distal spines; middle spine elongate; anterior distal spine thicker but shorter than 1st posterior distal spine; 2nd posterior distal spine thinnest and shortest.

Paraheterorhabdus compactus (Sars 1900) differs from the other species of the genus in several respects and a new subgenus is erected for it here.

SUBGENUS *PARAHETERORHABDUS* BRODSKY 1950

Paraheterorhabdus Brodsky 1950, p. 347.

Type by original designation: *Paraheterorhabdus robustus* (Farran 1908)

Diagnosis. Left caudal ramus (Fig.43-c) distinctly longer than right and completely fused with anal segment. Fourth marginal seta of left caudal ramus slightly thicker than other marginal setae and about as long as body. Antennule extending beyond posterior end of prosome. Masticatory edge of left Md (Fig.44-a) with 2 subequal teeth next to basal spine and ventralmost tooth at least twice as long as that of right Md (Fig.44-b). First inner lobe of Mx1 (Fig.44-c) with 12 setae. In coxa of Mxp (Fig.44-e), middle spine about as long as segment and longer than distal spines. Hooklike posterior process of P1 basis (Fig.44-g) relatively large.

List of species of the subgenus *Paraheterorhabdus*

Paraheterorhabdus vipera (Giesbrecht 1889)
Paraheterorhabdus robustus (Farran 1908)
Paraheterorhabdus farrani (Brady 1918)
Paraheterorhabdus longispinus (Davis 1949)
Paraheterorhabdus medianus (Park 1970)
Paraheterorhabdus compactoides Heptner 1971
Paraheterorhabdus illgi, new species

Remarks. Paraheterorhabdus compactoides Heptner 1971 was not found in the present study. It was originally described from both female (3.87-4.00 mm long) and male (3.12-3.50 mm long) specimens collected from the deep depths in the Kuril-Kamchatka Trench and has not been rediscovered. It is similar in habitus and body size to *P. robustus* but can be distinguished from this and all other species of the subgenus by the massive and relatively short ventralmost tooth of the left Md and the relatively long outer spines of the P1 exopod.

Key to species of the subgenus *Paraheterorhabdus*

Females

1a. Dorsally, posterolateral corners of prosome rounded (Fig.52-b) . . . . *P. medianus*
1b. Dorsally, posterolateral corners of prosome (Fig.43-c) angular 2

2a. Laterally, 2nd urosomal somite (Fig.48-b) with a swelling along ventral margin . . . . *P. farrani*
2b. Laterally, 2nd urosomal somite without such swelling 3

3a. Laterally, posterior slope of genital prominence concave (Fig.46-e) . . . . *P. longispinus*
3b. Laterally, posterior slope of genital prominence convex or straight 4

4a. Genital somite about as long as wide (Fig.43-c) 5
4b. Genital somite distinctly longer than wide (Fig.54-b) . . . . *P. vipera*

5a. Ventrally, genital operculum (Fig.43-f) wider than long . . . . *P. robustus*
5b. Ventrally, genital operculum (Fig.50-f) longer than wide . . . . *P. illgi*, n. sp.

Males

1a. Basis of right P5 (Figs.53-h, 55-g) with wide, rectangular inner lobe 2
1b. Basis of right P5 (Fig.45-e) with conical inner lobe 3

2a. Exopod of right P5 (Fig.53-e) with wide bilobed distal end; 2nd lobe of Mx2
(Fig.52-i) with a long seta . . . . *P. medianus*
2b. Exopod of right P5 (Fig.55-e) tapering distally into a single process; 2nd lobe
of Mx2 (Fig.54-i) with a short seta . . . . *P. vipera*

3a. Fifth legs with very thick exopods (Fig.49-d); medial projection of 2nd
exopodal segment of right P5 (Fig.49-g) with a lobe on distal side . . . . *P. farrani*
3b. Fifth legs with relatively slender exopods (Fig.45-d); medial projection of 2nd
exopodal segment of right P5 without a lobe on distal side 4

4a. Medial projection of 2nd exopodal segment of right P5 (Fig.51-f) fingerlike . . . . *P. illgi*, n. sp.
4b. Medial projection of 2nd exopodal segment of right P5 (Fig.45-j) toothlike 5

5a. In 3rd exopodal segment of left P5 (Fig.45-d), inner spine much higher in
position than outer spine . . . . *P. robustus*
5b. In 3rd exopodal segment of left P5 (Fig.47-d), inner spine about on the
same level as outer spine . . . . *P. longispinus*

Paraheterorhabdus (Paraheterorhabdus) robustus (Farran 1908)
Figures 43-45

Heterorhabdus robustus Farran 1908, p. 65, pl. 7, figs. 1-10.—Sars 1925, p. 224, pl. 61.
—Wilson 1950, p. 240.—Tanaka 1964, p. 18, fig. 182.
Heterorhabdus (Paraheterorhabdus) robustus; Vervoort 1965, p. 117.

Description of female. Prosome length, 2.04-2.80 mm; body length, 2.92-4.16 mm. Body relatively stout (Fig.43-a). Prosome approximately 2/3 length of body and about 1.9 times length of urosome. Laterally, prosome deepest at beginning of 1st pedigerous somite, typically with a length-depth ratio of 100:47. Cephalosome approximately 4/9 length of prosome, with a cervical groove halfway along dorsal margin, tapering anteriorly into a narrowly rounded forehead, produced ventrad into a

73

conical rostrum bearing 2 slender filaments (Fig.43-d). Laterally, posterior end of prosome rounded (Fig.43-b). Dorsally, forehead (Fig.43-e) somewhat truncated, with a low midanterior tubercular process bearing suprafrontal sensillae. Posterolateral corners of prosome (Fig.43-c) produced into bluntly angular processes extending over anterior 1/6 length of genital somite.

Urosome comparatively long, about 1/3 length of body. Dorsally, genital somite (Fig.43-c) with smoothly curved, symmetrical lateral margins, widest at middle, and about as long as wide. Laterally, genital somite (Fig.43-b) about as deep as long, with large genital prominence, which has only a slightly bulging anterior slope and a nearly straight posterior slope; both slopes are similar in length. Ventrally, genital operculum (Fig.43-f) typically with a width-length ratio of 100:76. Genital field with 4 pores on each side. Caudal rami relatively long. Length ratios of 4 urosomal somites and left caudal ramus, 26.7 : 18.2 : 15.9 : 11.4 : 27.8 = 100. Dorsal appendicular seta of left caudal ramus longer than that of right. Fourth marginal seta of left ramus unarmed and about as long as body. Third marginal seta of left ramus and 3rd and 4th marginal setae of right armed with spines in addition to normal setules.

Antennule reaching about posterior end of 2nd urosomal somite. Four lobes of 1st segment of antennule (Fig.43-h) with 1+2+2+2 setae and 0+1+1+1 aesthetes. Second to 19th segments each with 2 setae and 1 aesthete. In antenna (Fig.43-i), marginal setae of coxa and basis well developed and furnished with normal setules. First endopodal segment long, approximately 1.5 times length of basipod and as long as exopod, with 2 relatively short inner marginal setae and an elevated pore proximal to marginal setae. First exopodal segment with distomedial corner produced into a tubercular process.

Palp of Md (Fig.44-a) about 1/2 length of masticatory blade including teeth; exopod reaching distal end of endopod. In masticatory edge of left Md, 1st tooth from basal spine tricuspid; 2nd bicuspid and about as long as 1st; 3rd, ventralmost tooth approximately 2.8 times length of the corresponding tooth on right Md (Fig.44-b). Edge between 2nd and 3rd teeth slightly convex. In masticatory edge of right Md (Fig.44-b), 1st tooth from basal spine shortest, bicuspid; 2nd tricuspid, longer than either 1st or 3rd; 3rd single-pointed; 4th, ventralmost tooth extending to same level as 3rd and separated from it by a smoothly hollow gap. Ventralmost teeth in both mandibles with a well-developed longitudinal rib along ventral margin.

First inner lobe of Mx1 (Fig.44-c) with 2 anterior and 2 posterior submarginal and 8 marginal setae and an elevated pore on posterior surface close to proximomedial corner. Seta of 2nd inner lobe long, extending beyond distal end of 1st inner lobe by 1/2 its length. Measured along medial margins, exopod about as long as rest of appendage.

In Mx2 (Fig.44-d), 1st segment almost as long as rest of appendage; short seta of 1st lobe, setae of 2nd and 3rd lobes similar in length. Spine of 3rd lobe approximately 1.8 times length of seta. In 4th lobe, seta about 5/9 length of short spine; long spine about 1.2 times length of short spine. Fifth lobe about as long as segment itself, with 2 pores on posterior surface; distal 1/4 length of its spine unserrated. Sixth lobe reaching less than halfway to distal end of 5th; distal 1/4 length of its spine unserrated. Endopodal setae extending beyond distal end of 5th lobe by about 1/3 their length.

In coxa of Mxp (Fig.44-e), middle marginal spine about as long as segment; anterior distal spine about 3/4 length of 1st posterior distal spine and about 1.5 times length of 2nd posterior distal spine. Basis approximately 1.2 times length of coxa when measured along medial margins, with a large, smoothly convex swelling on lateral margin close to proximal end; 1st middle seta arising about halfway along medial margin, 2nd middle seta closer to 1st than 3rd; 2nd and 3rd middle setae each followed by a large elevated pore; anterior band of spinules extending to base of 3rd middle seta; medial margin not fringed with a row of setules.

In 1st leg (Figs.44-f, g), combined lengths of first 2 exopodal segments shorter than basipod but longer than endopod. Outer spine of 1st exopodal segment reaching about base of following spine on 2nd segment, which is far short of reaching base of following spine on 3rd segment. Second outer spine of 3rd segment longer than 1st; terminal spine approximately 3/4 length of exopod. Distolateral corners of all 3 endopodal segments pointed.

Measured along medial margins, basipod of P2 (Fig.44-h) approximately 1.4 times length of endopod and about 1.3 times combined lengths of first 2 exopodal segments. Third exopodal segment about as long as basipod; its terminal spine approximately 3/10 length of segment itself. Outer spines of exopod gradually decrease in size in order from 1st to last. Distolateral corners of first 2 endopodal segments produced into spiniform processes. Anteriorly, first 2 exopodal segments each with a pore; 3rd with 4 pores.

Third leg (Fig.44-i) similar to P2 except that 3rd exopodal segment comparatively long and wide, approximately 1.2 times length of basipod, its distal inner setae rather poorly developed, terminal spine only 1/7 length of segment itself. Anteriorly, 2nd exopodal segment with 2 pores, 3rd with 9 pores.

Fourth leg (Fig.44-j) similar to P2 but basis with a small outer seta, basipod approximately 1.5 times length of endopod and about 1.4 times combined lengths of first 2 exopodal segments. Third exopodal segment about as long as basipod, its terminal spine about 3/10 length of segment itself. Anteriorly, 2nd exopodal segment with 2 pores and 3rd with 9 pores.

In P5 (Fig.45-a), basipod about as long as combined lengths of first 2 exopodal segments and approximately 1.2 times length of endopod. Third exopodal segment about as long as basipod, its terminal spine slightly longer than 1/2 length of segment. Inner spine of 2nd exopodal segment only slightly shorter than 3rd exopodal segment but slightly longer than endopod. Distolateral corners of first 2 endopodal segments produced into spiniform processes. Anteriorly, 2nd and 3rd exopodal segments with a pore close to each outer spine.

Description of male. Prosome length, 2.00-2.72 mm; body length, 2.92-3.96 mm. Prosome similar in shape to that of female, about 2/3 length of body and about twice as long as urosome. Laterally, posterior margin of prosome rounded as in female. Dorsally, posterolateral corners of prosome (Fig.45-b) bluntly angular, covering about 1/4 length of genital somite. Urosome relatively long, corresponding to about 1/3 length of body. Length ratios of 5 urosomal somites and left caudal ramus, 12.3 : 16.2 : 17.5 : 15.5 : 11.3 : 27.2 = 100. Left caudal ramus distinctly longer than right, its dorsal appendicular seta about twice as long as that of right. Fourth marginal seta of left ramus unarmed and about as long as body. Third marginal seta of left ramus and 3rd and 4th marginal setae of right ramus armed with conspicuous spines in addition to normal setules as in female.

Right A1 as in female in segmentation and setation. In left, geniculated A1 (Fig.45-c), segments 1 and 2, segments 19-21, and segments 22 and 23 fused; setation as in right A1 except that segment 18 with middle seta missing, segment 19 with only an aesthete, segment 20 without setae or aesthetes. Other cephalosomal appendages and 1st to 4th legs as in female.

In 5th pair of legs (Figs.45-d, e), coxa with 6-7 pores on anterior surface; inner lobe of right basis rounded distally, pointing distomediad and reaching close to distal end of 1st endopodal segment. Posteriorly, right basis (Fig.45-f) with a band of setules on medial margin. Left basis with a nearly straight inner margin fringed with setules. Left endopod slightly longer than right, each with inner marginal seta on 2nd segment. In right exopod (Figs.45-g, h), outer spines small; terminal spine of 3rd segment smaller than outer spine; medial projection of 2nd segment (Fig.45-j) toothlike, approximately 2/3 length of segment, closer to distal end of segment than to proximal end, bifurcated distally, with short setules along proximal margin; 3rd segment approximately 3/4 combined lengths of 1st and 2nd, with 7 pores on posterior surface and 2 elevated pores on outer margin—1 next to outer spine and another next to terminal spine (Fig.45-g).

In left exopod (Figs.45-j-l), 2nd segment produced laterad into a large conical process terminated with a small outer spine pointing in the same direction as the process; 3rd segment approximately 9/10 combined lengths of 1st and 2nd; posteriorly, 3 pores close to inner spine, 1 elevated pore close to outer spine, and another elevated pore close to terminal spine; its terminal spine about 1/2 as long as segment itself; outer spine small, located about 2/3 length of segment from proximal end; inner spine finely serrated, approximately 2/5 length of terminal spine and located about 1/3 length of segment from proximal end.

Distribution. Since its original description from the Atlantic Slope of Ireland by Farran (1908), this species has been recorded widely in the Atlantic from the Gulf of Guinea (Vervoort 1965) and the Gulf of Mexico (Park 1970) to the waters south of Iceland (64°15'N) (Jespersen 1940). In the Pacific, the species has been recorded in the eastern Pacific from off Baja California and Colombia, and from the Philippines by Wilson (1950) and from the Izu region of Japan by Tanaka (1964). As already mentioned by Vervoort (1965), Farran's (1929) record of this species from the Antarctic may actually pertain to *P. farrani* (Brady 1918).

This species is represented in the present study by 97 specimens including 76 females and 21 males from 33 samples—8 taken in the Atlantic between 33°47'S and 63°24'N, 23 taken throughout the Pacific including the Malay Archipelago between 41°26'S and 34°33'N, 1 at 35°00'S, 60°00'E in the Indian Ocean, and 1 at 25°25'S, 38°14'E in the Mozambique Channel. These findings plus the previous records indicate that *P. robustus* is a circumglobal species that is displaced by *P. longispinus* in the northern Pacific and by *P. farrani* in the Antarctic. In the North Atlantic, however, the species ranges to high latitudes, apparently in the absence of closely related species.

Remarks. The original species description by Farran (1908) is based on female specimens 3.5-4.0 mm long and male specimens 3.4-3.7 mm long obtained from the Irish Atlantic Slope. It was redescribed by Sars (1925) from North Atlantic specimens. Two closely related species have since been described, *P. farrani* from the Antarctic and *P. longispinus* from the northwestern Pacific. A new species, *P. illgi*, described below is also similar to *P. robustus*. All of these species have been carefully examined in the present study and the most reliable characters by which *P. robustus* can be distinguished from these close relatives are as follows: 1) Genital somite of the female about as long as deep, posterior slope of genital prominence straight or slightly bulging but never concave, genital operculum wider than long, 2) second urosomal somite of the female with straight ventral margin, 3) long saberlike spines of 5th and 6th lobes of Mx2 each with an unserrated distal section about 1/4 length of respective spine, 4) basis of Mxp with medial margin not fringed with setules, and 5) in male 5th pair of legs, inner spine of left 3rd exopodal segment finely serrated, approximately 2/5 length of terminal spine and located about 1/3 length of segment from proximal end; medial projection of right 2nd exopodal segment toothlike, with nearly straight proximal margin and smoothly bulging distal margin.

Tanaka (1964) recorded *P. robustus* from the Izu region of Japan based on females 3.83-4.75 mm long and males 3.45-4.90 mm long. As the largest female specimen of this species found in the present study was 4.16 mm long and the largest male specimen was 3.96 mm long, his large specimens probably are *P. longispinus*, which in the northwestern Pacific has a size range of 4.8-5.0 mm in the female and 4.6-4.8 mm in the male (Brodsky 1950, as *P. robustoides*).

Paraheterorhabdus (Paraheterorhabdus) longispinus (Davis 1949)
Figures 46 and 47

Heterorhabdus longispinus Davis 1949, p. 58, pl. 10, figs. 126, 127; pl. 11. figs. 128-137.
Heterorhabdus (Paraheterorhabdus) robustoides Brodsky 1950, p. 347, fig. 243.
Heterorhabdus (Paraheterorhabdus) robustus; Heptner 1971, p. 156, fig. 39.

Description of female. Prosome length, 3.00-3.50 mm; body length, 4.50-5.33 mm. Similar in habitus to *H. robustus* but much larger than that species. Prosome approximately 2/3 length of body and about 1.9 times length of urosome. Laterally, prosome (Fig.46-a) deepest at beginning of 1st pedigerous somite, typically with a length-depth ratio of 100:45. Cephalosome approximately 4/9 length of prosome, with smoothly curved dorsal margin indented by a cervical groove at middle and with narrowly rounded forehead produced ventrad into a small rostrum bearing 2 relatively long filaments. Posterior margin of prosome rounded. Dorsally, forehead somewhat truncated, with a low midanterior tubercular process bearing suprafrontal sensillae. Posterolateral corners of prosome produced into angular processes covering about 1/7 length of genital somite.

Urosome approximately 1/3 length of body. Length ratios of 4 urosomal somites and left caudal ramus, 28.7 : 19.6 : 15.3 : 10.7 : 25.7 = 100. Dorsally, genital somite (Fig.46-c) with smoothly curved lateral margins, widest about 1/3 length from anterior end, typically with a length-width ratio of 100:96. Laterally, genital somite (Fig.46-b) a little longer than deep, with large genital prominence. Anterior slope of genital prominence slightly bulging and much shorter than posterior slope, which is distinctly concave (Fig.46-e). Ventrally, genital operculum (Fig.46-d) typically with a width-length ratio of 100:84. Genital field with about 13 pores on each side. Left caudal ramus distinctly longer than right. Fourth marginal seta of left ramus unarmed and about as long as body. Third marginal seta of left ramus and 3rd and 4th marginal setae of right armed with conspicuous spines in addition to normal setules.

Antennule reaching about posterior end of 2nd urosomal somite. All cephalosomal appendages similar in setation and other morphological details to those of *H. robustus* except for some minor features as follows: 1) In masticatory edge of left Md (Fig.46-g), 1st tooth from basal spine bicuspid and a little shorter than 2nd, which is tricuspid; ventralmost tooth relatively thick and about 2.5 times length of corresponding tooth of right Md (Fig.46-f), 2) in Mx2 (Fig.46-h), short seta of 1st lobe longer than setae of 2nd and 3rd lobes; spine of 3rd lobe 1.7 times length of seta; in 4th lobe, seta about 7/10 length of short spine; fifth lobe longer than segment itself, and 3) in coxa of Mxp (Fig.47-a), middle marginal spine about as long as segment; 1st posterior distal spine approximately 1.5 times length of anterior distal spine; in basis, 3rd middle seta followed by 2 elevated pores, instead of just 1.

First to 5th legs also similar to those of *H. robustus* except for some minor differences found in P5 exopod (Fig.47-b). Third exopodal segment of P5 a little longer than combined lengths of first 2 exopodal segments. Terminal spine of 3rd segment approximately 1/2 length of segment. Inner spine of 2nd segment about 6/7 length of 3rd segment. Anteriorly, 3rd segment with 3 large pores— 2 close to 2nd outer spine and 1 close to 3rd.

Description of male. Prosome length, 3.08-3.33 mm; body length, 4.64-5.08 mm. Prosome similar in shape to that of female and approximately 2/3 length of body. Urosome and left A1 similar to those of the male of *H. robustus*. All other cephalosomal appendages and 1st to 4th legs as in female.

Fifth pair of legs (Figs.47-c, d) similar to that of the male of *H. robustus* except for the following details: Coxa with 8-10 pores on anterior surface and 3 pores on posterior surface; inner lobe of right basis pointing in an almost straight distal direction and reaching short of distal end of 1st endopodal segment; left endopod only slightly shorter than right; in right exopod (Figs.47-e, f), terminal spine relatively well developed; medial projection of 2nd segment relatively short, approximately 1/2 length of segment itself; 3rd segment with about 10 pores on posterior surface in addition to 2 elevated pores on outer margin; in left exopod (Figs.47-g, h), lateral conical process of 2nd segment extending along distal margin of segment and comparatively long; 3rd segment with both outer and inner spines on about the same level in posterior half of segment, terminal spine slightly longer than 1/2 length of segment.

Distribution. Davis's (1949) original specimens were obtained in the northeastern Pacific and Brodsky's (1950) (as *Heterorhabdus robustoides*) were taken in the northwestern Pacific and Bering Sea. Subsequently, Heptner (1971) recorded the species (as *Heterorhabdus robustus*) from the Kuril-Kamchatka Trench. As mentioned above, the large specimens from the Izu region of Japan that Tanaka (1964) placed in *H. robustus* seem to be *P. longispinus*. In the present study, 58 specimens including 45 females and 13 males that are referable to *P. longispinus* were found in 5 samples—4 samples taken along the west coast of California between 32°26′N and 40°35′N and 1 sample obtained at 52°20′N, 155°16′W in the Gulf of Alaska. According to these findings and the previous records, this species seems to be endemic to the northern Pacific including the Bering Sea and ranges south along the west coast of North America down to about 32°N and along the east coast of Japan down to about 35°N.

Remarks. The original description of this species was based on male specimens 4.3 mm long captured in 2 hauls, respectively, from 500-0 m, 100 miles off Cape Flattery and from 1100-0 m, 60 miles off Cape Flattery, Washington, in the northeastern Pacific. Apparently unaware of Davis (1949),

Brodsky (1950) described the same species as *H. (Paraheterorhabdus) robustoides* from both female (4.8-5.0 mm long) and male (4.6-4.8 mm long) specimens collected in the northwestern Pacific. Brodsky (1950) distinguished his new species from *P. robustus* by its large size, relatively short anal segment, long caudal rami, and some morphological details of the male P5. However, the caudal ramus of *P. longispinus* is comparatively short. It is 25.7% length of the urosome in this species, while it is 27.2% length of the urosome in *P. robustus*.

Heptner (1971) considered *P. robustoides* Brodsky 1950 synonymous with *P. robustus*, but the former is definitely synonymous with *P. longispinus* (Davis 1949). The female of *P. longispinus* can be distinguished from that of *P. robustus* by the following features: 1) Genital somite longer than wide and deep and widest in anterior half, 2) laterally, posterior slope of genital prominence longer than anterior slope and distinctly hollowed inward, 3) ventralmost tooth of left Md about 2.5 times, instead of 3 times, as long as corresponding tooth of right Md, and 4) in Mx2, seta of 4th lobe relatively long; 5th lobe longer than segment itself.

The most reliable character by which the male 5th pair of legs of *P. longispinus* can be distinguished from that of the *P. robustus* male is the position of the inner and outer spines of the 3rd segment of the left exopod. In the former, both spines are about on the same level in the distal half of the segment; in the latter, the inner spine is on the proximal half and the outer spine is on the distal half of the segment.

Paraheterorhabdus (Paraheterorhabdus) farrani (Brady 1918)
Figures 48 and 49

Heterorhabdus farrani Brady 1918, p. 27, pl. 4, figs. 10-18.—Vervoort 1951, p. 128,
    figs. 70-77.—Vervoort 1957, p. 134, fig. 126.
Heterorhabdus austrinus, not of Giesbrecht; Wolfenden 1905, p. 13, pl. 4, fig. 9.
Alloiorhabdus austrinus, not of Giesbrecht; Wolfenden 1911, p. 304, pl. 35, figs. 3, 4.

    Description of female. Prosome length, 2.36-2.80 mm; body length, 3.52-4.24 mm. Similar in habitus to *P. robustus*. Prosome approximately 2/3 length of body and about 1.8 times length of urosome. Laterally, prosome (Fig.48-a) deepest at junction between cephalosome and 1st pedigerous somite, typically with a length-depth ratio of 100:44. Cephalosome approximately 4/9 length of prosome, with a cervical groove halfway along dorsal margin, and with narrowly rounded forehead produced ventrad into a small rostrum bearing 2 slender filaments. Posterior margin of prosome (Fig.48-b) produced in a triangular form with rounded posterior end. Dorsally, forehead somewhat truncated, with a low midanterior tubercular process bearing suprafrontal sensillae. Posterolateral corners of prosome (Fig.48-c) produced into angular processes covering about 1/7 length of genital somite.

    Urosome approximately 1/3 length of body. First 3 urosomal somites fringed with spinules along posterior margins. Length ratios of 4 urosomal somites and left caudal ramus, 29.3 : 17.8 : 15.1 : 9.7 : 28.1 = 100. Dorsally, genital somite (Fig.48-c) with smoothly curved lateral margins, widest about 2/5 length of somite from anterior end, typically with a length-width ratio of 100:89. Laterally, genital somite (Figs.48-b, d) only a little longer than deep, with nearly straight dorsal margin and large genital prominence ventrally. Anterior slope of genital prominence slightly bulging and posterior slope concave and about 1.4 times length of anterior slope. Ventrally, genital operculum (Fig.48-e) typically with a length-width ratio of 100:86. Genital field with about 7 pores on each lateral side and 3 pores on anterior side.

    Laterally, 2nd urosomal somite with posterior half of ventral margin produced like a nodular swelling. Left caudal ramus extending beyond posterior end of right by 1/9 its length as measured along medial margin. Dorsal appendicular seta of left ramus almost twice as long as that of right. Fourth marginal seta of left ramus unarmed and about as long as body. Third marginal seta of left ramus and 3rd and 4th marginal setae of right conspicuously armed with spines in addition to normal setules.

    Antennule reaching about posterior end of 3rd urosomal somite. In setation and other morphological details, all cephalosomal appendages similar to those of *P. robustus* and *P. longispinus* described above except for some minor differences as follows: In Md (Figs.48-f, g), ventralmost tooth of left masticatory edge about 3 times length of corresponding tooth of right; in Mx2 (Fig.48-h), short seta of 1st lobe shorter than setae on 2nd and 3rd lobes; spine of 3rd lobe approximately 1.5 times length of seta; in 4th lobe, seta about 3/4 length of short spine; 5th lobe longer than segment itself; in coxa of Mxp (Fig.49-a), middle marginal spine much longer than segment; 1st posterior distal spine about as long as segment and approximately 1.6 times length of anterior distal spine; in basis, 1st middle seta less than 1/2 length of 2nd; 2nd middle seta followed by 1 elevated pore and 3rd by 2 elevated pores.

Swimming legs similar to those of *P. robustus* but outer spines of all exopods better developed. In P1 exopod (Fig.49-b), outer spine of 2nd segment nearly as long as segment itself; 1st outer spine of 3rd segment about as long as outer spine of 2nd segment and a little longer than 2nd outer spine. In P5 exopod (Fig.49-c), outer spine of 2nd segment almost twice as long as those of 3rd segments. Terminal spine of 3rd segment approximately 3/5 length of segment itself. Inner spine of 2nd segment approximately 7/10 length of 3rd segment. Anteriorly, 2nd and 3rd segments with a pore close to each outer spine.

Description of male. Prosome length, 2.32-2.56 mm; body length, 3.40-3.92 mm. Prosome approximately 7/10 length of body, about 1.9 times length of urosome, and similar in shape to that of female. Laterally, prosome typically with a length-depth ratio of 100:44. Urosome approximately 1/3 length of body. Length ratio of 5 urosomal somites and left caudal ramus, 15.3 : 16.2 : 17.1 : 14.4 : 9.0 : 28.0 = 100. First to 4th urosomal somites armed with spinules along posterior margins. Caudal rami including their setation as in female.

Geniculated A1 as in male of *P. robustus*. All other cephalosomal appendages and 1st to 4th legs as in female. Fifth pair of legs (Fig.49-d) stout, with stubby exopods. Inner lobe of right basis (Fig.49-e) large, pointing in a distomedial direction. Both endopods of similar length, rather poorly developed, each with inner marginal seta on 2nd segment. In right exopod (Figs.49-g, h), 1st segment relatively small; 2nd segment large, about 1.5 times as deep (including medial projection) as long; medial projection of 2nd segment approximately 3/5 length of segment, bifurcated distally, and with a prominent tubercular outgrowth on distal margin; 3rd segment about as long as 2nd, with about 17 pores on posterior surface (Fig.49-i) and 1 elevated pore at distal end (Fig.49-h), its terminal spine needlelike, approximately 1/3 length of segment. In left exopod (Figs.49-j, k), 2nd segment produced laterad into a large conical process terminating with outer spine. Posteriorly, 3rd segment (Fig.49-l) with 6 pores close to inner spine and 1 large elevated pore next to terminal spine; inner spine large, approximately 3/5 length of segment, coarsely serrated along proximal margin; terminal spine equal to 7/10 length of segment.

Distribution. This species occurs exclusively in the Southern Ocean. Brady's (1918) original description of the species was based on specimens collected at 2 localities (64°34'S, 127°08'E and 64°34'S, 117°01'E) in the Antarctic south of Australia. Subsequently, the species was found by Vervoort (1951) in the Atlantic sector of the Antarctic (between 66°05'S-66°58'S and between 11°00'W-16°03'W) and by Vervoort (1957) at numerous localities in the subantarctic and Antarctic waters of the Indian Ocean and western Pacific sectors (between 44°05'S-66°46'S and between 45°32'E-178°29'E).

In the present study, 95 specimens including 46 females and 49 males were found in 6 samples—2 from 32°27'S, 08°49'E and 33°47'S, 15°39'E, respectively, in the southeastern Atlantic, 1 from 60°42'S, 59°48'W in the Drake Passage, 2 from 35°10'S, 74°08'W and 46°55'S, 75°50'W, respectively, off Chilean Coast, and 1 from 35°00'S, 60°00'E in the Indian Ocean.

Remarks. The male (4.0 mm long) identified by Wolfenden (1905) as *Heterorhabdus austrinus* and the female (4.35 mm long) identified by Wolfenden (1911) as *Alloiorhabdus austrinus* from 2 localities (65°03'S, 85°04'E and 65°16'S, 80°28'E) in the Indian Ocean sector of the Antarctic were regarded by Vervoort (1951) as being *P. farrani.*

The female of this species can readily be distinguished from those of the congeners by the 2nd urosomal segment, which in lateral view has a conspicuous nodular swelling on the posterior half of the ventral margin, the comparatively long ventralmost tooth of the left mandible, the long spines on the coxa of Mxp, and the relatively well developed exopodal outer spines of all swimming legs. The male can readily be identified by its 5th pair of legs, which is very stubby in addition to the following characteristics: Medial projection of 2nd exopodal segment of right leg has a prominent tubercular outgrowth on distal margin; inner spine of 3rd exopodal segment of left leg is large and coarsely serrated along proximal margin.

Paraheterorhabdus (Paraheterorhabdus) illgi, new species
Figures 50 and 51

Heterorhabdus medianus (male only) Park 1970, p. 525, figs. 262-264.

Type material. Holotype female (USNM 287023) and 10 paratype females (USNM 287024) found in IKMT sample from 4000-0 mwo, SIO Lusiad VII, station 79, 01°10'N, 11°36'W in the tropical Atlantic, July 6, 1963.

Description of female. Prosome length, 1.96-2.48 mm; body length, 2.84-3.56 mm. Body moderately stout. Prosome approximately 7/10 length of body and about 2.2 times length of urosome. Laterally, prosome (Fig.50-a) deepest at junction between cephalosome and 1st pedigerous somite, typically with a length-depth ratio of 100:48. Cephalosome slightly shorter than 1/2 length of prosome, with a cervical groove halfway along dorsal margin and a narrowly rounded forehead produced ventrad into a small rostrum bearing 2 filaments. Posterior margin of prosome broadly rounded. Dorsally, forehead (Fig.50-d) broadly rounded, with a low midanterior tubercular process bearing suprafrontal sensillae. Posterolateral corners of prosome (Fig.50-c) produced into angular processes covering about 1/7 length of genital somite.

Urosome approximately 3/10 length of body. Length ratios of 4 urosomal somites and left caudal ramus, 30.1 : 18.7 : 14.5 : 10.8 : 25.9 = 100. First 3 urosomal somites armed with spinules along posterior margins. Dorsally, genital somite (Fig.50-c) with smoothly curved lateral margins, widest about 1/3 length of somite from anterior end, typically with a length-width ratio of 100:92. Laterally, genital somite (Figs.50-b, e) about as long as deep, produced ventrad into a roughly triangular genital prominence with anterior and posterior slopes of similar length; anterior slope nearly straight but posterior slope only slightly bulging. Ventrally, genital operculum (Fig.50-f) relatively small, only 1/5 as wide as somite itself, a little longer than wide, typically with a length-width ratio of 100:95. Genital field with 5 pores on each side. Left caudal ramus longer than right by 1/7 its length as measured along medial margin. Dorsal appendicular seta of left caudal ramus twice length of that of right. Fourth marginal seta of left ramus about as long as body. Third marginal seta of left ramus and 3rd and 4th marginal setae of right armed with spines in addition to normal setules.

Antennule reaching about posterior end of 3rd urosomal somite. In setation and other morphological details, all cephalosomal appendages similar to those of *P. robustus* except for the following features: 1) In left Md (Fig.50-h), ventralmost tooth about 2 times length of corresponding tooth on right Md (Fig.50-g); masticatory edge between 2nd and 3rd, ventralmost teeth flat with a small round swelling close to 3rd, 2) in Mx2 (Fig.50-i), seta of 3rd lobe almost as long as spine; seta of 4th lobe about 7/10 length of short spine, which is about 5/9 length of long spine; 5th lobe longer than segment itself; spine on 5th lobe with distal half unserrated; spine on 6th lobe with distal 48% length unserrated, and 3) in coxa of Mxp (Fig.50-j), 1st posterior distal seta approximately 1.5 times length of anterior distal spine; basis of Mxp fringed with setules along proximal half of medial margin and with a large tubercular outgrowth on lateral margin close to proximal end.

Swimming legs also similar to those of *P. robustus* except for some minor differences as described below. In P1 exopod (Fig.51-a), 1st outer spine short of reaching base of 2nd, which is reaching about base of 3rd. Terminal spine of P2 exopod (Fig.51-b) approximately 4/9 length of segment. Third exopodal segment of P3 (Fig.51-c) relatively elongate, with 7 pores on anterior surface, its terminal spine approximately 1/7 length of segment. Third exopodal segment of P4 (Fig.51-d) with 7 pores on anterior surface, its terminal spine about 2/5 length of segment. In P5 (Fig.51-e), basipod about as long as endopod and much shorter than combined lengths of first 2 exopodal segments. Terminal spine of 3rd exopodal segment approximately 2/3 length of segment. Anteriorly, 3rd exopodal segment with 3 large pores.

Description of male. Prosome length, 2.12 mm; body length, 3.00 mm. Prosome about 7/10 length of body and 2.3 times length of urosome. Laterally, prosome widest at junction between cephalosome and 1st pedigerous somite, typically with a length-width ratio of 100:43. Cephalosome with a cervical groove halfway along dorsal margin. Dorsally, forehead with a low midanterior tubercular process bearing suprafrontal sensillae. Posterolateral corners of prosome produced into angular processes covering about 1/4 length of genital somite.

Urosome approximately 3/10 length of body. Five urosomal somites and left caudal ramus with length ratios of 18.4 : 14.7 : 16.2 : 13.2 : 10.3 : 27.2 = 100. Left caudal ramus distinctly longer than right. Fourth marginal seta of left ramus about as long as body. Third marginal seta of left ramus and 3rd and 4th marginal setae of right armed with spines in addition to regular setules.

Antennule reaching about posterior end of 3rd urosomal somite. In geniculated A1, first 2 segments fused, segments 19-21 fused, segments 22 and 23 fused, and similar in setation to geniculated A1 of *P. robustus* male. All other cephalosomal appendages and 1st to 4th legs as in female.

Fifth pair of legs including basal inner lobes (Figs.51-g, h) similar to that of *P. robustus* male but differs from it in the following aspects: Exopods (Fig.51-f) relatively slender; in right exopod, medial projection of 2nd segment elongate, fingerlike, nearly as long as segment, and slightly curved proximad; terminal spine of 3rd segment relatively well developed; in left exopod (Figs.51-i, j), lateral conical process of 2nd segment relatively small; inner and outer spines of 3rd segment located on the same level in distal half of segment; terminal spine of 3rd segment a little shorter than 1/2 length of segment.

Etymology. This species is named for Professor Paul L. Illg. It was his teaching and inspiration that led me acquire a lifelong interest in copepod taxonomy.

Distribution. The species was represented by a total of 53 specimens including 52 females and 1 male found in 16 samples—5 taken in the Atlantic between 33°47'S and 01°25'N, 6 in the eastern Pacific between 30°46'S and 22°35'N, 2 at 00°02'S, 165°44'W and 30°40'N, 154°55'W, respectively, in the central Pacific, 1 at 09°25'S, 159°06'E in the western Pacific, and 2 at 35°00'S, 60°00'E and 02°28'S, 86°23'E, respectively, in the Indian Ocean. These findings demonstrate that the species has a circumglobal distribution.

Remarks. The species can readily be distinguished from the congeners by the relatively short ventralmost tooth of the left Md, the 3rd lobe of Mx2 with a seta and spine of similar length, the 4th lobe with 2 spines of greatly different length, spines of the 5th and 6th lobes with long unserrated distal sections, the coxa of Mxp with a relatively long 1st posterior distal seta, and the basis of Mxp with a proximal medial margin fringed with setules. The female is very characteristic, having a small and narrow genital operculum and the male having an elongate and slightly curved medial projection on the 2nd segment of the right exopod. The male I (Park 1970, p.525) previously described as belonging to *Heterorhabdus medianus* is referable to this new species.

Paraheterorhabdus (Paraheterorhabdus) medianus (Park 1970)
Figures 52 and 53

Heterorhabdus medianus (female only) Park 1970, p. 525, figs. 253-264.
Heterochaeta vipera (male only); Giesbrecht 1892, p. 373, pl. 20, figs. 6, 32, 33.

Description of female. Prosome length, 1.64-1.88 mm; body length, 2.48-2.88 mm. Body (Fig.52-a) relatively elongate. Prosome approximately 2/3 length of body and about 1.9 times length of urosome. Laterally, prosome typically with a length-depth ratio of 100:41. Cephalosome approximately 1/2 as long as prosome, tapering into a rather narrowly rounded forehead produced ventrad into a small rostrum bearing 2 slender filaments (Fig.52-c), with a cervical groove halfway along dorsal margin. Prosome with a broadly rounded posterior margin (Fig.52-e). Dorsally, forehead (Fig.52-d) truncate, with a low midanterior tubercular process bearing suprafrontal sensillae. Posterolateral corners of prosome (Fig.52-b) rounded and only slightly produced posteriad.

Urosome approximately 1/3 length of body. Length ratios of 4 urosomal somites and left caudal ramus, 32.0 : 18.0 : 15.7 : 10.5 : 23.8 = 100. Dorsally, genital somite (Fig.52-b) with smoothly curved lateral margins, widest 2/3 length of somite from anterior end, typically with a length-width ratio of 100:76. Laterally, genital somite (Fig.52-e) longer than deep, typically with a length-depth ratio of 100:84. Genital prominence with posterior slope extending to posterior end of somite by merging imperceptibly into ventral margin of somite, nearly parallel to dorsal margin of somite, and about 1.4 times length of anterior slope. Ventrally, genital operculum (Fig.52-f) close to anterior end of somite, typically with a width-length ratio of 100:74, about 3/5 as wide as somite itself. Genital field with 5 or 6 pores on each side. Left caudal ramus distinctly longer than right, with dorsal appendicular seta about 1.5 times length of that of right ramus. Fourth marginal seta of left caudal ramus about as long as body.

Antennule reaching about posterior end of 3rd urosomal somite. All cephalosomal appendages similar to those of *P. robustus* except for the features described below. Ventralmost tooth of left Md (Fig.52-g) approximately 2.3 times length of corresponding tooth of right Md (Fig.52-h). In Mx2 (Fig.52-i), setae and spines on 2nd and 3rd lobes relatively well developed; 5th lobe longer than segment itself. In Mxp (Fig.53-a), 1st posterior distal spine about twice as long as anterior distal spine; 2nd posterior distal seta about as long as anterior distal spine; basis with relatively acute lateral swelling and fringed with a row of setules along proximal half of medial margin. Legs relatively slender. In P1 exopod (Fig.53-b), 1st outer spine a little short of reaching base of 2nd; last 3 outer spines of similar length. In P5 exopod (Fig.53-c), inner spine of 2nd segment longer than 3rd segment; terminal spine of 3rd segment approximately 5/9 length of segment.

Description of male. Prosome length, 1.68-1.88 mm; body length, 2.60-2.84 mm. Body elongate; prosome approximately 2/3 length of body and 2 times length of urosome. Dorsally, posterolateral corners of prosome (Fig.53-d) rounded as in female. Urosome approximately 1/3 length of body, its 5 segments and left caudal ramus with length ratios of 12.4 : 18.8 : 18.8 : 15.9 : 9.4 : 24.7 = 100. Left caudal ramus distinctly longer than right; setation in both rami as in female.

Left, geniculate A1 similar to that of *P. robustus* male. All other cephalosomal appendages and 1st to 4th legs as in female. In 5th pair of legs (Fig.53-e), inner lobe of right basis very large, oblong, protruding medially, with proximal margin fringed with setules (Fig.53-h). Left basis with nearly straight inner margin partially fringed with setules. Left endopod slightly longer than right. In right exopod (Fig.53-g), outer spines of first 2 segments relatively small, that of 3rd well developed and close to distal end of segment; medial projection of 2nd segment arising from proximal end of medial margin, rounded distally and about 1/2 length of segment; 3rd segment about as long as combined lengths of first 2 segments, with distolateral corner produced into a large angular process bearing 3 pores on its posterior surface, and with an large elevated pore next to terminal spine; terminal spine relatively well developed, approximately 2/5 length of segment.

In left exopod (Figs.53-e, f) 2nd segment with lateral margin produced into a conical process terminating with outer spine; 3rd segment produced distally into a conical process tipped with an elevated pore; inner spine of 3rd segment close to proximal end of segment and outer spine close to distal end; terminal spine a little shorter than segment.

Distribution. The specimens from which the species was originally described came from the Caribbean Sea and Gulf of Mexico. The male described by Giesbrecht (1892) as belonging to *P. vipera*, which was found in this study to be referable to *P. medianus*, was taken in the eastern tropical Pacific. In the present study 37 specimens including 26 females and 11 males were found in 20 samples—2 taken at 18°44'N, 10°15'W and 33°47'S, 15°39'E, respectively, in the Atlantic; 8 between 14°58'S and 29°35'N in the eastern Pacific; 8 between 24°33'S and 31°06'N in the central Pacific; 1 at 09°25'S, 159°06'E in the western Pacific; 1 at 02°28'S, 86°23'E in the Indian Ocean. These findings plus previous records establish its geographic range extending into all three major oceans.

Remarks. This species was originally described by Park (1970) from both female and male specimens collected from the Caribbean Sea and Gulf of Mexico. The male originally described by Park (1970) was found in this study to be referable to *P. illgi*, a new species described above. The correct male of *P. medianus* described above seems to be identical to the male Giesbrecht (1892) referred to *Heterochaeta vipera* Giesbrecht 1889 = *Paraheterorhabdus vipera*.

The species can be distinguished from the congeners by a relatively slender body, long urosome, and rounded posterolateral corners of the prosome in dorsal view. It is most closely related to *P. vipera* described below but can be distinguished from this by the seta of the 2nd lobe of Mx2 and the 1st posterior distal spine of the Mxp coxa, both of which are relatively well developed . The male of this species is further characterized by the 5th pair of legs, in which the inner lobe of the right basis is very large and oblong and the 3rd exopodal segment of the right leg with the distolateral corner produced as a large, conspicuous conical process.

Paraheterorhabdus (Paraheterorhabdus) vipera (Giesbrecht 1889)
Figures 54 and 55

Heterochaeta vipera (female only) Giesbrecht 1889, p. 812.—Giesbrecht 1892, p. 373, pl. 20,
    figs. 5, 12, 13, 18, 20, 27; pl. 39, fig. 41.
Heterorhabdus vipera (female only); Giesbrecht 1898, p. 116.—Wolfenden 1911, p. 303.—
    Farran 1926, p. 284.—Sewell 1932, p. 300.—Tanaka 1964, p. 17, fig. 181.
Heterorhabdus vipera; Park 1970, p. 523, figs. 245-252.
Heterorhabdus tenuis Tanaka 1964, p. 20, fig. 183.

Description of female. Prosome length, 1.60-2.04 mm; body length, 2.52-3.16 mm. Body slender (Fig.54-a). Prosome approximately 2/3 length of body and about 1.7 times length of urosome. Laterally, prosome typically with a length-depth ratio of 100:36; cephalosome approximately 4/9 length of prosome, with cervical groove halfway along dorsal margin; forehead (Fig.54-c) produced ventrad into a relatively long rostrum bearing 2 slender filaments; prosome produced posteriad into a roughly triangular process with rounded distal end (Fig.54-e). Dorsally, forehead (Fig.54-d) truncate, with a low midanterior tubercular process bearing suprafrontal sensillae; posterolateral corners of prosome (Fig.54-b) produced into angular processes covering about 1/7 length of genital somite.

Urosome almost 2/5 length of body. Length ratios of 4 urosomal somites and left caudal ramus, 30.2 : 18.6 : 16.3 : 11.1 : 23.8 = 100. Posterior margins finely serrated in first 3 urosomal somites. Dorsally, genital somite (Fig.54-b) with smoothly curved lateral margins, widest at middle, typically with a length-width ratio of 100:75. Laterally, genital somite (Fig.54-e) with a high genital prominent, typically with a length-depth ratio of 100:84. Posterior slope of genital prominent extending down to posterior end of somite, about 1.5 times length of anterior slope, only slightly bulging. Ventrally, genital operculum (Fig.54-f) about at middle of somite, typically with a width-length ratio of 100:90, about 1/2 as wide as somite itself. Genital field with 4 pores on each side. Left caudal ramus distinctly longer than right. Dorsal appendicular setae of both rami of similar length.

Antennule reaching about posterior end of 3rd urosomal somite. In morphological details, all cephalosomal appendages similar to those of *P. robustus* with the following exceptions: 1) In both mandibles, ventralmost teeth of masticatory edges relatively short, that of left Md (Fig.54-h) about 3 times length of corresponding tooth of right Md (Fig.54-g), 2) in Mx2 (Fig.54-i), seta on 2nd lobe relatively thick but short, densely bipectinate with stiff setules; 1st spine of 4th lobe equal to 65% length of 2nd spine; 5th lobe shorter than segment itself, and 3) in Mxp (Fig.54-j), middle marginal spine of coxa shorter than segment itself; 1st posterior distal spine only a little longer than anterior distal spine; 2nd posterior distal seta approximately 1/2 length of anterior distal spine; basis armed with a row of setules along proximal half of medial margin.

Swimming legs relatively slender but otherwise similar to those of *P. robustus* except for the following features: In P1 exopod (Fig.55-a), 1st outer spine far short of reaching base of following spine; in P5 (Fig.55-b), 3rd exopodal segment a little shorter than basipod; its terminal spine approximately 3/5 length of segment itself; inner spine of 2nd exopodal segment about as long as combined lengths of first 2 exopodal segments and approximately 1.2 times length of 3rd exopodal segment; inner marginal setae of first 2 endopodal segments armed with long setules on proximal half and short setules on distal half.

Description of male. Prosome length, 1.74-1.96 mm; body length, 2.92-3.04 mm. Body slender. Prosome approximately 2/3 length of body and about 1.7 times length of urosome. Dorsally, posterolateral corner of prosome (Fig.55-c) produced into angular process covering about 1/4 length of genital somite. Urosome approximately 1/3 length of body, its 5 segments and left caudal ramus with length ratios of 12.9 : 17.8 : 18.4 : 15.3 : 10.4 : 25.2 = 100. Left caudal ramus distinctly longer than right; dorsal appendicular setae of both rami of similar length; marginal setae of rami as in female.

Left, geniculate A1 similar in setation to *P. robustus.* In morphological details, all cephalosomal and 1st to 4th legs as in female.

In 5th pair of legs (Fig.55-d), inner lobe of right basis wide and relatively short, extending in a distomedial direction, with proximal margin fringed with setules (Fig.55-g). Left basis (Fig.55-h) with straight inner margin partially fringed with setules. In right exopod (Figs.55-e, f), outer spines relatively small; outer spine of 3rd segment in proximal half of segment and terminal spine very small; medial projection of 2nd segment digitiform and about 1/2 length of segment. In left exopod (Fig.55-i), 2nd segment with only moderately produced lateral margin bearing a small outer spine (Fig.55-j); 3rd segment with inner spine about halfway along medial margin and outer spine about 3/5 length of segment from proximal end; terminal spine approximately 4/5 length of segment.

Distribution. This species was originally described from the eastern tropical Pacific (between 99°-132°W and between 03°S-14°N) and has since been recorded from the Izu region of Japan by Tanaka (1964), the Indian Ocean by Sewell (1932), the southern Atlantic by Wolfenden (1911), and the Bay of Biscay by Farran (1926).

In the present study, the species is represented by 29 specimens including 24 females and 5 males found in 15 samples—1 taken in the South Atlantic at 18°44'S, 10°15'W; 8 in the eastern Pacific between 14°46'S and 31°38'N; 1 in the central Pacific at 10°05'N, 161°52'W; 3 in the western Pacific including the East China Sea between 09°25'S and 34°33'N; and 2 in the Indian Ocean at 02°28'S, 86°23'E and 25°25'S, 38°14'E (in the Mozambique Channel), respectively. According to these findings and the previous records, this species has a circumglobal distribution.

Remarks. The original description of this species was based on both female and male specimens, 2.8 mm and 2.6 mm long, respectively. The male that Giesbrecht (1892) described as *Heterochaeta vipera* was, however, found in this study to be referable to *P. medianus* Park 1970. The correct male of *P. vipera* was described for the first time by Park (1970). Tanaka (1964) described a new species, *Heterorhabdus tenuis,* based on a single male specimen obtained in Suruga Bay of Japan, which seems to be identical to the male here referred to *P. vipera.*

Paraheterorhabdus vipera resembles closely *P. medianus* but can be distinguished from it by the following characters: 1) Dorsally, posterolateral corners of prosome angular, 2) in Mx2, seta on 2nd lobe thick but short and densely bipectinate with stiff setules, and 3) in Mxp coxa, middle marginal spine shorter than segment, 1st posterior distal spine only a little longer than anterior distal spine, 2nd posterior distal seta about 1/2 length of anterior distal spine. The male is further characterized by the 5th pair of legs, in which the inner lobe of the right basis wide but relatively short; the medial projection of the 2nd segment of the right exopod fingerlike; the 3rd segment of the right exopod tapering distally, tipped with a small terminal spine, and with an outer spine close to the proximal end.

SUBGENUS *ANTIRHABDUS*, NEW SUBGENUS

Type by monotypy: *Paraheterorhabdus compactus* (Sars 1900)

Diagnosis. Left caudal ramus (Fig.56-c) only slightly longer than right and fully separated from anal segment. Fourth marginal seta of left caudal ramus unarmed, longer than other marginal setae, and about as long as urosome. Antennule a little short of reaching posterior end of prosome. Masticatory edge of left Md (Fig.57-a) with one small and 2 long teeth next to basal spine and ventral-most tooth similar in size and shape to that of right Md (Fig.57-b). First inner lobe of Mx1 (Fig.57-c) with 11 setae. In coxa of Mxp (Fig.57-e), middle marginal spine much shorter than segment and about as long as 1st posterior distal spine. Posterior hooklike process of P1 basis (Fig.57-f) elongate, taper-ing into a sharp point and only slightly curved.

Etymology. The subgeneric name is formed from the Greek *anti*, like, and *rhabdus,* rod, allud-ing to the nearly symmetrical caudal rami.

Paraheterorhabdus (Antirhabdus) compactus (Sars 1900)
Figures 56-58

Heterochaeta compacta Sars 1900, p. 83, pls. 24, 25.
Heterorhabdus compactus; Sars 1925, p. 226, pl. 62, figs. 1-8.—Farran 1929, p. 267.—
    Brodsky 1950, p. 348, fig. 244.—Tanaka 1964, p. 21, fig. 184.—Park 1970, p. 477.—
    Heptner 1971, p. 154, fig. 38.

Description of female. Prosome length, 1.72-2.48 mm; body length, 2.24-3.32 mm. Prosome almost 4/5 length of body and about 3 times length of urosome in small individuals (BL=2.52 mm), about 3/4 length of body and about 2.8 times length of urosome in large individuals (BL=3.08 mm). Laterally, prosome (Fig.56-a) deepest at junction between cephalosome and 1st pedigerous somite, typically with a length-depth ratio of 100:50. Cephalosome approximately 4/9 length of prosome, with a cervical groove halfway along dorsal margin, tapering anteriorly into a narrowly rounded fore-head, which is produced ventrad into a small rostrum bearing 2 relatively short filaments. Last pro-somal somite (Fig.56-b) somewhat triangular, with rounded posterior end. Dorsally, forehead (Fig.56-d) broadly rounded, with a low midanterior tubercular process bearing suprafrontal sensillae and continuing into rostrum. Posterolateral corners of prosome (Fig.56-c) produced posteriad into short angular processes covering 1/6 length of genital somite.

Urosome relatively short, 25% length of body in small individual (BL=2.52 mm) and 27% as long as body in large individual (BL=3.08 mm). Dorsally, genital somite with symmetrically swollen lateral margins, slightly wider than long in small individuals (Fig.56-c), but slightly longer than wide in large individuals (Fig.56-h). Laterally, genital somite (Fig.56-f) with smoothly curved ventral mar-gin, without a delimited genital prominence; as long as deep in small individuals (Fig.56-b) and slightly longer than deep in large individuals (Fig.56-g). Ventrally, genital operculum (Fig.56-e) typ-ically with a width-length ratio of 100:58 and about 4/9 as wide as somite itself. Genital field with 4 pores on each side. Length ratios of 4 urosomal somites and left caudal ramus are 38.2 : 17.8 : 14.0 : 8.9 : 21.1 = 100 in small individual (BL=2.52 mm) and 37.2 : 16.6 : 14.1 : 10.3 : 21.8 = 100 in large individual (BL=3.08 mm). Dorsally, anal segment with 2 pores on right side and 1 pore on left side; each caudal ramus with a pore close to anterior end. Dorsal appendicular setae of caudal rami similar in size. Fourth marginal seta of left caudal ramus slightly thinner than 3rd and about as long as uro-some. Third marginal seta of left caudal ramus and 3rd and 4th marginal setae of right ramus about 2/3 length of 4th marginal seta of left ramus and conspicuously armed with small spines in addition to regular setules.

Antennule a little short of reaching posterior end of prosome. Four lobes of 1st segment with 1+2+2+2 setae and 0+1+1+1 aesthetes. Second to 19th segments each with 2 setae and an aesthete. In A2 (Fig.56-i), marginal setae of coxa and basis all well developed, about as long as basipod; 2 mar-ginal setae of basis armed with stiff setules radiating nearly at right angles. First endopodal segment longer than basipod, nearly as long as exopod, and with 2 small unequal inner marginal setae and an elevated pore on inner margin proximal to marginal setae. Second endopodal segment with 8 (marginal) + 1 (posterior appendicular) setae on inner lobe and 6+1 setae on outer lobe. Exopod without setae on first 2 segments; inner marginal seta of last segment well developed.

In mandible (Fig.57-a), palp approximately 3/5 length of masticatory blade; endopod small, reaching only middle of exopod. In left masticatory edge, 1st tooth from basal spine about 1/2 length of 2nd. Second and 3rd teeth bicuspid and of equal length. Ventralmost tooth separated from other teeth by a smoothly curved concavity. Right masticatory edge (Fig.57-b) nearly symmetrical with left, except that first 2 teeth from basal spine bicuspid; 2nd longer than 1st; 3rd spiniform, terminated with a sharp point, about as long as 1st.

First inner lobe of Mx1 (Fig.57-c) with 2 anterior setae plus 9 marginal setae furnished with fine, short setules; 2nd with a long seta; outer lobe with 5 subequal setae; basis with a seta; endopod with 5 subequal setae; exopod about as long as rest of appendage, with 5 long setae plus a very small seta. In Mx2 (Fig.57-d), short seta of 1st lobe approximately 1/3 length of long seta; seta of 2nd lobe relatively thick and fringed with short setules, shorter than short seta of 1st lobe. Seta of 3rd lobe approximately 1/2 length of spine. Two spines of 4th lobe similar in length, and more than twice length of seta. Fifth lobe about as long as segment itself and two together about as long as 1st segment; its spine about 1.7 times as long as segment including lobe; distal 29% length of spine unserrated. Sixth lobe reaching about 1/3 way to distal end of 5th; its spine reaching about 3/4 length of spine on 5th lobe; distal 31% length of spine unserrated. Endopodal setae relatively short, reaching about distal end of 5th lobe.

Measured along lateral margins, basis of Mxp (Fig.57-e) nearly twice as long as coxa and approximately 1.3 times length of endopod. Anterior distal spine of coxa about as long as 2nd posterior distal spine and approximately 4/5 length of 1st posterior distal spine. First middle seta of basis arising about 2/3 length of segment from proximal end. Anteriorly, 2nd endopodal segment with 2 large pores; 3rd with 1 large pore.

In 1st leg (Fig.57-f), combined lengths of first 2 exopodal segments a little shorter than basipod but a little longer than endopod. Outer spine of 1st exopodal segment reaching about base of outer spine of 2nd segment, which is extending beyond base of following outer spine by about 1/4 its length. Two outer spines of 3rd segment of similar size and shorter than outer spine of 2nd segment. Terminal spine of 3rd segment about as long as combined lengths of first 2 segments. Distolateral corners of first 2 endopodal segments produced into spiniform processes.

In 2nd leg (Fig.57-g), basipod, combined lengths of first 2 exopodal segments, and 3rd exopodal segment all similar in length. Outer spines of first 2 exopodal segments similar in size. First outer spine of 3rd segment approximately 2/3 length of outer spine of 2nd segment. Three outer spines of 3rd segment gradually decreasing in size from 1st to 3rd. Terminal spine of 3rd segment about 1/3 length of segment itself. Endopod a little shorter than basipod, with distolateral corners of first 2 segments produced into sharp points. Anteriorly, first 2 exopodal segments each with a relatively small pore; 3rd segment with 4 large pores.

Basipod of P3 (Fig.57-h) about as long as combined lengths of first 2 exopodal segments and only slightly shorter than 3rd exopodal segment. Outer spines of first 2 exopodal segments similar in size and only slightly larger than 1st outer spine of 3rd segment; outer spines of 3rd segment gradually decreasing in size from 1st to 3rd as in P2. Terminal spine of 3rd segment slightly longer than 1/4 length of segment. Endopod approximately 2/3 length of 3rd exopodal segment; distolateral corners of all 3 segments produced into sharp points; 2nd outer marginal seta of 3rd segment preceded by a sharp spiniform process. Anteriorly, 1st and 2nd exopodal segments with 1 and 2 pores, respectively; 3rd segment with 8 pores.

In 4th leg (Fig.57-i), basipod about as long as combined lengths of first 2 exopodal segments, a little longer than 3rd exopodal segment and much longer than endopod. Outer spine of 1st exopodal segment larger than that of 2nd. Third with 3 outer spines similar in size and smaller than outer spine of 2nd. Terminal spine of 3rd segment approximately 1/3 length of segment. Distolateral corners of all 3 endopodal segments produced into sharp points; each of 2 outer marginal setae of 3rd segment preceded by a small spiniform process. Anteriorly, 1st and 2nd exopodal segments with 1 and 2 pores, respectively; 3rd segment with 8 pores.

Basipod of 5th leg (Fig.58-a) about as long as endopod and shorter than combined lengths of 1st and 2nd exopodal segments, which are about as long as 3rd exopodal segment. Outer spine of 1st exopodal segment relatively slender and longer than that of 2nd. Two outer spines of 3rd segment similar in size and smaller than outer spine of 2nd. Terminal spine of 3rd segment approximately 5/9 length of segment. Inner spine of 2nd segment about as long as endopod. Distolateral corners of 1st and 2nd endopodal segments produced into sharp points. Anteriorly, 1st exopodal segment with a small pore, 2nd and 3rd with a large pore close to each outer spine.

Description of male. Prosome length, 1.60-2.32 mm; body length, 2.08-3.20 mm. Prosome similar in shape to that of female, approximately 3/4 length of body and about 3 times length of urosome. Laterally, prosome about 4/9 as deep as long; cephalosome with a cervical groove halfway along dorsal margin, narrowly rounded forehead produced ventrad into a small rostrum bearing 2 rather short filaments. Dorsally, forehead broadly rounded with a low midanterior tubercular process, which extends ventrad into rostrum as in female; posterolateral corners of prosome (Fig.58-b) produced into triangular processes covering about 1/2 length of genital somite. Length ratios of 5 urosomal somites and left caudal ramus are 19.6 : 17.1 : 16.5 : 13.3 : 10.7 : 22.8 = 100. Dorsal appendicular seta of left ramus longer than that of right. Fourth marginal seta of left ramus about as long as urosome. Third marginal seta of left ramus and 3rd and 4th marginal setae of right about 2/3 length of 4th marginal seta of left ramus and armed with spinules in addition to regular setules. Dorsally, anal segment with a pore on each side; caudal ramus with 2 pores.

Right A1 similar in setation to that of female. Left, geniculated A1 with first 2 segments fused and 22nd and 23rd segment also fused. Other cephalosomal appendages and 1st to 4th legs as in female.

In 5th pair of legs (Fig.58-c), inner lobe of right basis rounded, extending directly distad, reaching middle of 1st endopodal segment. Inner margin of left basis straight, its distal 1/3 fringed with setules. Both endopods with an inner marginal seta on 2nd segment. In right exopod (Figs.58-d-f), outer spine of 2nd segment a little larger than that of 1st; 3rd segment almost as long as combined lengths of first 2 segments, with a toothlike process at distolateral end (Fig.58-f), with an outer spine halfway along lateral margin, terminal spine approximately 1/3 length of segment itself, and an elevated pore laterally close to distal end. Medial projection of 2nd segment closer to proximal end than to distal end of segment, more or less spiniform, almost as long as segment itself, with obtusely pointed distal end, and provided with short setules. Second segment of left exopod (Figs.58-g, h) with distolateral corner expanded into a large tubercular outgrowth; outer spine attached to distal side of outgrowth and pointing straight distad. Third segment almost as long as combined lengths of first 2 segments, with a small outer spine about 2/3 length of segment from proximal end, a relatively large inner spine halfway along inner margin, terminal spine approximately 2/3 length of segment, and an elevated pore on anterior surface close to distal end.

Distribution. This species is known to have a worldwide distribution from the Arctic Ocean (Sars 1900) to the Antarctic (Farran 1929). It is represented in the study by a total of 266 specimens including 252 females and 14 males found in 50 samples—6 taken in the Atlantic between 33°47'S and 01°25'N, 40 taken throughout the Pacific including the Malay Archipelago between 46°57'S and 40°35'N, and 4 taken in the Indian Ocean between 35°00'S and 17°05'S.

Remarks. This species seems to show a wide range of size variation. Sars (1900) originally described this species from female specimens 3.35 mm long collected from between 80°N and 85°N and between 79°E and 134°E in the Arctic Ocean. Subsequently, Sars (1925) redescribed the species from both female and male specimens taken in the North Atlantic, which were much smaller (2.30 mm long in the female) than those from the Arctic Ocean. The females and males Farran (1929) found in the Ross Sea of the Antarctic measured 3.2 and 2.88 mm long, respectively. Brodsky's (1950) specimens from the northwestern Pacific were 3.0-3.4 mm long in the female and 2.9-3.4 mm long in the male. Heptner's (1971) specimens from the Kuril-Kamchatka Trench were 2.90-3.50 mm long in the female and 2.95-3.60 mm long in the male. Tanaka's (1964) specimens from the Izu region of Japan measured only 2.43 mm long in the female and 2.59 mm long in the male.

In the present study, the body size of the female varied from 2.24 to 3.32 mm and that of the male from 2.08 to 3.20 mm. The largest individuals were found off the west coast of South Africa in the southern Atlantic, off southern Chile, and northern California in the Pacific. The smallest individuals were found in the equatorial waters of all three major oceans. However, it was not possible to subdivide the species into any groups that can be defined in terms of size as well as morphology.

This species can readily be recognized by its stout body, relatively short urosome, nearly symmetrical caudal rami (each separated from the anal segment), and short antennule. The female can also be easily recognized by its genital somite, which in lateral view is nearly square, without a protruding genital prominence. The male's 5th pair of legs is similar to that of the subgenus *Paraheterorhabdus* except that the outer spine of the 2nd exopodal segment of the left leg is relatively large and attached subterminally to a distolateral process of the segment. The same spine in the subgenus *Paraheterorhabdus* is relatively small and attached terminally to the distolateral process of the segment.

GENUS *HETERORHABDUS* GIESBRECHT 1898

Heterorhabdus Giesbrecht 1898, p. 113.
Heterochaeta Claus 1863, p. 180.
Alloiorhabdus Wolfenden 1911, p. 303.

Type by page priority: *Heterorhabdus spinifrons* (Claus 1863)

Diagnosis. Dorsally, forehead (Fig.59-e) with a midanterior tubercular process, which extends ventrad into rostrum. Sternite of 1st pedigerous somite (Fig.59-a) with a spiny papilla. Left caudal ramus (Fig.59-c) fused with anal segment and distinctly longer than right. In 1st segment of A1 (Fig.59-j), 1st lobe with a small seta, 2nd with 1 long and 1 very short setae and 1 aesthete, 3rd with 1 small and 2 medium-sized setae, and 4th with 1 medium-sized and 1 long setae and 1 aesthete. Left, geniculate A1 (Fig.61-c) of male with segments 22 and 23 fused. In A2 (Fig.60-a), first 2 exopodal segments without setae, last exopodal segment with a normally developed inner marginal seta, inner and outer lobes of 2nd endopodal segment with 8 (marginal)+1 (posterior appendicular) and 6+1 setae, respectively. In Md (Fig.60-c), basis with a seta, 1st endopodal segment with 1 very small and 2 well-developed setae, 2nd with 8 terminal and 1 anterior appendicular setae. Masticatory edge with basal spine followed by 2 teeth in left Md and by 3 teeth in right Md (Fig.60-e); ventralmost tooth separated from remaining teeth by a wide, flat gap. Ventralmost teeth of both mandibles symmetrical, large, falciform, each extending proximad, in the form of a reinforcement rib attached to the tooth, along ventral margin of mandibular blade to a distance about equal to length of tooth protruding beyond masticatory edge. Maxillule (Fig.60-f) with exopod greatly elongated and 2nd inner lobe bearing a relatively short seta. In Mx2 (Fig.60-g), 1st segment greatly elongated; 1st lobe small, papillate, with 2 unequal setae; 2nd represented by a short, stout spine attached directly to segment; 3rd represented by a small seta and a medium-sized stout spine, both arising directly from segment; 4th elongate, with a small posterior subterminal spine and 2 long, saberlike terminal spines of similar length; 5th lobe elongate, with a relatively short saberlike spine and a long falcate spine; 6th with a very small anterior seta and a long falcate terminal spine serrated with short spinules along proximal section of medial margin; endopod small, with 7 relatively short setae. Coxa of Mxp (Fig.60-h) with middle marginal spine developed into a long, saberlike spine reaching to distal end of basis; anterior distal spine large and stout and 2 posterior distal spines greatly reduced in size. Third exopodal segment of P3 broad (Fig.60-l), with small terminal spine.

Additional description of female. Rostrum with 2 slender filaments. First pedigerous somite fully separated from cephalosome; 4th and 5th pedigerous somites completely fused. Laterally, cephalosome with cervical groove halfway along dorsal margin; posterior margin of prosome rounded. Dorsally, posterolateral corner of prosome (Fig.59-c) rounded. Genital somite (Fig.59-b) with well-developed genital prominence. Fourth marginal seta of left caudal ramus unarmed and longer than body; other marginal setae typically plumose.

Antennule reaching or extending beyond posterior end of caudal ramus. In A2 (Fig.60-a), basipod shorter than 1st endopodal segment; endopod and exopod of similar length; setae on coxa and basis moderately developed; 1st endopodal segment with 2 rather poorly developed marginal setae on medial margin and a row of furcate spinules on anterior surface. In Md (Fig.60-c), mandibular blade a little longer than palp; endopod small, reaching distal end of exopod. In Mx1 (Fig.60-g), 1st inner lobe with 1 posterior submarginal, 9 marginal, and 2 anterior submarginal setae; 2nd inner lobe poorly developed, with a relatively small seta; outer lobe with 5 setae; 2nd inner lobe missing; basis normally with a seta; endopod very small, normally with 3 setae; exopod approximately 3/4 length of basipod, with 5 well-developed setae and 1 or 2 very small setae. Measured along lateral margins, 1st segment of Mx2 (Fig.60-g) approximately 1.6 times length of 2nd, which is nearly as long as 4th segment including 5th lobe; 6th lobe reaching about halfway to distal end of 5th lobe. In Mxp (Fig.60-h), coxa approximately 3/5 length of basis; endopod only slightly longer than or about equal to coxa.

Legs relatively slender. Coxa in 1st to 4th legs with a well-developed inner marginal seta. Basis of P1 with a well-developed inner marginal seta and a small hooklike process on posterior sur-

89

face close to lateral margin. Basis with a small outer seta only in P4 and P5. None of endopodal segments of P1 armed with setules or spinules on anterior surface that could be referred to von Vaupel-Klein's organ. Third endopodal segment with 7 setae in P2 and P4, 8 setae in P3. In P5 (Fig.61-a), 1st exopodal segment without inner seta, 2nd with a well-developed inner spine.

　　Additional description of male. Similar in habitus to female except that urosome 5-segmented. Left caudal ramus fused with anal segment and distinctly longer than right. Setae of caudal rami as in female. Left, geniculate A1 (Fig.61-c) with first 2 segments fused, segments 19-21 fused to form the 1st postgeniculate segment, and segments 22-23 fused. In morphological details, all other cephalosomal appendages and 1st to 4th legs as in female.

　　In 5th pair of legs (Fig.61-d), right basis with well-developed inner lobe, which in certain species is greatly elongated; left basis normally with a low, inconspicuous inner lobe. Both endopods with an inner marginal seta on 2nd segment and 6 marginal setae on 3rd; left endopod longer than right. In right exopod, 2nd segment with a medial projection of various shape; 3rd segment elongate, with an outer spine and a terminal spine. Left exopod with an outer spine on each segment; 3rd segment also has an inner spine and long terminal spine.

　　Remarks. The genus *Heterochaeta* was established by Claus (1863) to accommodate *H. spinifrons* and *H. papilliger.* As the name *Heterochaeta* was preoccupied, the genus was renamed *Heterorhabdus* by Giesbrecht (1898). *Alloiorhabdus* erected by Wolfenden (1911) for *Heterorhabdus austrinus* Giesbrecht 1902 (the type species by page priority) and a new species, *A. medius,* is synonymous with *Heterorhabdus.*

　　The most important characters shared by all species of this genus are a spiny papilla of the sternite of the 1st pedigerous somite and a saberlike or falciform spine on the coxa of Mxp. Also important for identification of this genus are Mx1, which has a small endopod bearing 3 or fewer setae and an elongate exopod, and Mx2, in which the 5th lobe has 2 large, saberlike or falciform spines. The male 5th pair of legs are characteristic in having elongate exopods and the terminal spine of the left exopod greatly elongated and completely fused with the segment.

　　The species of this genus can be separated into 4 species groups: *spinifrons, papilliger, fistulosus,* and *abyssalis.*

KEY TO SPECIES GROUPS OF *HETERORHABDUS*

Females and Males

1a. Middle coxal spine of Mxp smoothly curved (Fig.60-h), none
　　of the spines on 5th lobe of Mx2 (Fig.60-g) conspicuously
　　serrated with long spinules . . . . *spinifrons* species group
1b. Middle coxal spine of Mxp distinctly bent (Fig.73-a), 1 of the
　　spines on 5th lobe of Mx2 (Fig.72-i) conspicuously serrated
　　with long spinules 2

2a. Unserrated spine on 5th lobe of Mx2 (Fig.72-i) nearly as long as
　　or only a little shorter than serrated spine, basal lobe of male right
　　P5 (Fig.73-c) relatively short and arising close to distal end of segment 3
2b. Unserrated spine on 5th lobe of Mx2 (Fig.84-i) much shorter than
　　serrated spine; basal lobe of male right P5 (Fig.85-g) relatively
　　long and arising from middle of segment. . . . . *abyssalis* species group

3a. Dorsally, midanterior tubercular process of forehead pointed
　　(Fig.72-d), basis of Mxp without large elevated pores . . . . *papilliger* species group
3b. Dorsally, midanterior tubercular process of forehead (Fig.82-d)
　　somewhat truncate, basis of Mxp (Fig.82-h) with large elevated
　　pore on medial margin close to distal end . . . . *fistulosus* species group

SPINIFRONS SPECIES GROUP

Diagnosis. Midanterior tubercular process (Fig.59-e) of forehead angular or produced into a spiniform process. Small posterior subterminal spine of 4th lobe of Mx2 (Fig.60-g) about 1/4 length of long, saberlike spines; large falciform spine of 5th lobe not conspicuously serrated with spinules and small saberlike spine far short of reaching distal end of falciform spine; endopodal setae relatively poorly developed. In Mxp (Fig.60-h), middle marginal spine saberlike and smoothly curved, reaching distal end of basis at most. In male 5th pair of legs (Fig.61-d), inner lobe of right basis arising from distal part of medial margin of segment, relatively short and pointing in a distomedial direction; medial projection of 2nd exopodal segment of right P5 large, arising along proximal margin of segment, terminated with a bilobed structure consisting of a rounded plumose proximal lobe and a naked or partially plumose distal lobe, and a small spiniform process in between (Fig.61-h).

Composition. The following species are referred to this species group:

Heterorhabdus spinifrons (Claus 1863)
H. subspinifrons Tanaka 1964
H. caribbeanensis Park 1970
H. heterolobus, new species
H. quadrilobus, new species
H. ankylocolus, new species
H. insukae, new species

Key to species of *spinifrons* species group

Females

1a. Forehead (Fig.68-d) without a spiniform process . . . . *H. caribbeanensis*
1b. Forehead (Fig.59-d) with a spiniform process 2

2a. Anterior wall of genital prominence (Fig.66-c) with a pair of rounded lobes 3
2b. Anterior wall of genital prominence (Fig.59-b) without such lobes 5

3a. Genital prominence (Fig.64-b) reaching posterior end of somite;
    endopod of Mx1 (Fig.64-d) with 1 seta . . . . *H. subspinifrons*
3b. Genital prominence (Fig.65-a) short of reaching posterior end of somite;
    endopod of Mx1 with 3 setae 4

4a. Genital prominence (Fig.67-c) without posterior lobes . . . . *H. ankylocolus*, n. sp.
4b. Genital prominence (Fig.65-c) with posterior lobes . . . . *H. quadrilobus*, n. sp.

5a. Dorsally, genital somite (Fig.59-c) relatively short, with smoothly
    bulging lateral margins; Mx1 (Fig.60-f) with 1 seta on basis and
    3 setae on endopod 6
5b. Dorsally, genital somite (Fig.70-c) relatively long, with wavy lateral
    margins; Mx1 (Fig.70-h) with 0 seta on basis and 1 seta on endopod . . . . *H. insukae*, n. sp.

6a. Laterally, genital prominence (Fig.59-b) reaching posterior end of somite,
    without posterior lobes . . . . *H. spinifrons*
6b. Laterally, genital prominence (Fig.63-a) short of reaching posterior
    end of somite, with posterior lobes . . . . *H. heterolobus*, n. sp.

Males

1a. Forehead without a spiniform process . . . . *H. caribbeanensis*
1b. Forehead with a spiniform process 2

91

2a. In right P5 (Fig.71-c), basal inner lobe relatively short and conical; medial projection of 2nd exopodal segment (Fig.71-f) with large terminal spiniform process and poorly developed distal lobe . . . . *H. insukae*, n. sp.
2b. In right P5 (Fig.61-d), basal inner lobe relatively long and somewhat fingerlike; medial projection of 2nd exopodal segment (Fig.61-h) with small terminal spiniform process and relatively well developed distal lobe 3

3a. Basal inner lobe of right P5 (Fig.64-h) almost as long as segment, pointing straight distad; endopd of Mx1 (Fig.64-d) with 1 seta. . . . . *H. subspinifrons*
3b. Basal inner lobe of right P5 much shorter than segment, pointing distomediad; endopod of Mx1 with 3 seta 4

4a. Second exopodal segment of left P5 (Fig.61-i) with outer spine longer than conical process as measured along medial margins; terminal spine of 3rd exopodal segment of right P5 (Fig.61-e) about 1/6 length of segment . . . . *H. spinifrons*
4b. Second exopodal segment of left P5 (Fig.63-g) with outer spine equal to or shorter than conical process; terminal spine of 3rd exopodal segment of right P5 (Fig.63-h) about 1/3 length of segment. 5

5a. Lateral margin of 3rd exopodal segment of right P5 (Fig.67-h) strongly elbowed about 1/5 length of segment from distal end . . . . *H. ankylocolus*, n. sp.
5b. Lateral margin of 3rd exopodal segment of right P5 (Fig.66-e) only slightly or moderately elbowed about 1/5 length of segment from distal end. 6

6a. Basal inner lobe of right P5 (Fig.66-c) with a deep notch on lateral margin; in 3rd exopodal segment of right P5 (Fig.66-e), lateral margin moderately elbowed about 1/5 length of segment from distal end, and distal lobe elongate (Fig.66-f); in 2nd exopodal segment of left P5 (Fig.66-d), outer spine distinctly shorter than conical process. . . . . *H. quadrilobus*, n. sp.
6b. Basal inner lobe of right P5 (Fig.63-f) with a shallow notch on lateral margin; in 3rd exopodal segment of right P5 (Fig.63-h), lateral margin only slightly elbowed about 1/5 length of segment from distal end, and distal lobe relatively short (Fig.63-k); in 2nd exopodal segment of left P5 (Fig.63-g), outer spine about as long as conical process . . . . *H. heterolobus*, n. sp.

Heterorhabdus spinifrons (Claus 1863)
Figures 59-62

Heterochaeta spinifrons Claus 1863, p. 182, pl. 32, figs. 8, 9, 14, 16.—(male only) Giesbrecht 1892, p. 372, pl. 20, figs. 3, 11, 16, 19, 31; pl 39, fig. 51.
Heterorhabdus spinifrons; Giesbrecht 1898, p. 114.—Sars 1925, p. 227, pl. 62, figs. 9-12.

Description of female. Prosome length, 2.12-2.80 mm; body length, 3.04-4.00 mm. Body elongate (Fig.59-a). Prosome approximately 7/10 length of body and about 2.2 times length of urosome. Laterally, prosome deepest at joint between cephalosome and 1st pedigerous somite, typically with a length-depth ratio of 100:35. Cephalosome approximately 4/9 length of prosome, with a cervical groove halfway along dorsal margin. Midanterior tubercular process of forehead (Fig.59-e) produced into a spiniform process. Laterally, rostral lobe (Fig.59-d) poorly developed, with 2 slender filaments.

Posterior margin of prosome (Fig.59-b) broadly rounded. Dorsally, posterolateral corner of prosome (Fig.59-c) broadly rounded.

Urosome slightly shorter than 1/3 length of body. First 3 urosomal somites armed with spinules along posterior margins. Dorsally, genital somite (Fig.59-f) widest at middle, with smoothly curved lateral margins and typically with a length-width ratio of 100:88. Laterally, genital somite (Figs.59-g, h) typically with a length-depth ratio of 100:91, with genital prominence extending to posterior end of somite. Ventrally, genital operculum (Fig.59-I) typically with a length-width ratio of 77:100 and about 7/10 as wide as somite. Genital field with a row of about 15 pores on each side.

Length ratios of 4 urosomal somites and left caudal ramus, 37.0 : 17.3 : 13.6 : 9.9 : 22.2 =100. Left caudal ramus (Fig.59-c) extending beyond posterior end of right by 1/6 its length as measured along medial margin. Dorsal appendicular seta of left ramus nearly as long as ramus itself and about twice as long as corresponding seta on right ramus. Fourth marginal seta of left ramus unarmed and much longer than body.

Antennule extending beyond posterior end of caudal ramus by its last 3 segments. Not all aesthetes are clearly distinguishable from setae. Segments 2-19 each with 1 middle and 2 distal setae/aesthetes (Fig.59-j).

In antenna (Fig.60-a), basipod a little shorter than 1st endopodal segment; inner marginal setae of coxa and basis only moderately developed; endopod and exopod of similar length; 1st endopodal segment with rather poorly developed inner marginal setae and a row of furcate spinules on anterior surface (Fig.60-b); first 2 exopodal segments each with a small conical process at distomedial corner, without marginal setae; inner marginal seta of last exopodal segment normally developed.

Mandibular palp (Fig.60-c) with small endopod reaching about distal end of exopod. Basis well developed, with 1 relatively small seta. First endopodal segment with 1 small and 2 long setae; 2nd with 8 terminal setae plus 1 anterior appendicular seta. Mandibular blade 1.2 times length of palp. Two dorsal teeth of left Md (Fig.60-d) tricuspid; first dorsal tooth of right Md (Fig.60-e) bicuspid, 2nd tricuspid, and 3rd single-pointed. Masticatory edge between dorsal teeth and ventral falcate tooth more or less straight.

In maxillule (Fig.60-f), 1st inner lobes with 12 setae, 2nd inner lobe and basis each with 1 seta, endopod with 3 setae; all of these setae are rather poorly developed; outer lobe with 5 well-developed setae; exopod nearly as long as rest of appendage as measured along medial margins, with 5 well-developed setae plus 2 tiny setae.

First segment of Mx2 (Fig.60-g) nearly as long as rest of appendage. Long seta of 1st lobe about twice length of short seta. Spine of 2nd lobe stout, about as long as short seta of 1st lobe. Spine of 3rd lobe a little longer than segment itself and approximately 5/9 length of longest spine of 4th lobe; seta of 3rd lobe very fine and short. First saberlike spine of 4th lobe only slightly shorter than 2nd and about 4/5 length of large, falcate spine of 5th lobe; small spine of 4th lobe about as long as spine of 2nd lobe and about 1/5 length of longest spine of same lobe. Fifth lobe a little shorter than segment; falcate spine a little longer than longest spine of 4th lobe and proximal half of its medial margin fringed with extremely fine spinules visible only under a high magnification; saberlike spine of 5th lobe approximately 2/3 length of falcate spine. Sixth lobe reaching about halfway to distal end of 5th; falcate spine of 6th lobe a little longer but thinner than falcate spine of 5th lobe; proximal 4/7 serrated along medial margin; seta of 6th lobe very fine and short. Endopod very small, with 7 delicate setae barely reaching distal end of 5th lobe.

In coxa of Mxp (Fig.60-h), distal half of saberlike spine thickly armed with spinules; anterior distal spine of coxa well developed, approximately 1/2 length of segment and reaching 2/5 way to distal end of basis; 2 posterior distal spines greatly reduced, less than 1/2 length of anterior distal spine and very fine. Basis with a long band of spinules on anterior surface and about 1.4 times length of coxa as measured along medial margins; 3 middle marginal setae poorly developed and the first arising about 5/9 length of segment from proximal end. Endopod about as long as coxa.

First leg (Fig.60-i) slender. Measured along medial margins, basipod approximately 1.3 times length of endopod, which is a little shorter than combined lengths of first 2 exopodal segments. Inner marginal setae of coxa and basis well developed. Posterior hooklike process of basis relatively small (Fig.60-j). First 2 endopodal segments with distolateral corners pointed. First exopodal segment approximately 2/5 length of exopod; terminal spine of 3rd exopodal segment approximately 7/10 length of exopod. First outer spine of exopod a little short of reaching base of 2nd, which reaches about 1/2 way to base of 3rd. Third outer spine a little longer than 4th and reaching about 3/4 of the way to base of 4th.

In second leg (Fig.60-k), endopod 72% length of basipod, 85% combined lengths of first 2 exopodal segments, and 37% length of whole exopod. Distolateral corners of first 2 exopodal segments pointed. Third exopodal segment 57% length of whole exopod, its 1st outer spine arising 65% length of segment, as measured along medial margin, from proximal end; its terminal spine 28% length of segment. Anteriorly, 1st and 2nd exopodal segments each with a pore close to outer spine, 3rd with 4 pores.

Third leg (Fig.60-l) similar to P2 except for the following differences: 1) Third endopodal segment with 8 setae instead of 7, 2) third exopodal segment relatively wide, typically with a length-width ratio of 100:47, 60% length of whole exopod, its 1st outer spine arising about 3/5 length of segment from proximal end; its terminal spine less than 13% length of segment, and 3) anteriorly, 1st exopodal segment with 1 pore close to outer spine, 2nd with 2 pores close to outer spine, 3rd with 7 pores.

Fourth leg (Fig.60-m) also similar to P2 with the following exceptions: 1) Basis with a small outer seta, 2) endopod 80% length of basipod, 87% combined lengths of 1st and 2nd exopodal segments, and 38% length of whole exopod, 3) third exopodal segment 56% as long as whole exopod, its 1st outer spine arising 58% length of segment from proximal end; its terminal spine 23% as long as segment, and 4) anteriorly, 1st exopodal segment with 1 pore close to outer spine, 2nd with 2 pores close to outer spine, 3rd with 7 pores.

In 5th leg (Fig.61-a), basipod, endopod, 3rd exopodal segment, and inner spine of 2nd exopodal segment all similar in length. Endopod extending beyond distal end of 2nd exopodal segment. Distolateral corners of 1st and 2nd endopodal segments pointed; inner marginal setae provided with long setules for proximal halves and short setules for distal halves. Outer spines of exopod relatively small, all pointing in a distolateral direction. Third exopodal segment 46% as long as whole exopod, its 1st outer spine arising 63% length of segment from proximal end, its terminal spine 38% as long as segment. Anteriorly, 3rd exopodal segment with 1 pore.

Description of male. Prosome length 2.12-2.64 mm; body length, 3.00-3.84 mm. Similar in habitus to female except that urosome 5-segmented. Prosome 70% length of body and approximately 2.2 times length of urosome. Urosome approximately 3/10 length of body. First 4 urosomal somites (Fig.61-b) armed with spinules along posterior margins. Length ratios of 5 urosomal somites and left caudal ramus, 18.0 : 18.7 : 16.8 : 12.3 : 10.3 : 23.9 = 100. Left caudal ramus extending beyond posterior end of right by about 1/6 its length, with dorsal appendicular seta longer than corresponding seta of right; 4th marginal seta unarmed, thicker than other marginal setae and longer than body.

Left, geniculate A1 (Fig.61-c) with segments 1 and 2 fused, segments 19-21 fused to form 1st postgeniculate segment; segments 22 and 23 also fused. Other cephalosomal appendages and 1st to 4th legs as in female.

In 5th pair of legs (Fig.61-d), left endopod a little longer than right, both endopods with inner marginal seta on 2nd segment. Inner lobe of right basis more or less fingerlike, extending beyond distal end of 1st endopodal segment in a distomedial direction. In right exopod (Fig.61-e), combined lengths of 1st and 2nd segments about as long as 3rd . Anteriorly, proximolateral corner of 2nd segment with a low round ridge; medial projection (Fig.61-h) extending along proximal margin of segment; its proximal and distal lobes nearly symmetrical and its terminal spiniform process passing through distal lobe. Outer spine of 2nd segment about 2/3 length of segment from proximal end. Third segment smoothly curved (Fig.61-g), with an outer spine about 2/5 length of segment from proximal end; its distal lobe (Fig.61-f) extending about 1/3 way to distal end of terminal spine, which is about 1/6 length of segment. In left exopod (Fig.61-i), distolateral corner of 2nd segment produced into a long conical process bearing a large outer spine; measured along medial margins, outer spine approximately 1.8 times length of conical process. Third segment including terminal spine as long as

66% length of whole exopod, its inner spine about halfway along medial margin and outer spine close to distal end of segment.

Distribution. Since its original description from the Mediterranean Sea, *H. spinifrons* has been widely recorded throughout the world's oceans. In the Atlantic, the species has been found as far north as 62°45'N (Jespersen 1940) and as far south as 35°10'S (Wolfenden 1911). In the Indo-Pacific, the species has been recorded from off California (Esterly 1905, 1906), the Bering Sea (Wilson 1950), the Izu region of Japan (Tanaka 1964), throughout the tropical Pacific (Giesbrecht 1892, Wilson 1950, Grice 1962), the Malay Archipelago (Scott 1909, Wilson 1950), the South China Sea (Wilson 1950), off New Zealand and the Great Barrier Reef (Farran 1929, 1936), off Tasmania (Vervoort 1957), the Bay of Bengal (Sewell 1932), the Arabian Sea and Gulf of Aden (Sewell 1947). However, the Indo-Pacific records are believed to be a composite of *H. spinifrons, H. subspinifrons,* first described in 1964, and 3 new species (*H. heterolobus, H. ankylocolus,* and *H. quadrilobus*) described in this study, and it is not possible to determine to which species these records are referable.

Heterorhabdus spinifrons is represented by a total of 342 specimens including 244 females and 98 males found in 53 samples—10 taken in the Atlantic between 34°47'S and 38°53'N, 35 throughout the Pacific including the Malay Archipelago between 41°27'S and 40°35'N, and 8 in the Indian Ocean between 35°00'S and 12°40'S. These findings demonstrate that the species has a circumglobal distribution.

Remarks. This species was originally described from a female specimen 3.0 mm long obtained in the Mediterranean Sea (Messina, Sicily). Giesbrecht (1892) redescribed the species based on both female and male specimens collected from the equatorial Pacific (between 99°W-160°E, and between 3°S-14°N). According to Giesbrecht's figures of the male 5th pair of legs and female genital somite (pl. 20, fig. 31 and pl. 39, fig. 42), only his male seems to belong to *H. spinifrons,* and his female is referable to *H. heterolobus,* a new species described below. Sars (1925) redescribed the species from the North Atlantic with figures of the female genital somite and male 5th pair of legs, which are in full agreement with those of the species here referred to *H. spinifrons.* Sars's female specimen was 3.80 mm long.

The female of this species can be distinguished from those of the other species of the group by the spiniform process of the forehead, the relatively short genital somite with smoothly curved lateral margins, the genital prominence extending to the posterior end of the somite, and the absence of lobes on the anterior slope of the genital prominence. The male can be identified by its 5th pair of legs, in which the inner lobe of the right basis somewhat fingerlike, the 3rd segment of the right exopod smoothly curved, and the 2nd segment of the left exopod with its distolateral corner produced as a large conical process bearing a large outer spine.

The genital somite of the female varies considerably. Individuals from the Atlantic and the central and equatorial Pacific and Indian Oceans tended, as described above, to have a relatively elongate genital somite, with low lateral swellings and a relatively low genital prominence (typically length-width ratio of 100:88 and length-depth ratio of 100:91). Individuals (Figs.62-a-e) from the west coast of the Americas had relatively thick genital somites, typically with a length-width ratio of 100:90 and a length-depth ratio of 100:97. Dorsally, the lateral margins are strongly bulging. Laterally, the posterior slope of the genital prominence almost perpendicular to the longitudinal axis of the somite.

Heterorhabdus heterolobus, new species
Figure 63

Heterochaeta spinifrons (female only), not of Claus; Giesbrecht 1892, p. 372, pl. 20, figs. 1, 9, 24; pl. 39, figs. 42, 43, 52, 54.

Type material. Holotype female (USNM 287025), 8 paratype females (USNM 287027), allotype male (USNM 287026), and 4 paratype males (USNM 287028) found in IKMT sample from 3000-0 m, SIO Francis Drake III, station 4, 02°57'N, 80°49'W in the eastern tropical Pacific, May 8, 1975.

Description of female. Prosome length, 2.20-2.52 mm; body length, 3.16-3.56 mm. Very similar in habitus to *H. spinifrons*, with midanterior tubercular process of forehead produced as a spiniform process. Urosome approximately 3/10 length of body. First 3 urosomal somites (Fig.63-b) armed with spinules along posterior margins. Dorsally, genital somite widest at middle, with smoothly bulging lateral margins and typically with a length-width ratio of 100:80. Laterally, genital somite (Figs.63-a,c,d) typically with a length-depth ratio of 100:95, with high genital prominence and a low dorsal hump about 2/5 length of somite from anterior end; genital prominence with posterior slope nearly perpendicular to longitudinal axis of somite, with a distinct posterior lobe on each side and far short of reaching posterior end of somite; posterior lobe on left side larger. Ventrally, genital operculum (Fig.63-e) typically with a length-width ratio of 90:100 and about 2/3 as wide as somite. Genital field with a row of about 25 pores on each side.

Length ratios of 4 urosomal somites and left caudal ramus, 37.1 : 16.6 : 13.7 : 9.7 : 22.9 = 100. Left caudal ramus (Fig.63-b) extending beyond posterior end of right by 1/7 its length as measured along medial margin. Dorsal appendicular seta of left ramus approximately 1/2 length of ramus itself and about 1.3 times length of corresponding seta of right ramus. Fourth marginal seta of left ramus unarmed and much longer than body.

Antennule extending posterior end of caudal ramus by its last 3 segments. In anatomical details, all cephalosomal appendages and swimming legs similar to those of *H. spinifrons*.

Description of male. Prosome length, 2.08-2.32 mm; body length, 2.96-3.28 mm. Very similar in habitus to *H. spinifrons* male but prosome equal to 72% length of body. Fifth pair of legs (Fig.63-f) also similar to that of *H. spinifrons* male except for the features described below. Inner lobe of right basis (Fig.63-i) relatively long and extending more mediad. Inner lobe of left basis (Fig.63-j) relatively well developed. In right exopod (Fig.63-h), outer spine of 2nd segment relatively well developed and close to distal end of segment; lateral margin of 3rd segment, which appears as a longitudinal middle ridge bearing outer spine in posterior view, more strongly curved than medial margin of segment; terminal spine (Fig.63-k) relatively long, about 5 times length of distal lobe of segment and about 1/3 length of segment. Outer spine of 2nd exopodal segment of left P5 (Fig.63-g) relatively small, about as long as conical process of segment as measured along medial margins.

Etymology. The specific name refers to the unequal posterior lobes of the female genital prominence.

Distribution. This species is represented by a total of 90 specimens including 65 females and 25 males found in 13 samples all taken in the equatorial Pacific from the west coast of South America to the Malay Archipelago between 17°50'S and 04°59'N. Its distribution appears to be limited to the equatorial waters of the Pacific.

Remarks. According to Giesbrecht's figure (1892, pl. 39, fig. 42) of the genital somite in lateral view, the female he described as *Heterochaeta spinifrons* from the equatorial Pacific seems to be referable to this species.

This species is similar to *H. spinifrons* but can readily be distinguished from it by the following features: 1) In female, genital prominence far short of reaching posterior end of somite and with 2 distinct lobes along distal margin; and 2) in male 5th pair of legs, terminal spine of 3rd exopodal segment of right P5 relatively long, about 1/3 length of segment; outer spine of 2nd exopodal segment of left P5 relatively short, about as long as conical process of segment. This species is also very close to *H. quadrilobus* described below, but its female can be distinguished by the absence of rounded lobes on the anterior wall of the genital prominence. In the males, these 2 species are so close to each other that they can be identified only by careful observations of such structures as the basal inner lobe and 3rd exopodal segment of the right P5 and the lateral conical process and outer spine of the 2nd exopodal segment of the left P5 as described in the key. However, the species differ in distribution. *Heterorhabdus quadrilobus* occurs widely along the west coast of both North and South America between about 40°S and 40°N and in the equatorial waters its range extends westward almost to the middle of the Pacific. *Heterorhabdus heterolobus*, on the other hand, occurs exclusively in equatorial waters from the west coast of South America westward to the Malay Archipelago.

Heterorhabdus subspinifrons Tanaka 1964
Figure 64

Heterorhabdus subspinifrons Tanaka 1964, p. 14, fig. 180.

Description of female. Prosome length, 1.80-2.24 mm; body length, 2.56-3.20 mm. Very similar in habitus to *H. spinifrons*, with midanterior tubercular process of forehead produced into a spiniform process. Urosome 31% length of body. First 3 urosomal somites (Fig.64-a) armed with spinules along posterior margins. Dorsally, genital somite widest at middle, with smoothly bulging lateral margins and typically with a length-width ratio of 100:82. Laterally, genital somite (Fig.64-b) typically with a length-depth ratio of 100:85, with a low dorsal hump about 2/5 length of somite from anterior end and well-developed genital prominence. Anterior wall of genital prominence with a low, rounded lobe on each side. Laterally, posterior slope of genital prominence extending to posterior end of somite and slightly concave. Ventrally, genital operculum (Fig.64-c) typically with a length-width ratio of 72:100 and 72% as wide as somite itself. Genital field with a row of about 23 pores on each side.

Length ratios of 4 urosomal somites and left caudal ramus, 37.1 : 17.4 : 13.2 : 9.6 : 22.7 = 100. Left caudal ramus (Fig.64-a) extending beyond posterior end of right ramus by 1/6 its length as measured along medial margin. Dorsal appendicular seta of left ramus nearly as long as ramus itself and about 1.5 times length of corresponding seta of right ramus. Fourth marginal seta of left ramus unarmed and much longer than body.

Antennule reaching about posterior end of caudal ramus. In morphological details, all cephalosomal appendages and swimming legs similar to those of *H. spinifrons* except for the following features: 1) Endopod of Mx1 (Fig.64-d) very small, with a single seta, 2) in Mx2 (Fig.64-e), saberlike spine of 5th lobe relatively long, 78% as long as falcate spine, 3) in Mxp (Fig.64-f), anterior distal spine of coxa 43% as long as segment and reaching 2/7 way to distal end of basis; 1st posterior distal spine of coxa relatively long, about 1/2 length of anterior distal spine; 1st middle seta of basis arising about 3/5 length of segment from proximal end, and 4) in fifth leg (Fig.64-g), inner marginal setae of first 2 endopodal segments rather poorly developed; inner spine of 2nd exopodal segment relatively short, 84% as long as 3rd exopodal segment; terminal spine of 3rd exopodal segment 39% as long as segment; anteriorly, exopod without pores.

Description of male. Prosome length, 1.84-2.20 mm; body length, 2.60-3.16 mm. Very similar in habitus to *H. spinifrons* male. Prosome 68% length of body. Left, geniculate A1 as in *H. spinifrons* male. All other cephalosomal appendages and 1st to 4th swimming legs as in female. In 5th pair of legs, inner lobe of right basis (Fig.64-h) long, fingerlike, about as long as segment and pointing almost straight distad. Inner lobe of left basis (Fig.64-k) poorly developed. Posteriorly, lateral margin of right 3rd exopodal segment (Fig.64-i) runs longitudinally in the middle of segment and is strongly curved, with outer spine midway; terminal spine (Fig.64-j) relatively short, about 3 times length of distal lobe of segment and about 1/6 length of segment itself. Second exopodal segment of left P5 (Fig.64-k) with distolateral corner produced into a long conical process terminating with long outer spine. Measured along medial margins, outer spine a little longer than conical process.

Distribution. The species is represented in the study by 134 specimens including 115 females and 19 males found in 30 samples—3 taken in the South Atlantic between 33°47'S and 18°58'S, 20 taken throughout the Pacific from the west coast of the Americas to the Malay Archipelago and the South China Sea between 41°27'S and 35°01'N, and 7 taken in the Indian Ocean between 35°00'S and 17°03'S. These findings constitute the first record since the original discovery and establish a circumglobal distribution of the species.

Remarks. This species was originally described from both female and male specimens collected from deep waters in Sagami Bay, Japan. The female can be identified by the following features: Anterior wall of genital prominence with low, rounded lobes; posterior slope of genital prominence extends to posterior end of somite and is slightly concave; endopod of Mx1 with only 1 seta. The male can be distinguished from those of the congeners by the 5th pair of legs, in which the inner lobe of the right basis is relatively long, the 3rd exopodal segment of the right leg with a strongly curved lateral margin and short terminal spine, the lateral conical process of the 2nd exopodal segment of the left leg long and terminated with a long outer spine.

Heterorhabdus quadrilobus, new species
Figures 65 and 66

Type material. Holotype female (USNM 287029), 11 paratype females (USNM 287031), allotype male (USNM 287030), and 16 paratype males (USNM 287032) found in IKMT sample from 1000-0 mwo, SIO CalCOFI 7102G, station 120.62, 27°17'N, 116°16'W off San Diego in the eastern Pacific, April 1, 1971.

Description of female. Prosome length, 2.28-2.68 mm; body length, 3.20-3.84 mm. Very similar in habitus to *H. subspinifrons*, with midanterior tubercular process of forehead produced into a spiniform process and anterior wall of genital prominence with 2 lobular processes. Prosome 71% length of body. Urosome 30% length of body. First 3 urosomal somites armed with spinules along posterior margins. Dorsally, genital somite (Fig.65-b) widest at middle, with smoothly bulging lateral margins and typically with a length-width ratio of 100:83. Laterally, genital somite (Figs.65-a, c, d) typically with a length-depth ratio of 100:86, with a rather smoothly curved dorsal margin and well-developed genital prominence. Anterior wall of genital prominence smoothly curved outward, with a low, rounded lobe on each side; posterior wall nearly perpendicular to longitudinal axis of somite and far short of reaching posterior end of somite. Genital field with a distinct lobe on each side of posterior margin, which is clearly visible in lateral (Fig.65-c) as well as in ventral view (Fig.65-e). Ventrally, genital operculum (Fig.65-e) typically with a length-width ratio of 88:100 and 68% as wide as somite itself. Genital field with a row of about 30 small pores on each side.

Length ratios of 4 urosomal somites and left caudal ramus, 37.4 : 17.3 : 13.4 : 10.1 : 21.8 = 100. Left caudal ramus (Fig.65-b) extending beyond posterior end of right by about 1/6 its length as measured along medial margin. Dorsal appendicular seta of left ramus about as long as ramus itself and approximately 1.5 times length of corresponding seta of right ramus. Fourth marginal seta of left ramus thicker than other marginal setae, unarmed and much longer than body.

Antennule extending beyond posterior end of caudal ramus by its last 3 segments. In anatomical details, all cephalosomal appendages and swimming legs similar to those of *H. spinifrons* except that middle, saberlike spine of Mxp coxa (Fig.66-a) far short of reaching distal end of basis, anterior distal spine 42% as long as segment and reaching almost 1/3 way to distal end of basis, 1st posterior distal spine approximately 2/3 length of anterior distal spine.

Description of male. Prosome length, 2.16-2.60 mm; body length, 3.04-3.68 mm. Very similar in habitus to *H. spinifrons* male. Prosome 71% length of body. Left, geniculate A1 as in *H. spinifrons* male. All other cephalosomal appendages and 1st to 4th legs as in female.

In 5th pair of legs (Fig.66-b), inner lobe of right basis approximately 2/3 length of segment, extending in a distomedial direction, with a deep notch on lateral margin and a sigmoid row of setules along medial margin (Fig.66-c). Inner lobe of left basis (Fig.66-d) poorly developed. Posteriorly, 2nd exopodal segment of right P5 (Fig.66-e) with relatively well developed outer spine, approximately 1/2 length of segment and located close to distal end of segment; 3rd exopodal segment distinctly bent about 1/6 length of segment from distal end, and this flexion is clearly visible on lateral margin of segment that runs longitudinally along the middle of segment and bears an outer spine. Terminal spine (Fig.66-f) of 3rd exopodal segment approximately 1/3 length of segment and almost 4 times length of distal lobe of segment. In left exopod (Fig.66-d), 2nd segment with a long distolateral conical process terminating with outer spine. Measured along medial margins, outer spine approximately 2/3 length of conical process.

Etymology. The specific name refers to the four lobes on the genital somite.

Distribution. This species is represented by a total of 200 specimens including 119 females and 81 males found in 24 samples—17 taken along the west coast of the Americas between 39°26'S and

37°40'N and 7 taken in equatorial waters between 15°00'S and 10°05'N from the west coast of the Americas westward to 161°52'W. According to these findings, the species is mainly distributed along the west coast of both North and South America, and its range extends in equatorial waters from the American coast westward to the central Pacific.

Remarks. The female can be distinguished from those of the congeners by the following combination of characters: Frontal spiniform process of forehead; 2 low, rounded lobes on anterior wall of genital prominence and 2 posterior lobes of genital field; genital prominence far short of reaching posterior end of somite; and smoothly curved and relatively short saberlike middle spine of Mxp coxa. The male can be identified by the 5th pair of legs, in which the 3rd exopodal segment of the right leg is distinctly bent about 1/6 length of the segment from the distal end and has a relatively long terminal lobe and spine. In the left P5, the 2nd exopodal segment has a long distolateral conical process and relatively short outer spine.

Heterorhabdus ankylocolus, new species
Figure 67

Type material. Holotype female (USNM 287033), 2 paratype females (USNM 287035), allotype male (USNM 287034), and one paratype male (USNM 287036) found in IKMT sample from 3000-0 mwo, SIO Piquero V, station 7, 17°46'S, 110°19'W in the eastern tropical Pacific, March 29, 1969.

Description of female. Prosome length, 1.96-2.24 mm; body length, 2.80-3.24 mm. Very similar in habitus to *H. subspinifrons*, with midanterior tubercular process of forehead produced as a spiniform process and anterior wall of genital prominence with 2 lobular processes. Prosome 71% length of body. Urosome 30% length of body. First 3 urosomal somites armed with spinules along posterior margins. Dorsally, genital somite (Fig.67-b) widest at middle, with smoothly bulging lateral margins and typically with a length-width ratio of 100:87. Laterally, genital somite (Figs.67-a, c) typically with a length-depth ratio of 100:94, with a low, inconspicuous dorsal hump about 2/5 length of somite from anterior end and well-developed genital prominence. Anterior wall of genital prominence only slightly bulging, with a pair of asymmetrical rounded lobes; posterior wall smoothly curved outward and gradually sloping to merge onto somite some distance from its posterior end. Laterally or ventrally, genital field without posterior lobes. Ventrally, genital operculum (Fig.67-d) typically with a length-width ratio of 79:100 and 69% as wide as somite itself. Genital field with a row of about 30 small pores on each side.

Length ratios of 4 urosomal somites and left caudal ramus, 36.9 : 18.1 : 14.8 : 8.7 : 21.5 = 100. Left caudal ramus extending beyond posterior end of right by about 1/6 its length as measured along medial margin. Setae of caudal rami as in *H. subspinifrons*. Antennule reaching about posterior end of caudal ramus. All cephalosomal appendages and swimming legs similar to *H. spinifrons* except that anterior distal spine of Mxp coxa reaching about 1/3 way to distal end of basis

Description of male. Prosome length, 1.96-2.08 mm; body length, 2.72-2.92 mm. Very similar in habitus to male of *H. quadrilobus* described above, but can be distinguished from it by the following features of the 5th pair of legs: Anteriorly, inner lobe of right basis (Fig.67-g) with a relatively straight row of setules along medial margin and without a conspicuous lateral indentation; third exopodal segment of right P5 (Figs.67-h, i) strongly bent about 1/6 length of segment from distal end; this flexion is particularly pronounced on lateral margin of segment bearing outer spine; outer spine of 3rd exopodal segment relatively small; terminal spine approximately 1/3 length of segment and almost 4 times length of distal lobe (Fig.67-j) of segment as in *H. quadrilobus*; conical process of 2nd exopodal segment of left P5 (Figs.67-e, f) only a little longer than outer spine as measured along medial margins.

Etymology. The specific name *ankylocolus*, from Greek *ankylos* meaning bent and *kolon* denoting a leg, alludes to the conspicuously bent distal exopodal segment of the male right P5.

Distribution. This species seems to be an Indo-Pacific equatorial species. Sixty-seven speci-mens including 49 females and 18 males were found in 17 samples—13 taken throughout the equa-torial Pacific including the Malay Archipelago and South China Sea between 17°50′S and 17°44′N and 4 taken in the Indian Ocean between 25°42′S and 02°28′S.

Remarks. The female is readily distinguished by its genital somite, of which the posterior slope of the genital prominence in lateral view is smoothly bulging and gradually sloped to merge on the somite some distance from the posterior end of the somite; the genital prominence has lobes only on the anterior wall. The 5th pair of legs in the male is very characteristic with a strongly bent 3rd ex-opodal segment in the right leg.

Heterorhabdus caribbeanensis Park 1970
Figures 68 and 69

Heterorhabdus caribbeanensis Park 1970, p. 523, figs. 238-244.—Bradford 1971, p. 129, fig. 8.

Description of female. Prosome length, 1.30-1.80 mm; body length, 1.84-2.52 mm. Body elon-gate (Fig.68-a). Prosome 69% length of body and 2.1 times length of urosome. Laterally, prosome typically with a length-depth ratio of 100:39. Cephalosome 44% as long as prosome, with a cervical groove about halfway along dorsal margin. Laterally, forehead (Fig.68-d) ending in a wide angle and produced ventrad into a short rostrum bearing 2 slender filaments; posterior margin of prosome broadly rounded. Dorsally, forehead (Fig.68-c) broadly rounded, with a rather bluntly pointed mi-danterior tubercular process; posterolateral corner of prosome (Fig.68-b) rounded and slightly pro-duced posteriad, overlapping only about 1/20 length of genital somite.

Dorsally, genital somite (Fig.68-b) widest at middle, with smoothly and strongly bulging lat-eral margins and typically with a length-width ratio of 100:84. Laterally, genital somite (Fig.68-e) typically with a length-depth ratio of 100:86, with a low hump about halfway along dorsal margin and a low and long genital prominence, of which posterior slope broadly bulging and extending down to posterior end of somite. Ventrally, genital operculum (Fig.68-f) typically with a length-width ratio of 73:100 and 66% as wide as somite itself. Genital field with about 10 pores on each side.

First 3 urosomal somites (Fig.68-b) armed with spinules along posterior margins. Length ratios of 4 urosomal somites and left caudal ramus, 36.8 : 18.1 : 14.2 : 9.0 : 21.9 = 100. Left caudal ramus extending beyond posterior end of right by about 1/7 its length as measured along medial margin. Dorsal appendicular seta of left ramus a little longer than ramus and about twice as long as corre-sponding seta of right ramus. Fourth marginal seta of left ramus thicker than other marginal setae, un-armed and longer than body. Dorsally, caudal ramus with a pore.

Antennule extending a little beyond posterior end of prosome. In anatomical details, A1 and Md similar to those of *H. spinifrons*. Maxillule, Mx2, and Mxp also similar to those of *H. spinifrons* except for the following features: In maxillule (Fig.68-g), basis without setae, endopod with 2 setae; in maxilla (Fig.69-a), short seta of 1st lobe approximately 3/4 length of long seta; spine of 2nd lobe thin and short; spine of 3rd lobe a little longer than segment and 60% as long as longest spine of 4th lobe; first saberlike spine of 4th lobe only slightly shorter than 2nd and 81% as long as large, falcate spine of 5th lobe; small spine of 4th lobe approximately 1/4 length of longest spine of same lobe and nearly twice as long as spine of 2nd lobe; saberlike spine of 5th lobe 60% as long as falcate spine; in maxilliped (Fig.69-b), saberlike middle spine of coxa far short of reaching distal end of basis; ante-rior distal spine of coxa small, approximately 1/5 length of segment and about as long as 1st poste-rior distal spine; basis approximately 1.3 times length of coxa as measured along medial margins; first middle seta of basis located 51% length of segment from proximal end. Swimming legs similar to those of *H. spinifrons*.

Description of male. Prosome length, 1.34-1.70 mm; body length, 1.92-2.40 mm. Similar in habitus to female except that urosome 5-segmented. Prosome 71% length of body. Left, geniculate A1 similar to that of *H. spinifrons* male. All other cephalosomal appendages and 1st to 4th swimming legs as in female.

In 5th pair of legs (Fig.69-c), inner lobe of right basis large, about as long as segment itself; distal half of medial margin fringed with a row of long setules and proximal half with row of short and stiff setules. Inner lobe of left basis poorly developed, with 2 rows of relatively short setules. In right exopod (Fig.69-d), outer spine of 1st segment well developed, approximately 2/3 length of outer spine of 2nd segment. Third segment smoothly curved, only a little longer than combined lengths of first 2 segments, with outer spine nearly as long as outer spine of 2nd segment and located 35% length of segment from proximal end; terminal spine 43% as long as segment and 4.5 times as long as distal lobe of segment. In left exopod (Figs.69-e, f), outer spine of 1st segment only a little shorter than that of 2nd; lateral conical process of 2nd segment large, bearing a large outer spine, which is about 1.5 times as long as conical process as measured along medial margins; outer spine of 3rd segment well developed, longer than inner spine.

Distribution. Since the original description from the Caribbean Sea, this species has been recorded by Bradford (1971) from off New Zealand. It is represented in this study by a total of 66 specimens, 47 females and 19 males, from 21 samples—2 taken in the South Atlantic at 18°44'S, 10°15'W and 33°47'S, 15°39'E, respectively; 16 in the Pacific between 25°19'S and 34°33'N from the west coast of North America to Japan and New Guinea; and 3 in the central Indian Ocean between 35°00'S and 17°05'S. These findings plus the previous records demonstrate that the species has a circumglobal distribution.

Remarks. The species was originally described from a single female specimen 1.74 mm long obtained from 1900-980 m in the Caribbean Sea. The male was described for the first time by Bradford (1971) from 2 specimens from off New Zealand in the southwestern Pacific. This is the only species of this species group that has a midanterior tubercular process of the forehead not produced as a spiniform process.

Heterorhabdus insukae, new species
Figures 70 and 71

Type material. Holotype female (USNM 287037), 61 paratype females (USNM 287039), allotype male (USNM 287038), and 25 paratype males (USNM 287040) found in IKMT sample from 3000-0 mwo, SIO Piquero V, station 7, 17°46'S, 110°19'W in the eastern tropical Pacific, March 29, 1969.

Description of female. Prosome length, 1.52-1.80 mm; body length, 2.14-2.56 mm. Body somewhat robust (Fig.70-a). Prosome 71% length of body and about 2.4 times length of urosome. Laterally, prosome typically with a length-depth ratio of 100:37, with broadly rounded posterior margin. Cephalosome 42% length of prosome, with a cervical groove about halfway along dorsal margin. Laterally, forehead (Fig.70-d) produced anteriad into a spiniform process and produced ventrad into a short rostral lobe bearing 2 elongate filaments. Dorsally, forehead (Fig.70-e) broadly rounded, with a relatively tall midanterior tubercular process tapering into a sharp point. Posterolateral corner of prosome rounded, only slightly produced posteriad, overlapping only about 1/20 length of genital somite.

Genital somite (Fig.70-c) relatively long, 47% as long as whole urosome; in dorsal view, typically with a length-width ratio of 100:82, with wavy and somewhat asymmetrical lateral margins. Laterally, genital somite (Figs.70-b, f) typically with a length-depth ratio of 100:80, with a broadly bulging dorsal margin and large genital prominence, of which posterior wall smoothly curved outward and meeting ventral wall of somite some distance from its posterior end. Ventrally, genital operculum (Fig.70-g) close to anterior end of somite, typically with a length-width ratio of 76:100 and 74% as wide as somite itself. Genital field with a row of about 20 pores on each side.

First 3 urosomal somites (Fig.70-c) armed with spinules along posterior margins. Length ratios of 4 urosomal somites and left caudal ramus, 41.3 : 18.1 : 13.1 : 6.9 : 20.6 = 100. Left caudal ramus extending beyond posterior end of right by about 1/5 its length as measured along medial margin. Dorsal appendicular seta of left ramus about 1.5 times length of corresponding seta of right ramus. Fourth marginal seta of left ramus thicker than other marginal setae, unarmed, and longer than body.

Antennule extending about to posterior end of genital somite. In anatomical details all cephalosomal appendages similar to those of *H. spinifrons* except for the features described below. Basis of Mx1 (Fig.70-h) without setae, endopod with a single small seta. In Mx2 (Fig.70-i), short spine of 4th lobe longer than spine of 2nd lobe and approximately 1/4 length of longest spine of same lobe. Proximal 44% of falcate spine of 5th lobe serrated with extremely fine spinules along medial margin; saberlike spine 71% as long as falcate spine. Anterior distal spine of Mxp coxa (Fig.71-a) approximately 1/3 length of segment; 1st posterior distal spine only a little shorter than anterior distal spine. First middle seta of basis located about halfway along medial margin.

First to 4th swimming legs similar to those of *H. spinifrons*. In 5th leg (Fig.71-b), basipod shorter than 3rd exopodal segment. Inner seta of 2nd endopodal segment poorly developed, approximately 2/3 length of inner seta of 1st segment. In exopod, 1st outer spine pointing directly distad, all other outer spines pointing in a distolateral direction; inner spine of 2nd segment slightly shorter than 3rd segment; 1st outer spine of 3rd segment arising 62% length of segment from proximal end; terminal spine 46% as long as segment.

Description of male. Prosome length, 1.54-1.76 mm; body length, 2.12-2.48 mm. Similar in habitus to female except that urosome 5-segmented. Prosome 70% length of body. Left, geniculate A1 similar to that of *H. spinifrons*. All other cephalosomal appendages and 1st to 4th swimming legs as in female.

In 5th pair of legs (Fig.71-c), inner lobe of right basis relatively short (Fig.71d), approximately 1/3 length of segment and reaching about distal end of 1st endopodal segment. In right exopod (Figs.71-e, f), medial projection of 2nd segment extending in a proximomedial direction, with large terminal spiniform process and poorly developed distal lobe; outer spine relatively small and in posterior view, located about 1/3 length of segment from distal end. Third segment smoothly curved, with outer spine about 1/3 length of segment from proximal end; its terminal spine relatively short, approximately 1/4 length of segment and about 3 times as long as terminal lobe of segment. In left exopod (Figs.71-g, h), 1st outer spine larger than 2nd. Second segment with a long lateral conical process terminating with a small outer spine, which is about 2/3 length of conical process as measured along medial margin. Outer spine of 3rd segment shorter than inner spine.

Etymology. This species is named for my wife, Insuk, in appreciation of her encouragement and help throughout the course of this study.

Distribution. This species is represented in the study by a total of 318 specimens including 253 females and 65 males from 31 samples—1 taken in the South Atlantic at 33°47'S, 15°39'E; 24 throughout the Pacific between 17°50'S and 35°01'N and from the west coast of the Americas westward to the Malay Archipelago and East and South China Seas; 6 in the Indian Ocean between 25°50'S and 02°28'S. According to these findings, the species has a circumglobal distribution and is particularly common in equatorial waters.

Remarks. The female can be readily distinguished from all other members of the species group by its relatively long genital somite, which is almost 1/2 length of the whole urosome excluding the telescoped portion of each segment. The male's 5th pair of legs is very characteristic in having a large terminal spiniform process and a poorly developed distal lobe on the medial projection of the 2nd exopodal segment of the right P5. The maxillule of this species is different from those of the congeners in having no setae on the basis and a small single seta on the endopod.

PAPILLIGER SPECIES GROUP

Diagnosis. Midanterior tubercular process of forehead (Fig.72-d) angular or produced into a spiniform process. Small subterminal posterior spine of 4th lobe of Mx2 (Fig.72-i) longer than 1/2 length of large, saberlike spines of same lobe; large falcate spine of 5th lobe of Mx2 conspicuously serrated with spinules and small saberlike spine reaching close to distal end of falcate spine; endopodal setae relatively poorly developed. In Mxp (Fig.73-a), middle marginal spine falcate, distinctly bent at a point 2/7 length of spine from proximal end and extending beyond distal end of basis. In male 5th pair of legs (Fig.73-c), inner lobe of right basis arising from distal end of medial margin, shorter than segment and extending in a distomedial direction; medial projection of 2nd exopodal segment of right P5 (Fig.73-g) large, arising along proximal margin of segment, terminated with a bilobed structure consisting of a rounded plumose proximal lobe, a naked or partially plumose distal lobe, and a small terminal spiniform process in between.

Composition. The following species are referred to this species group:

Heterorhabdus papilliger (Claus 1863)

H. spinifer Park 1970

H. lobatus Bradford 1971

H. guineanensis, new species

H. prolatus, new species

Key to species of *papilliger* species group

Females

1a. Midanterior tubercular process of forehead (Fig.76-d) produced into a spiniform process; P5 (Fig.77-b) without inner marginal seta on 1st endopodal segment . . . .*H. spinifer*

1b. Midanterior tubercular process of forehead (Fig.72-e) bluntly pointed or angular and not produced into a spiniform process; P5 (Fig.73-a) with inner marginal seta on 1st endopodal segment 2

2a. Laterally, genital prominence (Fig.72-b) extending down to posterior end of somite 3

2b. Laterally, genital prominence (Fig.78-b) extending far short to posterior end of somite . . . .*H. prolatus*, n. sp.

3a. Laterally, genital prominence (Fig.74-b) somewhat truncate, with a steep anterior wall making an angle of about 75° with longitudinal axis of somite . . . .*H. lobatus*

3b. Laterally, genital prominence (Fig.72-b) more or less rounded, with a moderately sloping anterior wall making an angle of about 60° with longitudinal axis of somite 4

4a. Laterally, posteroventral margin of genital somite (Fig.72-g) highly inflated . . . .*H. guineanensis*, n. sp.

4b. Laterally, posteroventral margin of genital somite (Fig.75-c) nearly straight and not inflated . . . .*H. papilliger*

Males

1a. Midanterior tubercular process of forehead produced into a spiniform process; basal inner lobe of left P5 (Fig.77-g) large, rounded distally . . . . *H. spinifer*

1b. Midanterior tubercular process of forehead bluntly pointed or angular and not produced into a spiniform process; basal inner lobe of left P5 (Fig.73-c) rather poorly developed and if produced distally, is angular (Fig.79-d) 2

2a. In right P5, medial projection of 2nd exopodal segment (Fig.79-c) not terminally expanded, somewhat truncate, with nearly parallel proximal and distal sides . . . .*H. prolatus*, n. sp.

2b. In right P5, medial projection of 2nd exopodal segment (Fig.73-e) terminally expanded, with a large, rounded proximal lobe 3

3a.   In medial projection of 2nd exopodal segment of right P5 (Fig.73-g),
distal lobe below terminal spiniform process produced distad into
a large toothlike process and larger than proximal . . . . lobe *H. guineanensis*, n. sp.

3b.   In medial projection of 2nd exopodal segment of right P5
(Fig.74-k), distal lobe more or less rounded and similar in size
to proximal lobe . . . .*H. lobatus*

3c.   In medial projection of 2nd exopodal segment of right P5 (Fig.75-k),
distal lobe poorly developed and undifferentiated . . . .*H. papilliger*

Heterorhabdus guineanensis, new species
Figures 72 and 73

Type material. Holotype female (USNM 287041), 38 paratype females (USNM 287043), allotype male (USNM 287042), and 2 paratype males (USNM 287044) found in IKMT sample from 4000-
0 mwo, SIO Lusiad VII, station 79, 01°10'N, 11°36'W in the eastern tropical Atlantic, July 6, 1963.

Description of female. Prosome length, 1.36-1.66 mm; body length, 2.18-2.34 mm. Body elongate (Fig.72-a). Prosome length 71% length of body and about 2.4 times length of urosome. Laterally,
prosome deepest at junction between cephalosome and 1st pedigerous somite, typically with a length-
depth ratio of 100:36. Cephalosome 45% length of prosome, with cervical groove halfway along dorsal margin. Forehead (Fig.72-e) with widely angular anterior end and produced ventrad into a small
rostral lobe bearing 2 slender filaments. Posterior margin of prosome (Fig.72-b) broadly rounded.
Dorsally, forehead (Fig.72-d) with a tall midanterior tubercular process tapering into a blunt point.
Posterolateral corner of prosome broadly rounded.

Dorsally, genital somite (Fig.72-c) widest at middle, with smoothly bulging, symmetrical lateral margins and typically with a length-width ratio of 100:84. Laterally, genital somite (Figs.72-f, g)
about as deep as long, with a low hump halfway along dorsal margin and more or less rounded genital prominence. Anterior slope of genital prominence making an angle of about 60° with longitudinal
axis of somite; posterior slope extending down to posterior end of somite. Posteroventral corner of
genital somite greatly inflated, extending ventrad far beyond ventral margin of following somite. Ventrally, genital operculum (Fig.72-h) with poorly defined anterior margin, typically with a length-width
ratio of 84:100 and 75% as wide as somite itself. Genital field with 7 or 8 pores on each side.

Urosome 29% length of body. First 3 urosomal somites (Fig.72-c) armed with relatively large
spinules along posterior margins. Length ratios of 4 urosomal somites and left caudal ramus, 37.8 :
19.2 : 14.6 : 7.9 : 20.5 = 100. Left caudal ramus extending beyond posterior end of right by about 1/6
its length as measured along medial margin. Dorsal appendicular seta of left ramus nearly twice as
long as that of right. Fourth marginal seta of left ramus much thicker than other marginal setae, unarmed and longer than body.

Antennule extending about to posterior end of 3rd urosomal somite. In anatomical details,
cephalosomal appendages similar to those of *H. spinifrons* except for the features described below. In
Mx2 (Fig.72-i), 2nd seta of 1st lobe very small, about 1/5 length of 1st; subterminal posterior spine
of 4th lobe 63% as long as 2nd saberlike terminal spine, which is only slightly longer than 1st. Falcate spine of 5th lobe only slightly longer than saberlike spine, with proximal 55% of medial margin
serrated with spinules. Falcate spine of 6th lobe with proximal 64% of medial margin serrated with
spinules. In Mxp (Fig.73), middle marginal spine falcate, distinctly bent 2/7 length of spine from
proximal end and extending beyond distal end of basis by about 1/5 its length. Anterior distal spine
of coxa about as long as 1st posterior distal spine and approximately 1/3 length of segment. Basis approximately 1.5 times length of coxa, with 1st middle marginal seta located 51% length of segment
from proximal end.

First to 4th swimming legs similar to those of *H. spinifrons*. In 5th leg (Fig.73-b), basipod
about as long as combined lengths of first 2 exopodal segments. Endopod 81% as long as basipod,
with distolateral corners of first 2 segments each produced into a spiniform process; inner marginal

seta of 1st segment armed with long setules for proximal half and with short spinules for the rest. In exopod, inner spine of 2nd segment about as long as endopod; 3rd segment 44% as long as whole exopod, 93% as long as inner spine of 2nd segment, with 1st outer spine located 55% length of segment from proximal end; terminal spine 51% as long as segment; all outer spines relatively small and pointing in a distolateral direction. Anteriorly, 3rd segment with a pore.

Description of male. Prosome length, 1.44-1.46 mm; body length, 2.02-2.06 mm. Similar in habitus to female. Prosome 71% length of body and approximately 2.3 times length of urosome. Urosome 31% length of body. First 4 urosomal somites armed with spinules along posterior margins. Length ratios of 5 urosomal somites and left caudal ramus, 18.3 : 18.5 : 17.1 : 10.5 : 6.6 : 29.0 = 100. Left caudal ramus extending beyond posterior end of right by about 1/6 its length, with dorsal appendicular seta longer than that of right ramus; 4th marginal seta unarmed, thicker than other marginal setae and longer than body.

Left, geniculate A1 similar in morphological details to that of *H. spinifrons*. All other cephalosomal appendages and 1st to 4th legs as in female. In 5th pair of legs, left endopod a little longer than right (Fig.73-c). Inner lobe of right basis (Fig.73-f) arising from distomedial corner of segment, relatively wide, pointing in a distomedial direction and about 1/2 as long as segment itself; inner lobe of left basis (Fig.73-d) very low. In right exopod (Figs.73-d, e), medial projection of 2nd segment extending along proximal margin of segment, greatly enlarged terminally, with a round plumose anterior lobe and a large naked posterior lobe, which is produced distad into a large toothlike process (Fig.73-g); outer spines of 1st and 2nd segments relatively small; 3rd segment smoothly curved, with a short terminal lobe, a small outer spine a little proximal to midpoint of segment, and a very small terminal spine. In left exopod (Fig.73-h), 2nd segment with a large conical process arising from distolateral margin and terminated with a small outer spine, which is about 2/3 length of conical process as measured along medial margins; 3rd segment (Fig.73-e) tapering distally into a long spiniform process, about as long as 3rd segment of right exopod, with a small outer spine and long inner spine.

Etymology. The specific name *guineanensis* refers to the type locality of the species.

Distribution. This species is represented in the study by 50 specimens including 47 females and 3 males found in 5 samples all taken in the southeastern Atlantic between 15°39'E and 13°40'W and between 32°27'S and 01°10'N. Forty-two of the 50 specimens were found in one sample taken at 01°10'N, 11°36'W in the Gulf of Guinea. Apparently this species is endemic to the eastern Atlantic.

Remarks. This species is closely related to *H. papilliger* and *H. lobatus* described below but can easily be distinguished from them by the 5th pair of legs of the male, in which the medial projection of the 2nd exopodal segment of the right exopod is very characteristic in having a distal lobe produced into a large toothlike process. The female can be identified by the genital somite, which in lateral view has a nearly rounded genital prominence and greatly inflated posteroventral corner. In the present study, the species was found only in the southeastern Atlantic, while *H. papilliger* and *H. lobatus* occurred widely in all three major oceans.

Heterorhabdus lobatus Bradford 1971
Figure 74

Heterorhabdus lobatus Bradford 1971, p. 131, figs. 9, 10a-c.

Description of female. Prosome length, 1.40-1.92 mm; body length, 2.00-2.64 mm. Similar in habitus to *H. guineanensis* and *H. papilliger.* Dorsally, genital somite (Fig.74-a) widest at middle, with smoothly curved, symmetrical lateral margins and typically with a length-width ratio of 100:83. Laterally, genital somite (Figs.74b-d) more or less truncate and about as long as deep, with a low dorsal hump a little anterior to midpoint. Genital prominence high with a steep and slightly hollow anterior slope making an angle of about 75° with longitudinal axis of somite. Posteroventral corner of genital somite greatly inflated, extending posterior margin of somite ventrad to some distance from ventral margin of following urosomal segment.

Urosome (Figs.74-a, b) 30% length of body. First urosomal somites armed with spinules along posterior margins. Length ratios of 4 urosomal somites and left caudal ramus, 37.7 : 18.9 : 15.1 : 8.8 : 19.5 = 100. Dorsal appendicular caudal seta of left ramus longer than that of right ramus. Fourth marginal seta of left ramus thicker than other marginal setae and longer than body. Dorsally, anal segment with a pore on each side; each caudal ramus with a pore.

Antennule reaching about posterior end of 3rd urosomal somite. In anatomical details, all cephalosomal appendages and swimming legs similar to those of *H. guineanensis* described above.

Description of male. Prosome length, 1.36-1.92 mm; body length, 1.92-2.76 mm. Similar in habitus to female. Left, geniculate A1 similar to that of *H. spinifrons*. All other cephalosomal appendages and 1st to 4th legs as in female. In 5th pair of legs (Fig.74-e), inner lobe of right basis relatively narrow and short, arising from distomedial corner of segment and pointing in a distomedial direction (Fig.74-g). Inner lobe of left basis low, with distomedial corner produced into a short pointed process (Fig.74-f). In right exopod, medial projection of 2nd segment (Figs.74-h-k) distally enlarged, with a rounded, plumose proximal lobe, a more or less rounded, naked distal lobe, and a short terminal spiniform process. Distal lobe a little larger than proximal lobe but about similar in shape. Third segment (Fig.74-e) smoothly curved with very short terminal lobe and terminal spine and with outer spine a little proximal to midpoint of segment. In left exopod (Fig.74-f), 2nd segment with distolateral corner produced into a large conical process bearing a small outer spine. Measured along medial margins, outer spine approximately 1/2 length of conical process. Third segment (Fig.74-e) tapering into a long spiniform process, with a relatively small outer spine and long inner spine.

Distribution. This species is represented by a total of 437 specimens including 308 females and 129 males from 31 samples—2 taken from off the west coast of South Africa at 33°47'S, 15°39'E and 32°27'S, 08°49'E, respectively; 20 in coastal waters along the west coast of the Americas between 39°26'S and 40°35'N; 4 in the equatorial offshore waters in the eastern Pacific east of 112°57'W and between 00°55'S and 09°52'N; 1 off the east coast of North Island of New Zealand; 1 in the Tasman Sea at 49°39'S, 152°34'E; 1 in the southwestern Indian Ocean at 35°00'S, 60°00'E; 2 off the east coast of Africa at 25°42'S, 35°28'E and 25°25'S, 38°14'E, respectively. According to these findings and the original record by Bradford (1971), this species occurs mainly in the temperate coastal waters of the whole Southern Hemisphere. However, in the eastern Pacific, its range extends along the west coast of the Americas from about 40°S northward up to about 40°N and in the eastern tropical Pacific, this species is carried westward in the equatorial current from the American coast to about 113°W.

Remarks. This species was originally described from both female and male specimens taken in the waters around New Zealand between 53°26'S and 34°32'S and between 157°31'E and 179°18'W. It is very close in morphology to *H. papilliger* and *H. guineanensis* described above but can be distinguished from them in the female by the shape of the genital somite in lateral view and in the male by the shape of the medial projection of the 2nd exopodal segment of the right P5 as shown in the key.

Heterorhabdus papilliger (Claus 1863)
Figure 75

Heterochaeta papilligera Claus 1863, p. 182, pl. 32, figs. 10-13, 15.—Giesbrecht 1889,
    p. 811.—Giesbrecht 1892, p. 372, pl. 20, figs. 4, 7, 8, 10, 15, 17, 22, 23, 34, 35, 36;
    pl. 39, figs. 40, 53.
Heterorhabdus papilliger, Giesbrecht and Schmeil 1898, p. 114.—Sars 1925, p. 229, pl. 62,
    figs. 13-19.—Sewell 1932, p. 300, fig. 97.—Mori 1937, p. 73, pl. 37, figs. 7-13, pl. 38,
    figs. 1-4.—Tanaka 1964, p. 10.—Vervoort 1965, p. 120.—Park 1970, p. 477.

Description of female. Prosome length, 1.20-1.88 mm; body length, 1.66-2.66 mm. Similar in habitus to *H. guineanensis* and *H. lobatus* described above. Prosome 71% length of body and about 2.3 times length of urosome. Dorsally, genital somite (Fig.75-b) widest at middle, with smoothly bulging, symmetrical lateral margins and typically with a length-width ratio of 100:78. Laterally, genital somite (Figs.75-a, c) only slightly longer than deep, with a low dorsal hump about 3/7 length of somite from anterior end and a more or less rounded genital prominence, which has a nearly straight anterior slope making an angle of about 60° with longitudinal axis of somite and a posterior slope only slightly bulging and extending down to posterior end of somite (Fig.75-c).

Urosome (Figs.75-a, b) 31% length of body. First 3 urosomal somites armed with spinules along posterior margins. Length ratios of 4 urosomal somites and left caudal ramus, 37.7 : 19.5 : 14.5 : 8.2 : 20.1 = 100. Left caudal ramus extending beyond posterior end of right ramus by about 1/6 its length as measured along medial margin. Dorsal appendicular seta of left caudal ramus a little longer than that of right ramus. Fourth marginal seta of left ramus much thicker than other marginal setae and longer than body. Dorsally, anal segment with a pore on each side and each caudal ramus with a pore.

Antennule reaching about posterior end of 3rd urosomal somite. All cephalosomal appendages and swimming legs similar in anatomical details to those of *H. guineanensis* described above.

Description of male. Prosome length, 1.28-1.88 mm; body length, 1.80-2.64 mm. Similar in habitus to female. Left, geniculate A1 similar to that of *H. spinifrons*. All other cephalosomal appendages and 1st to 4th swimming legs as in female.

In 5th pair of legs (Fig.75-d), inner lobe of right basis relatively narrow, slightly shorter than 1/2 length of segment, and pointing in a distomedial direction (Fig.75-e). Inner lobe of left basis (Fig.75-f) low but clearly definable, and distally produced into a short process. In right exopod, medial projection of 2nd segment (Figs.75-h-k) with a large, rounded, plumose proximal lobe and without a definable distal lobe; whole distal margin of medial projection smoothly curved and inseparably merged into a relatively large terminal spiniform process (Fig.75-i); outer spine of 2nd segment relatively long and arising close to distal end of segment. Third segment (Fig.75-d) smoothly curved, about as long as combined lengths of first 2 segments; its outer spine small, located distal to midpoint of segment; terminal spine approximately 1/6 length of segment, and terminal lobe about 2/5 length of terminal spine. In left exopod (Fig.75-g), 2nd segment with a large lateral conical process terminating with small outer spine. Measured along medial margins, outer spine approximately 2/3 length of conical process. Third segment (Fig.75-d) tapering distally into a rather bluntly pointed spiniform process, with a small outer spine and a long inner spine.

Distribution. Since its original description by Claus (1863) from the Mediterranean Sea, *H. papilliger* has been recorded widely in all three major oceans (Vervoort 1965). Some of these records are believed to include species closely related to *H. papilliger* but unknown then.

The species is represented by 760 specimens including 470 females and 290 males from 39 samples—5 taken in the Atlantic between 33°47′S and 36°59′N; 29 throughout the Pacific from the west coast of the Americas westward to the Malay Archipelago between 31°47′S and 40°35′N; 5 in the Indian Ocean between 35°00′S and 12°31′S. These findings once again demonstrate that *H. papilliger* is a common circumglobal species occurring widely in tropical and temperate waters. In this study, as many as 214 specimens were found in a single sample taken from off Baja California.

Remarks. This species was originally described from the Mediterranean Sea. The species was redescribed by Giesbrecht (1892) from specimens obtained in the equatorial Pacific, Sars (1925) from specimens obtained in the North Atlantic, and Mori (1937) from specimens taken from off Japan. All of these previous descriptions seem to agree well with the specimens referred to this species in the present study. However, the species Grice (1962) described and figured as *H. papilliger* from the equatorial Pacific is obviously different in the shape of the female genital somite and the male 5th pair of legs from the material referred to this species here. According to his figures of the 5th pair of legs, Grice's male seems to belong to *H. prolatus*, a new species described below.

The female of *H. papilliger* can be recognized by the genital somite, which in lateral view has a more or less rounded genital prominence and an uninflated posteroventral margin and the male by the medial projection of the 2nd exopodal segment of the right P5, which has a poorly developed distal lobe.

Heterorhabdus spinifer Park 1970
Figures 76 and 77

Heterorhabdus spinifer Park 1970, p. 519, figs. 226-237.—Bradford 1971, p. 129.

Description of female. Prosome length, 1.14-1.34 mm; body length, 1.60-1.90 mm. Body elongate (Fig.76-a). Prosome 71% length of body and approximately 2.3 times length of urosome. Laterally, prosome deepest at joint between cephalosome and 1st pedigerous somite, typically with a length-depth ratio of 100:38. Cephalosome 48% length of prosome, with cervical groove halfway along dorsal margin. Forehead (Fig.76-d) produced anteriad into a small spiniform process and ventrad into a small rostrum bearing 2 slender filaments. Posterior margin of prosome broadly rounded (Fig.76-e). Dorsally, midanterior tubercular process of forehead (Fig.76-c) relatively tall, terminated with a spiniform process. Posterolateral corner of prosome broadly rounded (Fig.76-b).

Dorsally, genital somite (Fig.76-b) widest at middle, with smoothly curved, symmetrical lateral margins and typically with a length-width ratio of 100:80. Laterally, genital somite (Fig.76-e) typically with a length-depth ratio of 100:88, with a smoothly curved dorsal margin and more or less rounded genital prominence. Anterior slope of genital prominence with a low hump at middle, making an angle of about 53° with longitudinal axis of somite; posterior slope extending almost straight down to posterior end of somite. Ventrally, genital operculum (Fig.76-f) about as wide as long and 65% as wide as somite itself. Genital field with about 15 pores on each side.

First 3 urosomal somites (Fig.76-b) armed with spinules along posterior margins. Length ratios of 4 urosomal somites and left caudal ramus, 36.4 : 18.8 : 15.7 : 8.5 : 20.6 = 100. Left caudal ramus extending beyond posterior end of right ramus by about 1/5 its length as measured along medial margin. Dorsal appendicular seta of left ramus approximately 1.5 times length of that of right ramus. Fourth marginal seta of left ramus much thicker than other marginal setae and longer than body. Dorsally, anal segment with a pore on each side; each caudal ramus with 1 or 2 pores.

Antennule reaching close to posterior end of caudal ramus. In anatomical details, all cephalosomal appendages similar to those of *H. guineanensis* described above except that in Mx2 (Fig.76-g), falcate spine of 5th lobe with proximal 68% of medial margin serrated with spinules; falcate spine of 6th lobe with proximal 73% of medial margin serrated with spinules. Swimming legs similar to those of *H. guineanensis* except that in P1, 1st outer spine of exopod (Fig.77-a) relatively large, overreaching base of following outer spine and P5 (Fig.77-b) without inner marginal seta on 1st endopodal segment.

Description of male. Prosome length, 1.16-1.38 mm; body length, 1.64-2.02 mm. Similar in habitus to female. Prosome 68% length of body and approximately 2.2 times length of urosome. Urosome 32% length of body. First 4 urosomal somites armed with relatively large spinules along posterior margins.

Left, geniculate A1 similar in morphology to that of *H. spinifrons.* All other cephalosomal appendages and 1st to 4th legs as in female. In 5th pair of legs, inner lobe of right basis (Fig.77-c) elongate, only a little shorter than segment, extending in a distomedial direction. Inner lobe of left basis (Fig.77-d) well developed, relatively wide and rounded distally, extending beyond distal end of 1st endopodal segment by 1/2 its length. In right exopod (Figs.77-e, f), outer spine of 2nd segment small; medial projection somewhat truncated, with well-developed proximal lobe and terminal spiniform process but its distal lobe poorly developed and indistinguishable from the latter. Third segment smoothly curved, with outer spine about 2/5 length of segment from proximal end, terminal spine approximately 1/5 length of segment, and terminal lobe about 1/4 length of terminal spine. In left exopod (Figs.77-g, h), 2nd segment with a large lateral conical process terminating with a small outer spine. Measured along medial margins, outer spine a little longer than 1/2 length of conical process. Third segment about as long as combined lengths of first 2 segments, with falciform terminal spine, which is not tapering into a spiniform process; its outer spine approximately 2/3 length of inner spine.

Distribution. This species was originally described from the Caribbean Sea and Gulf of Mexico. Bradford (1971) found the species at 25°46'S, 170°16'W, off North Island, New Zealand. In the present study, it is represented by a total of 168 specimens including 136 females and 32 males from

18 samples—3 taken in the Atlantic, 1 at 38°53'N, 72°22'W off the east coast of North America and 2 at 33°47'S, 15°39'E and 30°09'S, 04°59'W, respectively, off the west coast of South Africa; 8 in the South Pacific between 25°19'S and 09°25'S and between 159°06'E and 106°06'W; 7 in the Indian Ocean between 35°00'S and 02°28'S from the east coast of Africa eastward to 86°23'E. The species therefore has a circumglobal distribution. In the Indo-Pacific, the species seems to be restricted mainly to the southern tropical regions, while it occurs widely in the Atlantic.

Remarks. This species can readily be distinguished from the other members of the species group by the presence of a spiniform process at the anterior end of the forehead, in the female by the absence of an inner marginal seta on the 1st endopodal segment of P5 and in the male by the large and distally rounded basal inner lobe and falciform exopodal terminal spine of the left P5.

Heterorhabdus prolatus, new species
Figures 78 and 79

Heterorhabdus papilliger (male only), not of Claus; Grice 1962, p. 222, pl. 24, figs. 13-16.

Type material. Holotype female (USNM 287045), 3 paratype females (USNM 287047), allotype male (USNM 287046), 5 paratype males (USNM 287048) found in IKMT sample from 2500-0 mwo, SIO Piquero V, station 3, 14°46'S, 93°37'W in the eastern tropical Pacific, March 18, 1969. Eight paratype females (USNM 287049) found in IKMT sample from 3000-0 mwo, SIO Piquero V, station 7, 17°46'S, 110°19'W in the eastern tropical Pacific, March 29, 1969.

Description of female. Prosome length, 1.28-1.56 mm; body length, 1.78-2.14 mm. Body elongate (Fig.78-a). Prosome length 70% length of body and approximately 2.3 times length of urosome. Laterally, prosome typically with a length-depth ratio of 100:37. Cephalosome 46% length of prosome, with a cervical groove halfway along dorsal margin. Forehead (Fig.78-d) angular and produced ventrad into a small rostrum bearing 2 elongate filaments. Posterior margin of prosome broadly rounded (Fig.78-b). Dorsally, midanterior tubercular process of forehead (Fig.78-e) tapering into a pointed tip. Posterolateral corner of prosome rounded (Fig.78-c).

Dorsally, genital somite (Fig.78-c) with somewhat asymmetrical lateral swellings, more or less parallel to each other, typically with a length-width ratio of 100:81. Laterally, genital somite (Figs.78-f, h) typically with a length-depth ratio of 100:94, with a smoothly curved dorsal margin and somewhat rounded genital prominence. Anterior slope of genital prominence straight or only slightly bulging, making an angle of about 55° with longitudinal axis of somite; posterior slope reaching far short of posterior end of somite. Ventrally, genital operculum (Fig.78-g) typically with a length-width ratio of 88:100 and 73% as wide as somite itself. Genital field with about 12 pores on each side.

First 3 urosomal somites (Figs.78-b, c) armed with a relatively large spinules along posterior margins. Length ratios of 4 urosomal somites and left caudal ramus, 38.8 : 18.7 : 13.7 : 8.8 : 20.0 = 100. Left caudal ramus extending beyond posterior end of right ramus by about 1/5 its length as measured along medial margin. Dorsal appendicular seta of left ramus almost twice as long as that of right ramus. Fourth marginal seta of left ramus much thicker than other marginal setae and longer than body. Dorsally, each caudal ramus with a relatively large pore.

Antennule reaching close to posterior end of caudal ramus. In anatomical details, all cephalosomal appendages similar to those of *H. spinifer* described above except that in Mx2 (Fig.79-a), falcate spine of 5th lobe with proximal 54% of medial margin serrated with spinules; falcate spine of 6th lobe with proximal 63% of medial margin serrated with spinules. All swimming legs also similar to those of *H. spinifer* except that P5 with an inner marginal seta on 1st endopodal segment.

Description of male. Prosome length, 1.26-1.52 mm; body length, 1.80-2.14 mm. Similar in habitus to female. Prosome 71% length of body; urosome 30% length of body. First 4 urosomal somites armed with relatively large spinules along posterior margins. Caudal rami and their setation as in female.

Left, geniculate A1 similar in morphology to that of *H. spinifrons*. All other cephalosomal appendages and 1st to 4th swimming legs as in female. In 5th pair of legs, inner lobe of right basis

(Fig.79-c) elongate, approximately 2/3 length of segment, and extending in a distomedial direction; inner lobe of left basis (Fig.79-d) relatively well developed, with distal margin high and nearly perpendicular to longitudinal axis of leg, with distomedial corner produced distad into a short spiniform process. In right exopod, 2nd segment with a characteristic medial projection (Fig.79-c), which is elongate, without terminal enlargement and with nearly parallel proximal and distal sides. Third segment smoothly curved, with outer spine about 2/5 length of segment from proximal end, terminal spine approximately 1/7 length of segment, and terminal lobe about 1/5 length of terminal spine. In left exopod (Fig.79-e), 2nd segment with a large lateral conical process tipped with a small outer spine. Measured along medial margins, outer spine approximately 2/3 length of conical process. Third segment with a relatively small outer spine and long inner spine, tapering distally into a relatively short, stout, and sharply pointed spine (Fig.79-d).

Etymology. The specific name *prolatus*, from the Latin meaning elongated, refers to the relatively long genital somite in the female.

Distribution. This species is represented by a total of 67 specimens including 55 females and 12 males found in 13 samples—10 taken widely in the equatorial waters of the Pacific including the Malay Archipelago between 17°46'S and 00°02'S and between 125°33'E and 93°37'W; 1 at 17°44'N, 119°55'E in the South China Sea; 2 at 25°42'S, 35°28'E and 02°28'S, 86°23'E, respectively, in the Indian Ocean. Apparently, it is an equatorial species of the Indo-Pacific.

Remarks. The female of this species can readily be distinguished from those of the other members of the species group by the genital somite, which in lateral view has a genital prominence far short of reaching the posterior end of the somite. The male can be identified mainly by the 5th pair of legs, in which the medial projection of the 2nd exopodal segment of the right P5 is elongate, without a terminal enlargement, and the basal inner lobe of the left P5 has a high distal margin, which is almost perpendicular to the longitudinal axis of the leg and its distomedial corner is produced into a spiniform process. Grice (1962) described a male as *H. papilliger* from the equatorial waters of the Pacific. According to his figures of the male 5th pair of legs, Grice's male seems to be referable to this new species.

FISTULOSUS SPECIES GROUP

Diagnosis. Midanterior tubercular process of forehead (Fig.82-d) low and somewhat truncated. Third marginal seta of left caudal ramus and 3rd and 4th marginal setae of right ramus armed with spines in addition to normal setules. Subterminal posterior spine of 4th lobe of Mx2 (Fig.80-j) longer than 1/2 length of large, saberlike spines of same lobe; falcate spine of 5th lobe of Mx2 conspicuously serrated with spinules and saberlike spine reaching close to distal end of falcate spine; endopodal setae relatively well developed. In Mxp (Fig.81-c), middle marginal spine of coxa falcate, strongly elbowed 1/4 length of spine from proximal end and extending beyond distal end of basis; basis (Figs.81-d, e) with 1 or 2 large, elevated pores distal to 3rd middle marginal seta. In male 5th pair of legs (Fig.81-f), inner lobe of right basis arising from distomedial corner of segment, shorter than segment and extending in a distomedial direction; medial projection of 2nd exopodal segment of right P5 about equal to or shorter than 1/2 length of segment, nearly conical, its proximal margin some distance from proximal end of segment.

Composition. The following species are referred to this species group:

Heterorhabdus fistulosus Tanaka 1964

H. egregius Heptner 1972

Remarks. This species group is intermediate between the *papilliger* species and *abyssalis* species groups, with the basal inner lobes of the male P5 of the *papilliger* type and the marginal setae of the caudal rami and the medial projection of the 2nd exopodal segment of the male right P5 of the *abyssalis* type. The Mx2 is intermediate with regard to the length of the saberlike spine of its 5th lobe.

Key to species of *fistulosus* species group

Females

1a. Dorsally, genital somite (Fig.80-c) widest in anterior half; dorsally as
well as laterally, genital prominence (Figs.80-g, h) far short of reaching
posterior end of somite . . . .*H. fistulosus*

1b. Dorsally, genital somite (Fig.82-b) widest in posterior half; dorsally as
well as laterally, genital prominence (Fig.82-e) extending down to
posterior end of somite . . . .*H. egregius*

Males

1a. Basal inner lobe of right P5 (Fig.81-f) elongate; lateral conical
process and outer spine of 2nd exopodal segment of left P5 (Fig.81-i)
elongate, their combined lengths approximately 2/3 length of segment. . . . .*H. fistulosus*

1b. Basal inner lobe of right P5 (Fig.83-d) wide; lateral conical process
and outer spine of 2nd exopodal segment of left P5 (Fig.83-i) short,
their combined lengths about 1/4 length of segment . . . .*H. egregius*

Heterorhabdus fistulosus Tanaka 1964
Figures 80 and 81

Heterorhabdus fistulosus Tanaka 1964, p. 7, fig. 177.—Heptner 1971, p. 149, fig. 36.

Description of female. Prosome length, 2.36-2.60 mm; body length, 3.40-3.72 mm. Body elongate (Fig.80-a). Prosome 70% length of body and approximately 2.2 times length of urosome. Laterally, prosome deepest at joint between cephalosome and 1st pedigerous somite, typically with a length-depth ratio of 100:38. Cephalosome 44% length of prosome, with cervical groove halfway along dorsal margin. Laterally, forehead with rounded anterior end and produced ventrad into a conical rostrum bearing 2 slender filaments; dorsally, forehead with a low, somewhat truncated midanterior tubercular process as in *H. egregius* (Fig.82-d). Laterally, posterior margin of prosome (Fig.80-b) broadly rounded; dorsally, posterolateral corner of prosome (Fig.80-c) also rounded, overlapping only about 1/10 length of genital segment.

Urosome 32% length of body. Genital somite with a tubercular outgrowth on dorsolateral side and close to posterior end. Its size varies from a small, low ridge (Fig.80-c) only visible by a careful observation to a highly prominent conical process (Figs.80-g-i). Dorsally, genital somite (Fig.80-c) widest in anterior half, with smoothly curved, nearly symmetrical lateral swellings not extending to posterior end of somite, and typically with a length-width ratio of 100:80. Laterally, genital somite (Figs.80-d, e) typically with a length-depth ratio of 100:84, with a low dorsal hump 45% length of somite from anterior end and more or less rounded genital prominence. Anterior wall of genital prominence slightly bulging and sloping at an angle of about 52° with reference to longitudinal axis of somite; posterior wall also only slightly bulging and short of reaching posterior end of somite. Ventrally, genital operculum (Fig.80-f) typically with a length-width ratio of 94:100 and about 3/4 as wide as somite itself. Genital field with 7-16 pores on left side and about 25 pores on right side.

First 3 urosomal somites (Figs.80-b, c) armed with spinules along posterior margins. Length ratios of 4 urosomal somites and left caudal ramus, 36.4 : 18.8 : 13.9 : 9.1 : 21.8 = 100. Left caudal ramus extending beyond posterior end of right ramus by about 1/6 its length as measured along medial margin. Dorsal appendicular seta of left ramus a little longer than that of right ramus. Fourth marginal seta of left ramus thicker than other marginal setae and longer than body. Third marginal seta of left ramus, 3rd and 4th marginal setae of right ramus armed with spines in addition to normal setules. Dorsally, each caudal ramus with 2 or 3 pores.

Antennule extending beyond posterior end of caudal ramus by its last 1 or 2 segments. In morphological details, A1, A2, Md, and Mx1 similar to those of *H. spinifrons*. In Mx2 (Fig.80-j), 2nd seta of 1st lobe approximately 1/3 length of 1st seta. Posterior subterminal spine of 4th lobe 68% as long as 2nd saberlike spine, which is only slightly longer than 1st. Saberlike spine of 5th lobe 83% as long as falcate spine. Proximal 68% of medial margin of falcate spine serrated with spinules. Falcate spine of 6th lobe with proximal 77% of medial margin serrated with spinules. In Mxp (Fig.81-c), middle marginal spine of coxa falcate, distinctly bent 1/4 length of spine from proximal end and extending beyond distal end of basis by about 1/5 its length; anterior distal spine of coxa a little shorter than 1st posterior distal spine and approximately 1/3 length of segment; 2nd posterior distal spine about 1/2 length of 1st. Basis approximately 1.5 times length of coxa, with 1st middle marginal seta about halfway along medial margin and separated from 2nd by a wide gap, and with 1 or 2 extraordinarily large elevated pores distal to 3rd marginal seta (Figs.81-d, e).

Swimming legs similar to those of *H. spinifrons* except for the following features: In P1 exopod (Fig.81-a), 1st outer spine extending beyond base of following outer spine by about 1/5 its length; each of 2nd and 3rd outer spines extending beyond base of following outer spine by about 1/4 its length; terminal spine of 3rd segment approximately 2.4 times length of segment; in P5 exopod (Fig.81-b), 3rd segment about as long as combined lengths of first 2 segments; 1st outer spine pointing straight distad, 2nd in a distomedial direction, 3rd and 4th in a distolateral direction; inner spine of 2nd segment 88% as long as 3rd segment; terminal spine of 3rd segment 54% as long as segment.

Description of male. Prosome length, 2.44-2.52 mm; body length, 3.40-3.56 mm. Similar in habitus to female. Prosome 76% length of body. Urosome 29% length of body. First 4 urosomal somites armed with spinules along posterior margins. Left, geniculate A1 similar to that of *H. spinifrons* male. All other cephalosomal appendages and 1st to 4th swimming legs as in female.

In 5th pair of legs (Fig.81-f), inner lobe of right basis arising from distomedial corner of segment, pointing nearly straight distad, conical, a little shorter than 1/2 length of segment. Inner lobe of left basis low, distally produced into a short process. Left endopod longer than right. In right exopod, medial projection (Figs.81-g, h) approximately 1/3 length of segment, conical, with proximal margin somewhat inflated at beginning and receded for some distance from proximal end of segment, with a plumose ridge running along proximal margin. Third segment (Fig.81-f) strongly curved, with outer spine about 1/5 length of segment from proximal end, terminal spine approximately 1/3 length of segment, and with a very small terminal lobe. In left exopod (Fig.81-i), 2nd segment with a greatly elongated lateral conical process bearing outer spine; combined lengths of conical process and outer spine approximately 2/3 length of segment; conical process a little longer than outer spine. Third segment (Fig.81-f) with a short outer and a long inner spine and tapering distally into a very long spine. Together with terminal spine, 3rd segment 73% as long as whole exopod.

Distribution. Since its original discovery in the Izu region of Japan, the species has been recorded from the Kuril-Kamchatka Trench by Heptner (1971). In the present study, it is represented by 40 specimens including 25 females and 15 males all found in a single sample taken at 40°29'N, 125°57'W off the coast of northern California. This species is apparently confined in distribution to the northern Pacific.

Remarks. The original description of this species was based on 2 females (3.39, 3.60 mm long, respectively) and 1 male (3.45 mm long) collected in the Izu region of Japan. Heptner (1971) found 2 different types of females, one with a tubercular outgrowth on the dorsolateral side of the genital somite and another without such an outgrowth and distinguished them as *H. fistulosus* f. *tuberculata* and *H. fistulosus* f. *typica*, respectively. In the present study, the tubercular outgrowth was found to vary from a small, low ridge only visible by a careful observation to a highly prominent conical process. Upon examination of all 25 females, with tubercular processes of varying size, from a single sample, no other morphological differences could be found. All 15 males in the sample also showed no morphological differences. It is therefore believed that all individuals found in the present study, regardless of the size of the tubercular process of the female genital somite, belong to the same species.

The female of this species can be distinguished by its genital somite, which is relatively long, widest in the anterior half, and with a genital prominence not extending to the posterior end of the somite. The male can be identified by the 5th pair of legs, in which the basal inner lobe of the right P5 is conical and the lateral conical process of the 2nd exopodal segment of the left P5 is greatly elongated.

Heterorhabdus egregius Heptner 1972
Figures 82 and 83

Heterorhabdus egregius Heptner 1972a, p. 1647, figs. 1, 2.

Description of female. Prosome length, 1.60-2.24 mm; body length, 2.24-3.12 mm. Similar in habitus to *H. fistulosus* described above. Prosome 72% length of body and approximately 2.5 times length of urosome. Laterally, prosome (Fig.82-a) deepest at junction between cephalosome and 1st pedigerous somite, typically with a length-depth ratio of 100:40. Cephalosome 49% length of prosome, with a cervical groove halfway along dorsal margin. Laterally, forehead (Fig.82-c) with rounded anterior end and produced ventrad into a conical rostrum bearing 2 elongate filaments. Dorsally, forehead (Fig.82-d) with a low, more or less truncate midanterior tubercular process. Laterally, posterior margin of prosome broadly rounded (Fig.82-e). Dorsally, posterolateral corner of prosome (Fig.82-b) rounded, overlapping about 1/15 length of genital somite.

Urosome 29% length of body. Dorsally, genital somite (Fig.82-b) widest in posterior half, with smoothly curved, symmetrical lateral swellings extending to posterior end of somite, and typically with a length-width ratio of 100:81. Laterally, genital somite (Fig.82-e) typically with a length-depth ratio of 100:85, with a low hump about halfway along dorsal margin and more or less rounded genital prominence. Anterior wall of genital prominence sloping at an angle of about 60° with reference to longitudinal axis of somite; posterior wall extending to posterior end of somite and smoothly bulging. Ventrally, genital operculum (Fig.82-f) typically with a length-width ratio of 82:100 and 67% as wide as somite itself. Genital field with 3-5 pores on left side and about 15 pores on right side.

First 3 urosomal somites (Fig.82-b) armed with spinules along posterior margins. Length ratios of 4 urosomal somites and left caudal ramus, 35.8 : 17.0 : 13.9 : 9.7 : 23.6 = 100. Left caudal ramus extending beyond posterior end of right by about 1/6 its length as measured along medial margin. Caudal rami with dorsal appendicular setae of similar length. Fourth marginal seta of left ramus thicker than other marginal setae and longer than body. Third marginal seta of left ramus and 3rd and 4th marginal setae of right ramus armed with spines in addition to normal setules.

Antennule extending beyond posterior end of caudal ramus by last 3 or 4 segments. All cephalosomal appendages similar in morphological details to those of *H. fistulosus* except for the features described below. In Mx2 (Fig.82-g), 2nd seta of 1st lobe approximately 1/2 length of 1st. Posterior subterminal spine of 4th lobe 63% as long as 2nd saberlike spine. Saberlike spine of 5th lobe 88% as long as falcate spine; falcate spine with proximal 64% of medial margin serrated with spinules. Falcate spine of 6th lobe with proximal 74% of medial margin serrated with spinules. In Mxp (Figs.82-h-k), basis with a small elevated pore in addition to 1 or 2 extraordinarily large elevated pores distal to 3rd middle marginal seta. Swimming legs similar to those of *H. fistulosus* except for some minor differences in P5 as follows: In exopod (Figs.83-a, b), 3rd segment shorter than combined length of first 2 segments; inner spine of 2nd segment 91% as long as 3rd segment; terminal spine of 3rd segment 60% as long as segment; anteriorly, 3rd segment with a pore.

Description of male. Prosome length, 1.64-2.24 mm; body length, 2.28-3.12 mm. Similar in habitus to female. Prosome 71% length of body. Urosome 29% length of body and its first 3 somites armed with spinules along posterior margins. Left, geniculate A1 similar to that of *H. spinifrons* male. All other cephalosomal appendages and 1st to 4th swimming legs as in female.

In 5th pair of legs, inner lobe of right basis (Figs.83-d, e) arising from distomedial corner of segment, wide, conical, distally rounded, approximately 1/3 length of segment. Inner lobe of left basis (Fig.83-f) low, produced distally into a small process. Left endopod (Fig.83-c) longer than right. In right exopod, medial projection (Figs.83-g, h) with proximal margin receded for some distance from proximal end of segment, about 2/5 as long as segment, conical, with a large lobular inflation on beginning half of proximal margin, and with a plumose ridge running along proximal margin. Third segment (Fig.83-c) strongly curved, with outer spine about 1/4 length of segment from proximal end, terminal spine approximately 1/2 length of segment, and with very small terminal lobe. In left exopod, 2nd segment (Fig.83-i) with a relatively short lateral conical process bearing outer spine; conical process and outer spine similar in length and their combined lengths approximately 1/3 length of segment. Third segment (Fig.83-c) with a small outer and a long inner spine and tapering distally into a very long spine. Third segment including terminal spine 69% as long as whole exopod.

Distribution. The species is represented by a total of 111 specimens including 81 females and 30 males from 21 samples—5 collected in the South Atlantic between 33°47'S and 18°30'S; 15 throughout the Pacific between 46°57'S and 31°13'N and from the west coast of the Americas westward to the Malay Archipelago; 1 at 35°00'S, 60°00'E in the Indian Ocean. These findings are the 2nd record since its original discovery in the Kuril-Kamchatka Trench and demonstrate that the species is distributed widely in all three major oceans.

Remarks. This species was originally described from females 2.50-2.75 mm long and males 2.50-2.65 mm long from the Kuril-Kamchatka Trench. The female can readily be distinguished from that of *H. fistulosus* by the genital somite, which is widest in the posterior half and the genital prominence of which is extending down to the posterior end of the somite. The male can be distinguished by the relatively short and wide basal inner lobe of the right P5 and the short lateral conical process and outer spine of the 2nd exopodal segment of the left P5.

ABYSSALIS SPECIES GROUP

Diagnosis. Dorsally, midanterior tubercular process of forehead (Fig.84-e) low and more or less truncated. Third marginal seta of left caudal ramus and 3rd and 4th marginal setae of right ramus armed with spines in addition to normal setules. Posterior subterminal spine of 4th lobe of Mx2 (Fig.84-i) longer than 1/2 length of large, saberlike spines of same lobe; falcate spine of 5th lobe of Mx2 conspicuously serrated with spinules, saberlike spine about equal to or less than 3/4 length of falcate spine; endopodal setae relatively well developed. In Mxp (Fig.84-j), middle marginal spine of coxa falcate, distinctly bent 1/4 length of spine from proximal end and extending far beyond distal end of basis. In male 5th pair of legs (Fig.84-g), inner lobe of right basis arising from middle or proximal half of segment, longer than segment and extending in a distolateral direction; medial projection of 2nd exopodal segment of right P5 relatively small, about equal to or shorter than 1/3 length of segment, nearly conical, its proximal margin receding for some distance from proximal end of segment.

Composition. The following species are referred to this species group:

Heterorhabdus norvegicus (Boeck 1872)
Heterorhabdus abyssalis (Giesbrecht 1889)
Heterorhabdus clausii (Giesbrecht 1889)
Heterorhabdus tanneri (Giesbrecht 1895)
Heterorhabdus austrinus Giesbrecht 1902
Heterorhabdus pustulifer Farran 1929
Heterorhabdus pacificus, Brodsky 1950
Heterorhabdus spinosus, Bradford 1971
Heterorhabdus habrosomus, new species
Heterorhabdus prolixus, new species
Heterorhabdus tuberculus, new species
Heterorhabdus oikoumenikis, new species
Heterorhabdus longisegmentus, new species
Heterorhabdus americanus, new species
Heterorhabdus cohibilis, new species
Heterorhabdus paraspinosus, new species
Heterorhabdus confusibilis, new species

Remarks. The females of this species group are similar to those of the *fistulosus* species group but distinguished from them by a relatively short saberlike spine of the 5th lobe of Mx2 and the absence of extraordinarily large elevated pores on the basis of Mxp. The males are different from those of all other species groups of the genus in having a greatly elongated inner basal lobe of the right P5.

Key to species of *abyssalis* species group

Females

1a. Genital somite (Fig.92-b) with a conical projection on dorsal side . . . . *H. pustulifer*
1b. Genital somite without such a conical projection 2

2a. Ventrally, genital somite (Figs.114-g, h) with a small conical process
bearing concentric ridges posterior to genital operculum . . . . *H. norvegicus*
2b. Ventrally, genital somite without such a process 3

3a. Laterally, genital somite (Fig.102-a) with a tubercular outgrowth
posteriorly on dorsal margin . . . . *H. paraspinosus*, n. sp.
3b. Laterally, genital somite without such a tubercular outgrowth 4

4a. Laterally, genital somite (Figs.104-f, 112-e) with a tubercular
outgrowth posteriorly on ventral margin 5
4b. Laterally, genital somite without such a tubercular outgrowth 6

5a. Laterally, genital operculum (Fig.104-b) reaching close to posterior
end of somite; endemic to the Antarctic . . . . *H. austrinus*
5b. Laterally, genital operculum (Fig.112-d) short of reaching posterior
end of somite; endemic to the northern Pacific . . . . *H. tanneri*

6a. Laterally, genital operculum (Figs.100-c, 106-b) reaching close to
posterior end of somite 7
6b. Laterally, genital operculum (Fig.84-b) far short of reaching posterior
end of somite 8

7a. Laterally, genital flange (Fig.100-f) short and high, distinctly separate
from posterior ventral margin of somite; posterior subterminal spine of
4th lobe of Mx2 (Fig.101-a) about 1/5 length of 2nd saberlike spine . . . . *H. spinosus*
7b. Laterally, genital flange (Fig.106-d) elongate and low, separate from
posterior ventral margin of somite by a shallow notch; posterior
subterminal spine of 4th lobe of Mx2 (Fig.106-h) almost 1/2 length
of 2nd saberlike spine . . . . *H. abyssalis*

8a. Laterally, right and left genital flanges (Figs.90-e, f) asymmetrical 9
8b. Laterally, right and left genital flanges symmetrical 11

9a. Laterally, left genital flange (Fig.90-e) high, like a tubercular
outgrowth . . . . *H. tuberculus*, n. sp.
9b. Laterally, left genital flange low, smoothly arched 10

10a. Laterally, genital somite (Fig.86-e) relatively elongate, typically
with a length-depth ratio of 100:83; endemic to coastal waters
of the eastern Pacific . . . . *H. prolixus*, n. sp.
10b. Laterally, genital somite (Fig.88-d) relatively short, typically with
a length-depth ratio of 100:89; endemic to equatorial oceanic waters
of the Pacific . . . . *H. clausii*

11a. Laterally, genital prominence (Fig.94-b) gradually merging onto ventral
wall of somite 12
11b. Laterally, genital prominence (Fig.108-f) distinctly separate from
ventral wall of somite 13

12a. Dorsally, genital somite (Fig.94-c) only slightly asymmetrical ....*H. oikoumenikis*, n. sp.
12b. Dorsally, genital somite (Fig.96-d) strongly asymmetrical ....*H. longisegmentus*, n. sp.

13a. Dorsally as well as ventrally, lateral swellings of genital somite
(Fig.108-b) reaching posterior end of somite....................................14
13b. Dorsally as well as ventrally, lateral swellings of genital somite
(Fig.84-c) short of reaching posterior end of somite............................16

14a. Laterally, genital prominence (Fig.108-f) extending close to posterior
end of somite; genital somite with a conspicuous dorsal hump ....*H. pacificus*
14b. Laterally, genital prominence far short of extending to posterior
end of somite; genital somite with a low dorsal hump...........................15

15a. Laterally, genital operculum (Fig.97-c) reaching posterior end
of genital prominence ....*H. americanus*, n. sp.
15b. Laterally, genital operculum (Fig.99-d) far short of reaching
posterior end of genital prominence................................*H. cohibilis*, n. sp.

16a. Laterally, genital prominence (Fig.110-e) reaching close to
posterior end of somite; posterior subterminal spine of 4th
lobe of Mx2 (Fig.110-h) shorter than 1/2 length of saberlike spine .....*H. confusibilis*, n. sp.
16b. Laterally, genital prominence (Fig.84-f) far short of reaching
posterior end of somite; posterior subterminal spine of 4th
lobe of Mx2 (Fig.84-i) longer than 1/2 length of saberlike spine ....*H. habrosomus*, n. sp.

Males

1a. Basis of left P5 (Fig.89-e) with a well-developed inner lobe 2
1b. Basis of left P5 (Fig.85-g) without a well-developed inner lobe.......................4

2a. Basal inner lobe of left P5 (Fig.89-e) wide...............................*H. clausii*
2b. Basal inner lobe of left P5 conical or short, fingerlike 3

3a. Basal inner lobe of left P5 (Fig.87-f) distally rounded ....*H. prolixus*, n. sp.
3b. Basal inner lobe of left P5 (Fig.91-f) distally truncate................*H. tuberculus*, n. sp.

4a. Basal inner lobe of right P5 (Fig.101-j) armed with thick
lancetlike bristles ....*H. spinosus*
4b. Basal inner lobe of right P5 armed with normal bristles............................5

5a. Third exopodal segment of right P5 (Fig.115-e) with a long
terminal lobe, about 2/3 length of terminal spine........................*H. norvegicus*
5b. Third exopodal segment of right P5 with a short terminal lobe,
equal to or less than 1/3 length of terminal spine................................6

6a. Basal lobe of right P5 (Fig.113-a) arising from anterior side of segment ....*H. tanneri*
6b. Basal lobe of right P5 arising from medial or anteromedial side of segment 7

7a. Second exopodal segment of left P5 (Fig.85-i) with outer spine borne
on a conical process 8
7b. Second exopodal segment of left P5 (Fig.103-h) with outer spine not
borne on a conical process 12

8a. Basal lobe of right P5 (Fig.85-g) sigmoid ....*H. habrosomus*, n. sp.
8b. Basal lobe of right P5 nearly straight or slightly curved............................9

9a. Basal lobe of right P5 (Fig.105-f) with a single band of bristles......................10
9b. Basal lobe of right P5 (Fig.95-i) with an additional row of bristles 11

10a. Basal lobe of right P5 (Fig.93-c) with wide beginning; 3rd exopodal
     segment of right P5 with terminal spine about 2/3 length of segment . . . . *H. pustulifer*
10b. Basal lobe of right P5 (Fig.105-c) with narrow beginning; 3rd exopodal
     segment of right P5 with terminal spine about 1/4 length of segment . . . . *H. austrinus*

11a. Third exopodal segment of right P5 (Fig.95-c) relatively short
     and curved . . . . *H. oikoumenikis*, n. sp.
11b. Third exopodal segment of right P5 (Fig.96-f) relatively long
     and straight . . . . *H. longisegmentus*, n. sp.

12a. Third exopodal segment of right P5 (Fig.103-d) with terminal
     spine about 1/3 length of segment . . . . *H. paraspinosus*, n. sp.
12b. Third exopodal segment of right P5 with terminal spine longer
     than 1/2 length of segment 13

13a. Anteriorly and when tilted clockwise, basal lobe of right P5
     (Fig.109-b) recurved without folding 14
13b. Anteriorly and when tilted clockwise, basal lobe of right P5
     (Fig.107-d) recurved by folding 15

14a. Lateral margin of basal lobe of right P5 (Fig.109-b) forming
     a narrow, rounded arch with medial margin of segment. . . . . *H. pacificus*
14b. Lateral margin of basal lobe of right P5 (Fig.111-f) forming
     a narrow angle with medial margin of segment. . . . . *H. confusibilis*, n. sp.

15a. Anteriorly, basal lobe of right P5 (Fig.107-d) strongly folded . . . . *H. abyssalis*
15b. Anteriorly, basal lobe of right P5 (Fig.98-e) slightly folded 16

16a. Basal lobe of right P5 (Fig.98-e) conspicuously inflated distally . . . . *H. americanus*, n. sp.
16b. Basal lobe of right P5 (Fig.99-g) not conspicuously inflated distally . . . . . . *H. cohibilis*, n. sp.

Heterorhabdus habrosomus, new species
Figures 84 and 85

Heterorhabdus abyssalis, not of Giesbrecht; Tanaka 1964, p. 2, fig. 175, male only.—Park 1968,
     p. 561, pl. 11, figs. 1-6.
Heterorhabdus proximus, not of Davis; Bradford 1971, p. 126, figs. 6d-f.

Type material. Holotype female (USNM 287050), 52 paratype females (USNM 287052), allotype male (USNM 287051), and 2 paratype males (USNM 287053) found in IKMT sample from
3000-0 mwo, SIO Indopac I-VIII, station 1, 34°33'N, 141°55'E, off Tokyo, April 29, 1976.

Description of female. Prosome length, 1.52-1.88 mm; body length, 2.16-2.56 mm. Body elongate (Fig.84-a). Prosome 73% length of body and approximately 2.4 times length of urosome. Laterally, prosome deepest at junction between cephalosome and 1st pedigerous somite, typically with a
length-depth ratio of 100:38. Cephalosome 46% length of prosome, with cervical groove halfway
along dorsal margin. Forehead (Fig.84-d) rounded, produced ventrad into a relatively long conical
rostrum bearing 2 slender filaments. Posterior margin of prosome (Fig.84-b) broadly rounded. Dorsally, midanterior tubercular process of forehead (Fig.84-e) more or less truncate, bearing
suprafrontal sensillae. Posterolateral corner of prosome (Fig.84-c) rounded, overlapping anterior 1/6
length of genital somite.

Dorsally, genital somite (Fig.84-c) widest at middle, typically with a length-width ratio of
100:75, with smoothly curved lateral swellings extending posteriad to 87% length of somite from anterior end. Laterally, genital somite (Figs.84-f, g) longer than deep, typically with a length-depth ratio of
100:89, with a low dorsal hump slightly anterior to midpoint and more or less rounded genital prominence extending far short to posterior end of somite. Genital flange smoothly arched. Genital operculum

extending to top of genital flange. Ventrally, genital operculum (Fig.84-h) about as long as wide and 62% as wide as somite. Genital field with about 12 pores on right side and 2-6 pores on left side.

First 3 urosomal somites (Figs.84-b, c) armed with spinules along posterior margins. Length ratios of 4 urosomal somites and left caudal ramus, 39.1 : 17.4 : 13.1 : 8.7 : 21.7 = 100. Left caudal ramus extending beyond posterior end of right by 1/7 its length as measured along medial margin. Dorsal appendicular seta of left ramus approximately 1.5 times as long as that of right ramus. Fourth marginal seta of left ramus much thicker than other marginal setae and longer than body. Third marginal seta of left ramus and 3rd and 4th marginal setae of right ramus armed with spines in addition to normal setules.

Antennule extending beyond posterior end of caudal ramus by its last 1 or 2 segments. Antennule, A2, Md, and Mx1 similar in anatomical details to those of *H. spinifrons*. In Mx2 (Fig.84-i), 2nd seta of 1st lobe approximately 1/2 length of 1st. Spine of 2nd lobe thick, about as long as 2nd seta of 1st lobe. Small, posterior spine of 3rd lobe approximately 1/4 length of anterior spine. Posterior subterminal spine of 4th lobe 58% as long as 2nd saberlike spine, which is about as long as 1st. Falcate spine of 5th lobe approximately 1.3 times length of saberlike spine, with proximal 63% of medial margin serrated with spinules. Falcate spine of 6th lobe with proximal 72% of medial margin serrated with spinules. In Mxp (Fig.84-j), middle marginal spine of coxa falcate, distinctly bent 1/4 length of spine from proximal end and extending beyond distal end of basis by about 1/4 its length. Anterior distal spine of coxa thick, a little longer than 1st posterior distal spine, about twice as long as 2nd posterior distal spine, and approximately 2/7 length of segment. Basis approximately 1.6 times length of coxa, with 1st middle seta 57% length of segment from proximal end.

In P1 exopod (Fig.85-a), outer spine of 1st segment extending a little beyond distal end of 2nd segment. Outer spine of 2nd segment barely reaching base of 1st outer spine of 3rd segment, which is about as long as preceding spine and reaches base of following spine. Third exopodal segment approximately 4/5 length of 1st, with terminal spine about 3/4 length of whole exopod.

Exopod of P2 (Fig.85-b) with relatively long 3rd segment, which is 59% length of whole exopod as measured along medial margin. First 4 outer spines well developed, about twice as long as 5th. All 3 outer spines of 3rd segment in distal half of segment. Terminal spine of 3rd segment 29% as long as segment. Anteriorly, 1st segment with 1 pore, 3rd with 4 pores.

In P3 exopod (Fig.85-c), 3rd segment 63% as long as whole exopod, widened, typically with a length-width ratio of 100:40. First 3 outer spines of similar size, about twice as long as 5th outer spine. Third segment with all outer spines in distal half and terminal spine approximately 1/7 length of segment. Anteriorly, 1st segment with 1 pore, 2nd with 2, and 3rd with 7.

Exopod of P4 (Fig.85-d) elongate, with 3rd segment making up 60% length of whole exopod. First 2 outer spines of similar size and larger than last 3, which are of similar size. Terminal spine of 3rd segment 28% as long as segment. Anteriorly, 1st segment with 1 pore, 2nd with 2, 3rd with 7.

Basipod of P5 (Figs.85-e, f) a little shorter than endopod, which is barely reaching distal end of 2nd exopodal segment. First 2 endopodal segments each with distolateral corner produced into spiniform process. First 2 exopodal segments each with distolateral corner divided into 2 spiniform processes, together forming a socket for outer spine. Outer spine of 2nd segment largest of all outer spines and pointing in a distomedial direction along posterior surface of segment. Inner spine of 2nd segment about as long as 3rd segment. Third segment distinctly shorter than combined lengths of first 2 segments, with 1st outer spine 61% length of segment from proximal end; its terminal spine approximately 1/2 as long as segment. Anteriorly, exopod with 1 pore on 3rd segment.

Description of male. Prosome length, 1.54-1.76 mm; body length, 2.14-2.48 mm. Similar in habitus to female. Prosome 73% length of body and approximately 2.6 times length of urosome. Urosome 28% length of body. First 4 urosomal somites armed with spinules along posterior margins. Caudal rami and their setation as in female.

Left, geniculate A1 similar in anatomical details to that of *H. spinifrons*. All other cephalosomal appendages and 1st to 4th legs as in female. In 5th pair of legs (Fig.85-g), inner lobe of right basis arising from middle of anteromedial margin of segment, longer than segment, characteristically sigmoid, with a row of stiff setules proximally in addition to a band of setules along its whole medial margin. Left basis (Fig.85-h) with smoothly curved medial margin, its distal 2/3 fringed with setules. Left endopod longer than right, each with inner marginal seta on 2nd segment. In right exopod (Fig.85-j), 1st segment approximately 1/2 length of 2nd. Medial projection of 2nd segment triangular, its proximal margin set off some distance from proximal end of segment and furnished with setules and its distal margin arising from middle of segment. Outer spine of 2nd segment about at middle of segment. Third segment including terminal lobe approximately 1.7 times length of 2nd, smoothly curved, with outer spine, which is moved mediad by the rotation of exopod, about 1/4 length of segment from proximal end. Terminal spine of 3rd segment approximately 2/3 length of segment.

In left exopod (Fig.85-g), 1st segment approximately 2/3 length of 2nd, with a rounded tubercular swelling proximally on medial margin. Second segment (Fig.85-i) with a long lateral, conical process terminating with outer spine. Measured along medial margin, conical process approximately 1/3 length of segment and about as long as outer spine. Third segment with well-developed inner and outer spines and tapering distally into a long spiniform process. Together with distal spiniform process, 3rd segment making up 70% length of whole exopod.

Etymology. The specific name *habrosomus*, from Greek *habros* meaning graceful and *soma* meaning body, alludes to the gracile habitus.

Distribution. This species has been recorded as *H. abyssalis* from the Izu region of Japan (Tanaka 1964) and the central North Pacific (Park 1968) and as *H. proximus* from the southwestern Pacific (Bradford 1971). In this study, the species is represented by a total of 132 specimens including 110 females and 22 males found in 21 samples—6 taken in the South Atlantic between 33°47′S and 01°25′N; 11 in the central and west Pacific from 155°W westward to the East China Sea and between 41°27′S (off New Zealand) and 35°01′N (off central Japan); 4 in the western Indian Ocean between 35°00′S and 23°23′S. These findings seem to demonstrate a circumglobal distribution of the species. However, it was not found in the whole east Pacific, where more samples were examined than any other parts of the world's oceans.

Remarks. This species is unusual in having a sigmoid basal inner lobe on the male right P5. The male of this species was reported for the first time by Tanaka (1964) from off the east coast of central Japan and both the female and male were described by Park (1968) from the central North Pacific; both authors referred the species to *H. abyssalis* (Giesbrecht 1889). The male 5th pair of legs as figured by Tanaka (1964) and Park (1968) is, however, obviously different from the figure given for *H. abyssalis* by Giesbrecht (1892). Subsequently, Bradford (1971) found in the southwestern Pacific *H. abyssalis* of Tanaka (1964) and Park (1968) and identified it as *H. proximus* Davis 1949.

Heterorhabdus abyssalis of Tanaka (1964) and Park (1968) and *H. proximus* of Bradford (1971) are the same and differ from both *H. abyssalis* (Giesbrecht 1889) and *H. proximus* Davis 1949, and therefore are described here as *H. habrosomus,* new species. This new species differs from *H. proximus* as originally described by Davis (1949) in the following aspects: 1) *Heterorhabdus proximus* is much larger (3.2-3.7 mm long in the female and 3.2-3.4 mm long in the male) than *H. habrosomus* (2.16-2.56 mm long in the female and 2.14-2.48 mm long in the male), 2) in the male 5th pair of legs of *H. proximus*, the basal inner lobe of the right P5 is short, reaching about the distal end of the segment, the terminal spine of the right exopod very short, the terminal spine of the left exopod short, not tapering into a long, needle-like spine; in the male 5th pair of legs of *H. habrosomus*, on the other hand, the basal inner lobe of the right P5 is long, extending beyond the distal end of the segment by 2/3 its length, the terminal spine of the right exopod long, the terminal spine of the left exopod long and tapering into a long needle-like spine, 3) *Heterorhabdus proximus* was found along the northern Pacific coast of North America and in the Gulf of Alaska, while *H. habrosomus* was completely absent from this area, although many samples from the area were examined in this study. *Heterorhabdus proximus* Davis 1949 is a synonym of *H. tanneri* (Giesbrecht 1895) (see below), which commonly occurs in the same area and is similar in body size to *H. proximus* but was not reported by Davis (1949).

Heterorhabdus prolixus, new species
Figures 86 and 87

Type material. Holotype female (USNM 287054), 22 paratype females (USNM 287056), allotype male (USNM 287055), and 44 paratype males (USNM 287057) found in IKMT sample from 1000-0 mwo, SIO Piquero IV, station 1, 31°47'S, 75°03'W off Chile in the southeastern Pacific, February 28, 1969.

Description of female. Prosome length, 1.68-2.08 mm; body length, 2.36-2.92 mm. Body (Fig.86-a) elongate and similar in habitus to *H. habrosomus*. Prosome 71% length of body and approximately 2.3 times length of urosome. Laterally, prosome typically with a length-depth ratio of 100:37. Cephalosome 45% length of prosome.

Dorsally, genital somite (Fig.86-b) typically with a length-width ratio of 100:67, widest at middle, asymmetrical, with smoothly curved swelling on left side and low and somewhat flattened swelling on right side. On each side, genital swelling extending posteriad to 87% length of somite from anterior end. Laterally, genital somite (Figs.86-e-h) typically with a length-depth ratio of 100:84. Dorsal margin of genital somite sigmoid, with a large, rounded hump at middle followed by a smoothly curved hollow. Genital prominence merging into ventral wall of somite posteriorly without exhibiting a clear junction. Genital operculum reaching about 3/4 length of somite from anterior end. Left genital flange curved out under genital operculum and extending as a wavy ridge close to posterior end of somite. Viewed from left side, right genital flange (Fig.86-e) extending beyond left genital flange. Viewed from right side, right genital flange low and flattened (Fig.86-f) but emerging, when genital operculum is open (Fig.86-h), as a distinct, bilobed ridge ending some distance before posterior end of somite. Ventrally, genital operculum (Fig.87-a) asymmetrical and slightly slanted, typically with a length-width ratio of 100:89 and 66% as wide as somite. Genital field with 7-9 pores on each side. Several individuals with abnormal genital somites were found among normal females (Figs.87-b, c). They were recognized as conspecific because of the identity in body size and all other anatomical details including certain characteristic features of the genital somite, such as genital flanges.

Length ratios of 4 urosomal somites (Fig.86-b) and left caudal ramus, 37.8 : 17.7 : 15.2 : 8.6 : 20. 7 = 100. Dorsal appendicular seta of left caudal ramus twice as long as that of right ramus. Anteriorly, caudal rami each with 2 pores.

Antennule extending beyond posterior end of caudal ramus by its last 3 segments. All cephalosomal appendages and swimming legs similar to those of *H. habrosomus* described above.

Description of male. Prosome length, 1.64-2.08 mm; body length, 2.28-2.88 mm. Similar in habitus to female. Prosome 73% length of body and approximately 2.7 times length of urosome. Urosome 27% length of body. Caudal rami and their setation as in female. Left, geniculate A1 similar in anatomical details to that of *H. spinifrons*. All other cephalosomal appendages and first 4 pairs of swimming legs as in female.

In 5th pair of legs (Fig.87-d), inner lobe of right basis arising from middle of anteromedial side of segment, longer than segment, extending beyond distal end of segment by its distal half, and with a row of setules proximally in addition to normal band of setules along its whole medial margin. Anteriorly and when tilted clockwise, inner lobe of right basis (Fig.87-e) initially narrow and folded at its proximal lateral margin. Left basis (Fig.87-f) with a well-developed conical inner lobe, distally rounded, with a row of setules proximally in addition to a band of setules along medial margin. In right exopod, 1st segment approximately 2/3 length of 2nd. Anteriorly, 2nd segment (Fig.87-g) with a conspicuous, large rounded tubercular swelling close to proximal end, a small rounded process at proximomedial corner, and a relatively small medial projection. Proximal margin of medial projection sloping at about 35° with reference to longitudinal axis of segment and its distal margin forming a right angle with medial margin of segment. Third segment smoothly curved, about twice as long as 2nd, with outer spine about 1/4 length of segment from proximal end, terminal spine approximately 1/3 length of segment, and terminal lobe (Fig.87-i) approximately 1/7 length of terminal spine.

In left exopod (Fig.87-h), 2nd segment with a lateral projection terminating with outer spine. Measured along medial margin, lateral projection approximately 1/6 length of segment and a little shorter than outer spine. Third segment with well-developed inner and outer spines and tapering distally into a long spiniform process. Together with distal spiniform process, 3rd segment making up 67% length of whole exopod.

Etymology. The specific name *prolixus*, from the Latin meaning stretched out or long, refers to the relatively long antennule.

Distribution. This species is represented by a total of 265 specimens including 133 females and 132 males from 15 samples taken along the west coast of the Americas between 39°26'S and 27°17'N and in equatorial waters between the west coast of South America and 107°56'W. Apparently it is endemic to the coastal waters of the east Pacific but may be carried westward to some distance in the equatorial currents.

Remarks. This species is almost identical in anatomical details of the appendages other than the male P5 to *H. habrosomus* described above. The female of this species can be identified by the shape of the genital flanges and the male by the shape of the basal inner lobe of the left P5.

In 4 out of a total of 15 samples yielding this new species, 11 abnormal females were found among 97 normal females. The abnormality occurred exclusively in the shape of the genital somite, and all abnormal forms were unique, although some differed from each other in a very minor feature. Bradford (1971) noted similar variations in the shape and length of the female genital somites of certain *Heterorhabdus* species and she attributed the variations to the different contractile states of the longitudinal muscles in the somite.

Heterorhabdus clausii (Giesbrecht 1889)
Figures 88 and 89

Heterochaeta clausii Giesbrecht 1889, p. 812.—Giesbrecht 1892, p. 372, pl. 20. figs. 2, 28, 37, 38.
Heterorhabdus clausii; Giesbrecht 1898, p. 115.

Description of female. Prosome length, 1.66-1.92 mm; body length, 2.32-2.64 mm. Very similar in habitus to *H. prolixus* described above. Prosome 72% length of body and approximately 2.5 times length of urosome. Laterally, prosome (Fig.88-a) typically with a length-depth ratio of 100:36. Cephalosome 43% length of prosome.

Dorsally, genital somite (Fig.88-c) typically with a length-width ratio of 100:72, widest at middle, with smoothly bulging lateral swellings, slightly asymmetrical, with left margin bulging slightly more than right. On either side, lateral genital swelling short of reaching posterior end of somite. Laterally, genital somite (Figs.88-d, e) typically with a length-depth ratio of 100:91, with a large, rounded hump halfway along dorsal margin followed by a smoothly curved concave margin. Genital prominence high, with nearly rounded outline. Genital operculum reaching 84% length of somite from anterior end. Left genital flange smoothly arched beside genital operculum, extending posteriorly to about the same level as genital operculum. Viewed from left side, right genital flange (Fig.88-d) extending beyond left genital flange. Viewed from right side, right genital flange (Fig.88-e) with an undulating outline, which has an outward curve, at level of posterior end of genital operculum, followed by a small, deep hollow. Shape of genital flange not significantly altered when genital operculum opens (Figs.89-a, c). Ventrally, genital operculum (Fig.89-b) almost symmetrical but slightly slanted, typically with a length-width ratio of 100:84 and 79% as wide as somite. Genital field with about 9 pores on right side and normally 2 pores on left side.

Length ratios of 4 urosomal somites (Figs.88-b, c) and left caudal ramus, 37.3 : 17.1 : 14.5 : 8.8 : 22.3 = 100. Dorsal appendicular seta of left caudal ramus approximately 1.5 times length of that of right ramus. Each caudal ramus with 2 pores on anterior surface.

Antennule extending beyond posterior end of caudal ramus by its last 1 or 2 segments. All cephalosomal appendages and swimming legs similar to those of *H. habrosomus* described above.

Description of male. Prosome length, 1.60-1.76 mm; body length, 2.24-2.40 mm. Similar in habitus to female. Prosome 73% length of body and approximately 2.6 times length of urosome. Urosome 28% length of body. Caudal rami and their setation as in female. Left, geniculate A1 similar in anatomical details to that of *H. spinifrons.* All other cephalosomal appendages and first 4 pairs of swimming legs as in female.

Fifth pair of legs (Fig.89-d) practically identical to that of *H. prolixus* described above except for shape of basal inner lobe of left P5 (Fig.89-e). Anteriorly, basal inner lobe of left P5 relatively wide and rounded distally; its proximal margin together with proximomedial margin of segment forming a smooth, shallow inward curve; its lateral margin about 1/4 length of segment and forming a right angle with distomedial margin of segment; its medial margin nearly parallel to lateral margin of segment.

Distribution. This species was originally described from the eastern equatorial Pacific (99°W-108°W, 0°-3°S). In the present study it is represented by 54 specimens including 47 females and 7 males from 6 samples all taken in the eastern and central equatorial waters of the Pacific between 17°50'S and 10°05'N and between 164°14'W and 110°18'W. Apparently this is an eastern equatorial Pacific species.

Remarks. This species was originally described briefly as *Heterochaeta clausii* by Giesbrecht (1889) from the eastern equatorial Pacific and redescribed by Giesbrecht (1892) with figures of P5 of the female and the left A1 and 5th pair of legs of the male. The female P5 figured by Giesbrecht (1892) has an inner marginal seta on the 1st exopodal segment. However, the seta is absent in all species of the family Heterorhabdidae according to the information available in the literature and not found in any species of the family examined in the present study. The male 5th pair of legs figured for *H. clausii* by Giesbrecht (1892) is highly characteristic in having a well-developed inner lobe on the left basis. In the present study 3 species were found to have a well-developed basal inner lobe on the male left P5 and to be similar in body size to *H. clausii* as described by Giesbrecht (1892). One of the 3 species was identified with *H. clausii* because the 5th pair of legs of the male was exactly identical to the figure given by Giesbrecht (1892) and was the most common in the equatorial Pacific, the type locality of *H. clausii.* The remaining 2 are described as new species, *H. prolixus* and *H. tuberculus.*

Besides its type locality in the equatorial Pacific, *H. clausii* has been recorded from off San Diego (Esterly 1905), the Malay Archipelago (Scott 1909), the Indian Ocean (Sewell 1932), the Gulf of Guinea (Vervoort 1965), and the South Atlantic (Wolfenden 1911). However, none of these records outside the type locality is accompanied by reliable species descriptions, and their validity therefore cannot be confirmed.

Heterorhabdus clausii can be distinguished from *H. prolixus* and *H. tuberculus,* the most closely related species, by the relatively short genital somite and characteristic genital flanges in the female and by the shape of the basal inner lobe of the left P5 in the male. They also have different distributions. *Heterorhabdus clausii* was found in this study exclusively in the equatorial waters of the eastern and central Pacific, *H. prolixus* in the coastal waters of the eastern Pacific, and *H. tuberculus* widely in the whole Indo-Pacific.

Heterorhabdus tuberculus, new species
Figures 90 and 91

Type material. Holotype female (USNM 287058) and 2 paratype females (USNM 287060) found in IKMT sample from 3000-0 mwo, SIO Piquero V, station 7, 17°46'S, 110°19'W in the eastern tropical Pacific, March 29, 1969. Thirty-eight paratype females (USNM 287061), allotype male (USNM 287059), and 35 paratype males (USNM 287062) found in IKMT sample from 1000-0 mwo, SIO Cal-COFI 7102G, station 120.55, 27°23'N, 116°12'W off San Diego in the eastern Pacific, April 2, 1971.

Description of female. Prosome length, 1.60-2.04 mm; body length, 2.24-2.84 mm. Similar in habitus to *H. prolixus* described above. Prosome 72% length of body and approximately 2.5 times length of urosome. Laterally, prosome (Fig.90-a) typically with a length-depth ratio of 100:37. Cephalosome 45% length of prosome.

Dorsally, genital somite (Fig.90-b) typically with a length-width ratio of 100:76, widest at middle, with smoothly curved lateral swellings, slightly asymmetrical, with left side bulging a little higher than right. Genital swelling on right side extending to posterior end of somite, while it is a little short of reaching posterior end of somite on left side. Laterally, genital somite (Figs.90-e-h) typically with a length-depth ratio of 100:89, with a large, rounded dorsal hump a short distance anterior to middle of somite followed by a smoothly curved hollow. Genital prominence relatively high. Genital operculum with rounded outline, reaching 65% length of somite from anterior end. Left genital flange rounded, tubercular, with gradually sloping anterior margin and steep posterior margin meeting ventral wall of somite at the same level as posterior end of genital operculum and followed by a low ridge extending longitudinally halfway to posterior end of somite. Viewed from left side, right genital flange (Fig.90-e) extending beyond left genital flange. Viewed from right side, right genital flange (Fig.90-f) low, extending down to posterior end of somite. Shape of genital flanges not significantly altered by opening of genital operculum (Fig.90-h). Ventrally, genital operculum (Fig.90-d) about as long as wide, nearly symmetrical, slightly slanted. Genital field with about 10 pores on right side and normally only 1 pore on left side.

Several females with abnormal genital somites were found along with normal individuals (Figs.91-a-d); they differed from the normal and from each other mainly in shape of the dorsal wall. All of them, however, had the same characteristic left genital flange.

Length ratios of 4 urosomal somites (Figs.90-b, c) and left caudal ramus, 38.9 : 16.5 : 12.9 : 8.2 : 23.5 = 100. Dorsal appendicular seta of left caudal ramus approximately 1.5 times longer than that of right ramus. Anteriorly, each caudal ramus with 2 pores.

Antennule extending beyond posterior end of caudal ramus by its last 1 or 2 segments. All cephalosomal appendages and swimming legs similar to those of *H. habrosomus* described above.

Description of male. Prosome length, 1.60-1.92 mm; body length, 2.24-2.68 mm. Similar in habitus to female. Prosome 71% length of body and approximately 2.4 times length of urosome. Urosome 30% length of body. Caudal rami and their setation as in female. Left, geniculate A1 similar in anatomical details to that of *H. spinifrons*. All other cephalosomal appendages and first 4 pairs of swimming legs as in female.

Fifth pair of legs (Fig.91-e) closely resembles those of both *H. prolixus* and *H. clausii* described above except for shape of basal lobe of left P5. Anteriorly, basal lobe fingerlike, extending distad, slightly curved laterad, with weakly bilobed or somewhat truncate distal end, fringed along medial margin with 2 separate rows of setules.

Etymology. The specific name *tuberculus* refers to the left genital flange of the female, which looks like a tubercule.

Distribution. This species is represented by 370 specimens including 213 females and 157 males from 26 samples—18 taken in the eastern Pacific between 17°50'S and 38°40'N; 4 in the central Pacific between 04°57'S and 15°00'N; 1 at 04°40'S, 125°33'E in the Malay Archipelago; 1 at 17°44'N, 119°55'E in the East China Sea; 1 at 02°28'S, 86°23'E in the eastern Indian Ocean; 1 at 25°42'S, 35°28'E in the Mozambique Channel. This seems to be an Indo-Pacific species.

Remarks. Heterorhabdus prolixus, H. clausii, and *H. tuberculus* are morphologically very close to one another and can be distinguished from one another by carefully comparing the shape of the female genital somite and the male 5th pair of legs. The genital somites in *H. prolixus* and *H. tuberculus* are relatively long compared to *H. clausii*. The genital somite of *H. prolixus* can be characterized by a strongly sigmoid dorsal margin and a long and low left genital flange, while that of *H. tuberculus* has a weakly sigmoid dorsal margin and a high and tubercular left genital flange. The males of these 3 species are even closer anatomically than the females. They can be distinguished by the shape of the basal lobe of the left P5. Otherwise, all appendages are practically identical in all 3 species except that antennules seem to be relatively longer in *H. prolixus* than in the other 2 species.

In 9 of the 26 samples yielding *H. tuberculus*, 1 out of about 8 females showed some form of abnormality in shape of the genital somite, mainly in contour of its dorsal wall. Similar abnormalities were also found in *H. prolixus*, described above.

Heterorhabdus pustulifer Farran 1929
Figures 92 and 93

Heterorhabdus pustulifer Farran 1929, p. 266, fig. 27.—Vervoort 1951, p. 128.—Vervoort 1957, p. 132, figs. 123, 124.

Description of female. Prosome length, 2.16-2.28 mm; body length, 3.04-3.24 mm. Prosome 72% length of body and approximately 2.5 times length of urosome. Laterally, prosome (Fig.92-a) typically with a length-depth ratio of 100:40. Cephalosome 43% length of prosome.

Dorsally, genital somite (Fig.92-c) typically with a length-width ratio of 100:75, nearly symmetrical, widest at middle, with lateral swellings extending more or less to posterior end of somite. Laterally, genital somite (Fig.92-e) with a conical spiny process halfway along dorsal margin and rounded genital prominence, typically with a length-depth ratio of 100:92 including dorsal conical process. Genital flange high and rounded; its posterior margin concave, merging with ventral wall of somite. Genital operculum reaching halfway along posterior slope of genital flange or 85% length of somite from anterior end. Ventrally, genital operculum (Fig.92-d) wider than long, typically with a length-width ratio of 92:100 and 80% as wide as somite. Genital field with about 18 pores on each side.

Length ratios of 4 urosomal somites (Figs.92-b, c) and left caudal ramus, 37.4 : 16.6 : 15.9 : 8.6 : 21.5 = 100. Dorsal appendicular seta of left caudal ramus approximately 1.5 times length of that of right ramus. Anteriorly, anal segment with 2 pores on each side, each caudal ramus with 3 pores.

Antennule reaching about posterior end of 3rd urosomal somite. In anatomical details, all cephalosomal appendages and swimming legs similar to those of *H. habrosomus* except for the following features: In Mx2 (Fig.92-f), posterior subterminal spine of 4th lobe relatively long, 66% as long as 2nd saberlike spine; in P1 exopod (Fig.93-a), 1st outer spine extending beyond distal end of segment by about 1/5 its length; in P5 (Fig.93-b), 3rd exopodal segment about as long as combined lengths of first 2 segments; outer spine of 2nd segment pointing straight distad.

Description of male. Prosome length, 2.16-2.20 mm; body length, 3.04-3.16 mm. Similar in habitus to female. Prosome 71% length of body and approximately 2.4 times length of urosome. Urosome 30% length of body. Caudal rami and their setation as in female. Left, geniculate A1 similar in morphology to that of *H. spinifrons*. All other cephalosomal appendages and first 4 pairs of legs as in female.

In 5th pair of legs (Fig.93-c), basal inner lobe of right leg with a wide initial section (which is actually an extension of the whole length of medial margin of segment) followed by distal section that extends in a distolateral direction. The distal section is about 1/2 as wide as the initial section, more or less inflated distally and fringed with a single band of setules (Fig.93-d). Left basis with convex medial margin fringed with a single band of setules. In right exopod, 2nd segment (Fig.93-e) approximately 1.5 times length of 1st, with a low tubercular process proximally on medial margin, followed by well-developed medial projection; its proximal edge plumose, straight, nearly perpendicular to medial margin of segment and its tip pointing proximad. Outer spine of 2nd segment close to distal end of segment. Third segment (Fig.93-c) including terminal lobe about twice as long as 2nd, smoothly curved, with outer spine 27% length of segment from proximal end and terminal spine approximately 3/5 length of segment. Terminal lobe (Fig.93-g) approximately 1/6 length of terminal spine.

In left exopod (Fig.93-f), 2nd segment approximately 1.5 times length of 1st, with a short lateral conical process terminating with outer spine. Measured along medial margins, conical process about 1/2 length of outer spine. Third segment with well-developed inner and outer spines and tapering distally into a long spiniform process. Together with distal spiniform process, 3rd segment approximately 7/10 length of whole exopod.

Distribution. This species has a circumantarctic distribution. It was originally described from female specimens captured from 1000-0 m at 71°41'S, 166°47'W and from 600-0 m at 71°49'S, 167°32'W in the Pacific sector of the Antarctic. Vervoort (1951) found 2 females of this species in a sample taken from 900-0 m at 66°S, 11°W in the Atlantic sector of the Antarctic. Vervoort (1957) found both the females and males at 9 stations between 66°30'S and 47°43'S and between 54°16'E and 164°29'E in the Indian Ocean sector of the Antarctic. In the present study, the species is represented by 9 females and 5 males all taken at 60°42'S, 59°48'W in the Drake Passage.

Remarks. The original description of the species by Farran (1929) was based on female specimens 2.98-3.04 mm long. Vervoort (1957) reported on both the female and male but no description of the male was given. The male is described here for the first time. The female of this species is similar in shape of the genital somite and in details of the appendages to that of *H. habrosomus* described above, but can be readily distinguished from it by the dorsal conical process of the genital somite and the relatively short A1. The male can be distinguished from those of the other members of the species group by the shape of the basal inner lobe of the right P5, which is unusually wide proximally.

Heterorhabdus oikoumenikis, new species
Figures 94 and 95

Type material. Holotype female (USNM 287063), 30 paratype females (USNM 287065), allotype male (USNM 287064), and 10 paratype males (USNM 287066) found in IKMT sample from 4000-0 mwo, SIO Lusiad VII, station 79, 01°10'N, 11°36'W in the eastern tropical Atlantic, July 6, 1963.

Description of female. Prosome length, 1.88-2.48 mm; body length, 2.60-3.52 mm. Body elongate (Fig.94-a). Prosome 71% length of body and approximately 2.4 times length of urosome. Laterally, prosome typically with a length-depth ratio of 100:36. Cephalosome 45% length of prosome, with broadly rounded forehead (Fig.94-d) produced ventrad into a conical rostrum bearing slender filaments. Dorsally, midanterior tubercular process of forehead (Fig.94-e) rather rounded, with suprafrontal sensillae. Posterolateral corner of prosome more or less truncate, extending over anterior 1/8 length of genital somite.

Dorsally, genital somite (Fig.94-c) typically with a length-width ratio of 100:77, widest at middle, somewhat asymmetrical with a broad swelling on right side and a slightly sigmoid swelling on left side. Laterally, genital somite (Figs.94-g, h) typically with a length-depth ratio of 100:90, with a low dorsal hump a short distance anterior to midpoint and more or less rounded genital prominence. Posterior margin of genital prominence straight from posterior end of genital operculum down to posterior margin of somite. Genital operculum reaching 60% length of somite from anterior end or peak of genital flange. Ventrally, genital operculum (Fig.94-f) slightly slanted, wider than long, typically with a length-width ratio of 90:100 and 63% as wide as somite. Genital field with about 12 pores on right side and normally 2 pores on left side.

Urosome (Figs.94-b, c) 30% length of body. Length ratios of 4 urosomal somites and left caudal ramus, 37.6 : 17.6 : 13.9 : 9.1 : 21.8 = 100. Dorsal appendicular seta of left caudal ramus about twice as long as that of right ramus.

Antennule extending beyond posterior end of caudal ramus by its last 3 or 4 segments. All cephalosomal appendages similar in morphology to those of *H. habrosomus* described above except for the following features of Mx2 (Fig.94-i): Posterior subterminal spine of 4th lobe 40% as long as 2nd saberlike spine; falcate spine of 5th lobe with proximal half of medial margin serrated with spinules and approximately 1.2 times length of saberlike spine.

Swimming legs also similar to those of *H. habrosomus* except that basipod of P5 (Figs.95-a, b) about as long as endopod and terminal spine of 3rd exopodal segment 58% as long as segment.

Description of male. Prosome length, 1.76-2.36 mm; body length, 2.40-3.28 mm. Similar in habitus to female. Prosome 72% length of body and approximately 2.5 times length of urosome. Urosome 28% length of body. Caudal rami and their setation as in female. Left, geniculate A1 similar in morphology to that of *H. spinifrons*. All other cephalosomal appendages and first 4 pairs of legs as in female.

In 5th pair of legs (Fig.95-c), basal inner lobe of right leg arising from proximal end of medial margin, turning backward by folding its lateral margin slightly over medial edge of somite, extending nearly straight distad when segment lies flat, with a short row of setules proximally in addition to normal band of setules fringing most of its medial margin (Figs.95-f-i). Although somewhat variable, basal inner lobe of right leg as measured along lateral margin normally shorter than segment. Medial margin of left basis (Fig.95-c) smoothly convex, with an additional row of setules proximally.

In right exopod (Figs.95-c, d), 2nd segment approximately 1.5 times length of 1st, with medial projection arising a short distance from proximal end of segment. Plumose proximal margin of medial projection (Fig.95-e) sloping at an angle of about 45° with respect to longitudinal axis of segment and distal margin at a right angle. Third segment approximately 1.8 times length of 2nd, smoothly curved, with outer spine 30% length of segment from proximal end and terminal spine about 1/2 length of segment. Terminal lobe (Fig.95-k) relatively short, approximately 1/10 length of terminal spine.

In left exopod (Figs.95-c, j), 2nd segment approximately 1.5 times length of 1st, with a long lateral process terminating with outer spine. Measured along medial margins, lateral process about 2/5 length of segment and about 1.9 times length of outer spine. Third segment with well-developed inner and outer spines, tapering distally into a long spiniform process. Together with distal spiniform process, 3rd segment making up 67% length of whole exopod.

Etymology. The specific name *oikoumenikis*, Greek, meaning worldwide, alludes to a world-wide distribution of the species. A noun in apposition.

Distribution. This species has an extremely wide distribution. A total of 355 specimens including 261 females and 94 males were found in 45 samples—12 taken in the Atlantic between 35°19'S and 63°24'N; 32 throughout the Pacific between 41°27'S and 40°35'N and from the west coast of the Americas westward to the Malay Archipelago and East and South China Seas; 1 at 28°05'S, 66°05'E in the Indian Ocean.

Remarks. The female of this species can be distinguished from the other members of the species group by the genital somite, which in lateral view has a straight posterior ventral margin extending from the posterior end of the genital operculum down to the posterior end of the somite and in dorsal view has a slightly sigmoid left lateral margin. The male can be identified mainly by the following features: Basal inner lobe of right P5 relatively short, straight, with an additional row of setules; 3rd exopodal segment with terminal spine about 1/2 length of segment and terminal lobe about 1/10 length of terminal spine; 2nd exopodal segment of left P5 with a long lateral process, almost twice as long as outer spine.

Heterorhabdus longisegmentus, new species
Figure 96

Type material. Holotype female (USNM 287067), 35 paratype females (USNM 287069), allotype male (USNM 287068), and 20 paratype males (USNM 287070) found in IKMT sample from 4000-0 mwo, SIO Lusiad VII, station 11, 33°47'S, 15°39'E in the southeastern Atlantic, June 4-5, 1963.

Description of female. Prosome length, 1.88-2.32 mm; body length, 2.64-3.24 mm. Similar in habitus to *H. oikoumenikis* described above. Prosome 72% length of body and approximately 2.5 times length of urosome.

Dorsally, genital somite (Fig.96-d) typically with a length-width ratio of 100:78, widest at middle, strongly asymmetrical; its right side with a low, flat-topped swelling short of extending to posterior end of somite; its left side somewhat sigmoid, with a high, tubercular swelling followed by a deep hollow. Laterally, genital somite (Fig.96-b) typically with a length-depth ratio of 100:83, with a low dorsal hump a short distance anterior to midpoint and more or less rounded genital prominence. Posterior slope of genital prominence extending nearly straight from posterior edge of genital operculum to posterior end of somite. Genital operculum reaching 62% length of somite from anterior end. Ventrally, genital operculum (Fig.96-c) somewhat slanted, wider than long, typically with a length-width ratio of 82:100 and 61% as wide as somite. Genital field with about 13 pores on right side and normally 2 pores on left side.

Urosome (Figs.96-a, d) 29% length of body. Length ratios of 4 urosomal somites and left caudal ramus, 39.0 : 16.9 : 14.5 : 8.7 : 20.9 = 100. Dorsal appendicular seta of left caudal ramus only a little longer than that of right ramus. Anteriorly, each caudal ramus with 3 pores.

Antennule extending beyond posterior end of caudal ramus by its last 3 segments. All cephalosomal appendages and swimming legs similar to those of *H. oikoumenikis* described above.

Description of male. Prosome length, 1.80-2.08 mm; body length, 2.52-2.88 mm. Prosome 72% length of body and approximately 2.5 times length of urosome. Urosome 29% length of body. Caudal rami and their setation as in female. Left, geniculate A1 similar in morphology to that of *H. spinifrons.* All other cephalosomal appendages and first 4 pairs of legs as in female.

Fifth pair of legs (Figs.96-e, f) similar to that of *H. oikoumenikis* male except for the following features: In right exopod (Figs.96-g-j), medial projection of 2nd segment relatively small, third segment including terminal lobe approximately 2.4 times length of 2nd segment, straight, with terminal lobe and spine turned about 50° mediad; its outer spine 20% length of segment from proximal end; its terminal spine approximately 1/3 length of segment and about 3 times length of terminal lobe; in the left exopod, lateral process of 2nd segment (Fig.96-k) relatively thick but relatively short—approximately 1/4 length of segment and about as long as outer spine.

Etymology. The specific name *longisegmentus* refers to the relatively long 3rd exopodal segment of the male right 5th leg.

Distribution. This species occurs only in the subtropical and temperate regions of the Southern Hemisphere. A total of 161 specimens including 111 females and 50 males were found in 12 samples—4 taken in the southeastern Atlantic between 33°47'S and 30°23'S and between 02°55'W and 15°39'E; 2 in the southwestern Pacific at 41°26'S, 176°04'E and 09°25'S, 159°06'E, respectively; 6 in the southwestern Indian Ocean between 35°00'S and 17°03'S and between the east coast of Africa and 66°05'E.

Remarks. This species is very close in habitus and details of the appendages to *H. oikoumenikis* described above but can be readily distinguished from it, in the female, by the genital somite, which in dorsal view is strongly asymmetrical and in the male, by the exopod of the right P5, which is highly characteristic with an unusually long and straight 3rd segment. This species also differs in distribution from *H. oikoumenikis*; the former occurs exclusively in the subtropical and temperate regions of the Southern Hemisphere and the latter throughout all three major oceans.

Heterorhabdus americanus, new species
Figures 97 and 98

Type material. Holotype female (USNM 287071), 11 paratype females (USNM 287073), allotype male (USNM 287072), and 6 paratype males (USNM 287074) found in IKMT sample from 1130-0 m, SIO Southtow IV, MV 72, station II-29, 16°09'S, 75°42'W in the eastern tropical Pacific, May 8, 1972.

Description of female. Prosome length, 2.16-2.52 mm; body length, 3.12-3.52 mm. Body elongate (Fig.97-a). Prosome 70% length of body and approximately 2.2 times length of urosome. Laterally, prosome typically with a length-depth ratio of 100:40. Cephalosome 46% length of prosome, with broadly rounded forehead (Fig.97-d) produced ventrad into a conical rostrum bearing 2 slender filaments. Dorsally, midanterior tubercular process of forehead more or less truncate, with suprafrontal sensillae. Posterolateral corner of prosome (Fig.97-b) broadly rounded, overlapping anterior 1/10 length of genital somite.

Dorsally, genital somite (Fig.97-b) typically with a length-width ratio of 100:69, widest at middle, more or less asymmetrical with a broad lateral swelling extending to posterior end of somite on right side and a somewhat wavy lateral swelling ending some distance before posterior end of somite on left side. Laterally, genital somite (Figs.97-f, g) typically with a length-depth ratio of 100:82; its dorsal margin sigmoid, with a broad dorsal hump followed by a smoothly curved concave margin; its genital prominence smoothly rounded. Genital flange high and rounded, merging into ventral wall of somite by a deep hollow, together forming a sigmoid outline. Genital operculum reaching center of concavity at junction between genital flange and ventral wall of somite or close to it. Ventrally, genital operculum (Fig.97-e) longer than wide, typically with a length-width ratio of 100:92 and 76% as wide as somite itself.

Urosome (Figs.97-b, c) 31% length of body. Length ratios of 4 urosomal somites and left caudal ramus, 39.6 : 18.0 : 13.4 : 8.1 : 20.9 = 100. Dorsal appendicular seta of left caudal ramus approximately 1.5 times length of that of right ramus.

Antennule extending beyond posterior end of caudal ramus by its last 2 segments. All cephalosomal appendages similar to those of *H. habrosomus* described above except for the following features of Mx2 (Fig.97-h): Posterior subterminal spine of 4th lobe 46% as long as 2nd saberlike spine; falcate spine of 5th lobe with proximal 56% of medial margin serrated with spinules; falcate spine of 6th lobe with proximal 70% of medial margin serrated with spinules.

Swimming legs also similar to those of *H. habrosomus* except that in P1 exopod (Fig.98-a), outer spine of 1st segment relatively long, 3rd segment approximately 5/6 length of 1st, and terminal spine of 3rd segment 90% as long as whole exopod; and in P5 (Fig.98-b), outer spine of 2nd exopodal segment relatively small and terminal spine of 3rd exopodal segment 69% as long as segment.

Description of male. Prosome length, 2.16-2.40 mm; body length, 2.96-3.36 mm. Prosome 72% length of body and approximately 2.5 times length of urosome. Urosome 28% length of body. Caudal rami and their setation as in female. Left, geniculate A1 similar in morphology to that of *H. spinifrons*. All other cephalosomal appendages and first 4 pairs of legs as in female.

In right P5, basal inner lobe (Figs.98-e-g) arising close to proximal end, tapering mediad and then bent backward, greatly inflated distally, and fringed with a single band of setules; its lateral margin folded over medial margin of segment. Second exopodal segment (Fig.98-c) approximately 1.5 times length of 1st, with a moderately developed medial projection, of which the plumose proximal margin is nearly perpendicular to longitudinal axis of segment (Figs.98-h, i). Third exopodal segment smoothly curved, approximately 2.3 times length of 2nd, with outer spine 32% length of segment from proximal end, terminal spine 69% as long as segment, terminal lobe (Fig.98-d) approximately 1/11 length of terminal spine. In left P5, basis (Fig.98-e) with a smoothly bulging inner margin fringed with a single band of setules. Second exopodal segment approximately 1.5 times length of 1st, with outer spine not borne on a conical process and similar in size to outer spine of 1st segment (Fig.98-j). Third segment with well-developed inner and outer spines, tapering distally into a long spiniform process (Fig.98-c). Together with spiniform process, 3rd segment making up 70% length of whole exopod.

Etymology. The specific name *americanus* refers to the distribution of the species.

Distribution. The species is represented by 94 specimens including 61 females and 33 males from 18 samples all taken in the east Pacific between 20°20'S and 31°38'N and from the west coast of the Americas westward to 118°53'W. These findings seem to indicate that the species is endemic to the coastal area of the eastern tropical Pacific.

Remarks. The female genital somite of this species is characteristic in having a moderately sigmoid dorsal margin, a strongly sigmoid ventral margin formed by the genital flange together with the posterior ventral wall and a long genital operculum reaching the junction between the genital flange and the posterior ventral wall. The male can be identified by the combination of the following characters of the 5th pair of legs: Inner lobe of right basis with a narrow beginning, turned backward by folding over, and greatly inflated distally; medial projection of 2nd right exopodal segment with proximal margin nearly perpendicular to longitudinal axis of segment; 3rd right exopodal segment with a long terminal spine; outer spine of 2nd left exopodal segment not borne on a conical process.

Heterorhabdus cohibilis, new species
Figure 99

Type material. Holotype female (USNM 287075), 13 paratype females (USNM 287077), allotype male (USNM 287076), and 4 paratype males (USNM 287078) found in IKMT sample from 3000-0 mwo, SIO Aries I, station 3, 00°55'N, 112°50'W in the eastern tropical Pacific, November 22, 1970.

Description of female. Prosome length, 2.00-2.52 mm; body length, 2.80-3.60 mm. Similar in habitus (Fig.99-a) to *H. americanus* described above. Prosome 71% length of body and approximately 2.4 times length of urosome. Laterally, prosome typically with a length-depth ratio of 100:37. Cephalosome 45% length of prosome, with broadly rounded forehead produced ventrad into a conical rostrum carrying 2 slender filaments. Dorsally, midanterior tubercular process of forehead somewhat truncate. Posterolateral corner of prosome (Fig.99-b) rounded, overlapping anterior 1/10 length of genital somite.

Dorsally, genital somite (Fig.99-b) typically with a length-width ratio of 100:78, widest at middle, only slightly asymmetrical, with a broadly curved right margin and a slightly sigmoid left margin. Laterally, genital somite (Fig.99-d) typically with a length-depth ratio of 100:82, with a sigmoid dorsal margin formed by a broadly rounded dorsal hump followed by a shallow concave margin and a rounded genital prominence. Genital flange large, rounded, with posterior edge meeting ventral wall in a wide angle 13% length of somite from posterior end. Genital operculum reaching about halfway along posterior slope of genital flange. Ventrally, genital operculum (Fig.99-c) only slightly wider than long, typically with a length-width ratio of 96:100 and 72% as wide as somite itself.

Urosome (Fig.99-b) 29% length of body. Length ratios of 4 urosomal somites and left caudal ramus, 37.7 : 18.8 : 14.1 : 8.8 : 20.6 = 100. Dorsal appendicular seta of left caudal ramus approximately 1.5 times length of that of right ramus. Anteriorly, caudal ramus with 2 pores.

Antennule extending beyond posterior end of caudal ramus by its last 2 segments. All cephalosomal appendages and swimming legs similar in anatomical details to those of *H. americanus* described above.

Description of male. Prosome length, 1.96-2.40 mm; body length, 2.68-3.32 mm. Prosome 73% length of body and approximately 2.6 times length of urosome. Urosome 28% length of body. Caudal rami and their setation as in female. Left, geniculate A1 similar in morphology to that of *H. spinifrons*. All other cephalosomal appendages and first 4 pairs of legs as in female.

Basal inner lobe of right P5 (Figs.99-e-g) arising from proximal half of medial margin of segment, initially tapering mediad before recurving as lateral margin is folded over medial margin of segment, only slightly inflated distally, and fringed with a single band of setules. In right exopod (Fig.99-e), 1st and 2nd segments each with a pore on anterior surface close to proximal end. Second segment approximately 1.5 times length of 1st, with a moderately developed medial projection, proximal margin of which nearly perpendicular to longitudinal axis of segment. Third segment smoothly curved, about 2 times length of 2nd, with outer spine 28% length of segment from proximal end, terminal spine 65% as long as segment, terminal lobe approximately 1/11 length of terminal spine.

In left P5 (Figs.99-h, i), basis with smoothly bulging medial margin armed with a single band of setules. First 2 exopodal segments with well-developed outer spines of similar length. Second exopodal segment approximately 1.3 times length of 1st, with outer spine not borne on a conical process. Third segment with well-developed inner and outer spines, tapering distally into a long spiniform process. Together with spiniform process, 3rd exopodal segment making up 71% length of whole exopod.

Etymology. The specific name *cohibilis*, from the Latin meaning short, refers to the relatively short genital operculum.

Distribution. This species is represented by 78 specimens including 54 females and 24 males from 6 samples—3 taken in the eastern equatorial Pacific between 00°55'S and 02°35'N and between 112°57'W and 112°36'W; 1 at 09°25'S, 159°06'E in the western equatorial Pacific; 1 at 04°40'S, 125°33'E in the Malay Archipelago; 1 at 25°35'N, 124°30'E in the East China Sea.

Remarks. This species is very close in morphology to *H. americanus* described above but can be distinguished from it in the female by the relatively short genital operculum, which in lateral view is far short of reaching the posterior edge of the genital flange, while the genital operculum in *H. americanus* is extending to the posterior edge of the genital flange. The male can be identified by the basal inner lobe of the right P5, which is only slightly inflated. Furthermore, the two species have different ranges. *Heterorhabdus cohibilis* occurs widely in the equatorial Pacific including the Malay Archipelago, while *H. americanus* was found only in coastal waters of the eastern tropical Pacific.

Heterorhabdus spinosus Bradford 1971
Figures 100 and 101

Heterorhabdus spinosus Bradford 1971, p. 121, figs. 1, 2g-k, 3c, 4c.

Description of female. Prosome length, 2.72-3.08 mm; body length, 3.84-4.40 mm. Prosome 71% length of body and approximately 2.3 times length of urosome. Laterally, prosome (Fig.100-a) typically with a length-depth ratio of 100:38. Cephalosome 45% length of prosome, with broadly rounded forehead (Fig.100-d) produced ventrad into a conical rostrum bearing 2 slender filaments. Dorsally, forehead (Fig.100-e) somewhat truncate; midanterior tubercular process more or less rounded. Posterolateral corner of prosome (Fig.100-b) broadly rounded, overlapping anterior 1/10 length of genital somite.

Dorsally, genital somite (Fig.100-b) widest close to anterior end, typically with a length-width ratio of 100:75 and nearly symmetrical lateral swellings; anterior slopes of lateral swellings relatively steep, while posterior slopes gradually extend to posterior end of somite. Laterally, genital somite (Figs.100-f, g) typically with a length-depth ratio of 100:92, with a strongly sigmoid dorsal margin and rounded genital prominence. Genital flange high and rounded when genital operculum is closed (Fig.100-f), with posterior edge meeting ventral wall of somite about 1/6 length of somite from posterior end. When genital operculum is fully open, genital flange extends ventrad into an angular process (Fig.100-h). Ventral wall of somite posterior to genital flange bulging ventrad into a rounded lobe, which is clearly visible when genital operculum is open. When closed, genital operculum extending almost to posterior end of somite (Fig.100-c). Ventrally, genital operculum (Fig.100-i) longer than wide, typically with a length-width ratio of 100:88 and 79% as wide as somite itself. Genital field with a cluster of about 12 pores on each side.

Urosome (Figs.100-b, c) 31% length of body. Length ratios of 4 urosomal somites and left caudal ramus, 37.1 : 18.3 : 13.7 : 8.6 : 22.3 = 100. Left caudal ramus extending beyond posterior end of right ramus by 1/6 its length as measured along medial margins. Dorsal appendicular seta of left caudal ramus a little longer than that of right ramus. Anteriorly, each caudal ramus with 3 pores.

Antennule extending beyond posterior end of caudal ramus by its last 2 segments. All cephalosomal appendages similar to those of *H. habrosomus* except for the following features of Mx2 (Fig.101-a): Posterior subterminal spine of 4th lobe approximately 1/5 length of 2nd saberlike spine, which is a little longer than 1st; falcate spine of 5th lobe approximately 1.3 times length of saberlike spine, with proximal 47% of medial margin finely serrated with short spinules; falcate spine of 6th lobe with proximal 63% of medial margin serrated with relatively long spinules.

Swimming legs also similar to those of *H. habrosomus* except for P5 (Figs.101-b, c), in which basipod about as long as endopod and terminal spine of 3rd exopodal segment 52% as long as segment.

Description of male. Prosome length, 2.40-2.76 mm; body length, 3.32-3.80 mm. Prosome 72% length of body and approximately 2.5 times length of urosome. Urosome 29% length of body. Caudal rami and their setation as in female. Left, geniculate A1 similar in morphology to that of *H. spinifrons*. All other cephalosomal appendages and first 4 pairs of legs as in female.

In 5th pair of legs, basal inner lobe of right leg (Figs.101-d, h) arising from proximal half of segment, tapering mediad for some distance, recurved distally by folding lateral margin. Medial margin of lobe, together with medial margin of segment, forming a smoothly curved arch (Fig.101-i). Lateral margin of lobe nearly straight. Medial margin of lobe armed with thick lancetlike bristles or spinules (Fig.101-j). In right exopod, 2nd segment (Figs.101-k, l) approximately 1.5 times length of 1st, with moderately developed medial projection and outer spine only a short distance distal to midpoint. Proximal margin of medial projection arising about 1/4 length of segment from proximal end and nearly perpendicular to longitudinal axis of segment. Third segment (Fig.101-d) smoothly curved, about twice the length of 2nd, with outer spine 30% length of segment from proximal end, terminal spine 54% as long as segment, terminal lobe (Fig.101-e) approximately 1/4 length of terminal spine.

Basis of left P5 (Fig.101-f) with smoothly inflated inner margin, of which distal half is armed with somewhat lancetlike bristles. In exopod, 2nd segment (Fig.101-g) almost twice as long as 1st, with a short lateral process terminating with outer spine. Third segment with well-developed inner and outer spines, distally tapering into a long spiniform process. Together with spiniform process, 3rd segment making up 67% length of whole exopod.

Distribution. This species is represented in the study by 343 specimens including 203 females and 140 males found in 9 samples—3 from off the west coast of South Africa in the southeastern Atlantic between 33°47'S and 31°09'S; 3 from off the Chilean coast in the southeastern Pacific between 46°55'S and 35°10'S; 1 from off the east coast of New Zealand at 41°26'S, 176°04'E; 2 in the Tasman Sea at 49°39'S, 152°34'E and 40°06'S, 152°00'E, respectively. According to these findings and the previous record, this species has a circumsubantarctic distribution.

Remarks. This species was originally described from females 3.4-4.2 mm long and males 3.1-3.7 mm long from the southwestern Pacific between 45°55'S and 34°33'S and 157°32'E and 175°14'E. The female can be distinguished from the other members of the species group by a long genital operculum, strongly sigmoid dorsal wall of the genital somite, a lobular outgrowth on the distoventral margin of the genital somite, and a short posterior subterminal spine of the 4th lobe of Mx2. The male 5th pair of legs is unique in that the basal inner lobe of the right leg is armed with thick lancetlike bristles.

Heterorhabdus paraspinosus, new species
Figures 102 and 103

Type material. Holotype female (USNM 287079), 17 paratype females (USNM 287081), allotype male (USNM 287080), and 8 paratype males (USNM 287082) found in IKMT sample from 6000-0 mwo, SIO Lusiad VII, station 14, 32°27'S, 08°49'E in the southeastern Atlantic, June 6, 1963.

Description of female. Prosome length, 2.60-3.00 mm; body length, 3.64-4.40 mm. Similar in habitus to *H. spinosus*. Prosome 69% length of body and approximately 2.2 times length of urosome.

Dorsally, genital somite (Fig.102-b) widest at middle, typically with a length-width ratio of 100:72, strongly asymmetrical, with broadly inflated left side and relatively low, smooth swelling followed by a tubercular outgrowth on right side. Genital somite in dorsal view varies among specimens to a certain extent, mainly in the shape of the lateral margins, but related variations were not found in lateral view. One example of such variation is shown in Figure 102-c, in which the left swelling is short of reaching posterior end of somite, right swelling is relatively short, and distal lobular outgrowth of right side is not as pronounced.

Laterally, genital somite (Figs.102-d-f) typically with a length-depth ratio of 100:91, with a large rounded hump at middle and rounded tubercular outgrowth posteriorly on dorsal margin; genital prominence large and rounded. Viewed from right side and when tilted counterclockwise, dorsal margin of genital somite (Fig.102-e) is distinctly concave, ending in a tubercular outgrowth of posterior end. Genital flange broadly rounded; its posterior edge meeting ventral wall of somite 12% length of somite from posterior end. Genital operculum extending to junction between genital flange and posterior ventral wall of somite.

Ventrally, genital operculum (Fig.102-g) nearly symmetrical, typically with a length-width ratio of 100:97 and 78% as wide as somite itself. Genital field with a cluster of about 12 pores on each side.

Urosome (Figs.102-a, b) 32% length of body. Length ratios of 4 urosomal somites and left caudal ramus, 38.9 : 17.4 : 14.0 : 8.4 : 21.3 = 100. Left caudal ramus extending beyond posterior end of right by 1/6 its length as measured along medial margins. Dorsal appendicular seta of left caudal ramus a little longer than that of right ramus. Each caudal ramus with 3 pores on anterior surface.

Antennule extending beyond posterior end of caudal ramus by its last 2 segments. All cephalosomal appendages and swimming legs similar in morphology to those of *H. spinosus* except for the following features of Mx2 (Fig.103-a): Posterior subterminal spine of 4th lobe 29% as long as 2nd saberlike spine; falcate spine of 5th lobe with proximal 48% of medial margin finely serrated with short spinules; falcate spine of 6th lobe with proximal 66% of medial margin serrated with relatively long spinules.

Description of male. Prosome length, 2.44-2.76 mm; body length, 3.44-3.96 mm. Similar in habitus to *H. spinosus.* Prosome 71% length of body and approximately 2.4 times length of urosome. Urosome 30% length of body. Caudal rami and their setation as in female. Left, geniculate A1 similar in morphology to that of *H. spinifrons.* All other cephalosomal appendages and first 4 pairs of legs as in female.

Fifth pair of legs similar to that of *H. spinosus* except for the following features: In basal inner lobe of right leg (Figs.103-b-d), lateral margin folded over to a lesser extent, medial margin with a hump proximally, and armed along medial margin with a band of normal setules; third exopodal segment of right leg with outer spine 34% length of segment from proximal end, terminal spine 27% length of segment, and terminal lobe (Fig.103-e) approximately 1/3 length of terminal spine; in left exopod (Figs.103-g, h), lateral process bearing outer spine of 2nd segment very low; 3rd segment together with distal spiniform process making up 64% length of whole exopod.

Etymology. The specific name, formed from the Greek prefix *para-* meaning near, alludes to the apparent relationship of this new species with *H. spinosus.*

Distribution. This species is represented by 64 specimens including 38 females and 26 males found in 5 samples—3 taken from off the west coast of South Africa in the southeastern Atlantic between 33°47'S and 31°09'S; 1 from off the Chilean coast at 46°55'S, 75°50'W; 1 in the Tasman Sea at 49°39'S, 152°34'E. This species seems to be restricted in distribution to the coastal waters of the subantarctic coastal waters.

Remarks. This species is almost identical in body size and distribution to *H. spinosus* Bradford 1971, but can be distinguished from it in the female by the genital somite, which in lateral view has a relatively short genital operculum, a tubercular outgrowth posteriorly on the dorsal margin, and no tubercular outgrowth posteriorly on the ventral wall and in the male by the right P5; its basal lobe is furnished with normal setules instead of lancetlike bristles, and its 3rd exopodal segment has a short terminal spine. The cephalosomal appendages of these species are almost identical except that in Mx2, the posterior subterminal spine of the 4th lobe is much longer in this species.

Heterorhabdus austrinus Giesbrecht 1902
Figures 104 and 105

Heterorhabdus austrinus Giesbrecht 1902, p. 28, pl. 6, figs. 1-9.—Brady 1918, p. 27, pl. 4, figs. 1-9.—Farran 1929, p. 265, fig. 26.—Vervoort 1951, p. 127.—Vervoort 1957, p. 131.

Description of female. Prosome length, 2.04-2.56 mm; body length, 2.84-3.68 mm. Body slender (Fig.104-a). Prosome 72% length of body and approximately 2.4 times length of urosome. Laterally, prosome typically with a length-depth ratio of 100:36. Cephalosome 45% length of prosome, with broadly rounded forehead (Fig.104-d) tapering ventrad into a relatively large conical rostrum provided with 2 elongate filaments. Dorsally, forehead (Fig.104-e) somewhat truncate. Midanterior tubercular process also somewhat truncate, bearing suprafrontal sensillae. Posterolateral corner of prosome (Fig.104-c) rounded, overlapping about anterior 1/7 length of genital somite.

Dorsally, genital somite (Fig.104-c) nearly symmetrical, with bilobed lateral swellings extending posteriad to 88% length of somite from anterior end, widest at anterior pair of lobes or 26% length of somite from anterior end, typically with a length-width ratio of 100:72. Posterior end of left margin of genital somite slightly produced. Laterally, genital somite (Figs.104-b, f) typically with a length-depth ratio of 100:76. Dorsal margin sigmoid, with a broadly rounded hump at middle followed by a smooth hollow. Whole ventral margin a smoothly curved arch, with genital operculum reaching close to posterior end of somite when closed. Genital flange a low triangular form followed by a large tubercular process arising from posterior end of ventral margin. This posterior tubercular process is easily seen when genital operculum is open (Fig.104-f). Ventrally, genital operculum (Fig.104-g) much longer than wide, typically with a length-width ratio of 100:73 and 78% as wide as somite itself. Genital field with a cluster of about 20 pores on each side.

Urosome (Figs.104-b, c) about 30% length of body. Length ratios of 4 urosomal somites and left caudal ramus, 39.4 : 17.6 : 14.5 : 7.9 : 20.6 = 100. Left caudal ramus extending beyond posterior end of right by 1/9 its length as measured along medial margin. Dorsal appendicular seta of left caudal ramus approximately 1.5 times length of that of right ramus. Anteriorly, each caudal ramus with 4 pores.

Antennule reaching about posterior end of caudal ramus. All cephalosomal appendages similar in morphology to those of *H. habrosomus* described above except for the following features of Mx2 (Fig.104-h): Spine of 2nd lobe relatively thin and short, posterior subterminal spine of 4th lobe 51% as long as 2nd saberlike spine, which is a little longer than 1st saberlike spine; falcate spine of 5th lobe approximately 1.3 times length of saberlike spine, with proximal 54% of medial margin serrated with spinules; falcate spine of 6th lobe with proximal 67% of medial margin serrated with spinules.

Swimming legs also similar in morphology to those of *H. habrosomus* with the following exceptions: In P1 exopod (Fig.105-a), 2nd outer spine longer than 3rd; in P5 (Fig.105-b), basipod about as long as endopod, inner spine of 2nd segment a little shorter than 3rd segment, terminal spine of 3rd segment 51% as long as segment.

Description of male. Prosome length, 1.88-2.28 mm; body length, 2.64-3.28 mm. Prosome 69% length of body and approximately 2.2 times length of urosome. Urosome 32% length of body. Caudal rami and their setation as in female. Left, geniculate A1 similar in morphology to that of *H. spinifrons*. All other cephalosomal appendages and first 4 pairs of legs as in female.

In 5th pair of legs, basal inner lobe of right leg (Figs.105-d-f) elongate, arising from proximal end of segment, tapering mediad into a relatively long and narrow neck and then turning backward making a wide rounded curve by folding over lateral margin, armed along medial margin with a single band of setules. In right exopod, 2nd segment (Figs.105-i, j) a little longer than 1st, with a triangular process proximal to medial projection, which is also somewhat triangular. Third segment (Fig.105-c) only slightly curved, with outer spine 38% length of segment from proximal end, terminal spine approximately 1/5 length of segment, terminal lobe (Fig.105-g) 27% as long as terminal spine.

In left exopod, 2nd segment (Fig.105-h) approximately 1.4 times length of 1st, with outer spine borne on a long lateral conical process. Measured along medial margins, conical process approximately 1/4 length of segment and about 1/2 length of outer spine. Third segment (Fig.105-c) with well-developed inner and outer spines, tapering distally into a long spiniform process. Together with distal spiniform process, 3rd segment making up 63% length of whole exopod.

Distribution. Since its original discovery in the Pacific sector of the Antarctic by Giesbrecht (1902), this species has been recorded throughout the circumantarctic waters and found as far north as 44°05'S in the Tasman Sea (Vervoort 1951, 1957). In the present study, a total of 40 specimens including 28 females and 12 males were found in samples taken in the following 5 localities: 33°47'S, 15°39'E off the west coast of South Africa; 60°42'S, 59°48'W in the Drake Passage; 46°55'S, 75°50'W off the Chilean coast; 41°26'S, 176°04'E off the east coast of New Zealand; 49°39'S, 152°34'E in the Tasman Sea.

Remarks. This species was originally described from a female specimen 3.4 mm long collected in the Pacific sector of the Antarctic (70°12'S, 84°03'W). Farran (1929) noted a wide range of size variation in both the female and male.

The female of this species can be easily distinguished from the other members of the species group by the genital somite, which in lateral view has a smoothly arched ventral margin, a genital operculum reaching almost the posterior end of the somite, and a conspicuous tubercular process at the posterior end of the ventral margin. The 5th pair of legs of the male is characteristic in having a right basal lobe turn backward in a broad smooth curve, a triangular process proximal to the medial projection of the 2nd exopodal segment of the right leg, a short terminal spine of the 3rd exopodal segment, and a long conical lateral process of the 2nd exopodal segment of the left leg.

Heterorhabdus abyssalis (Giesbrecht 1889)
Figures 106 and 107

Heterochaeta abyssalis Giesbrecht 1889, p. 812.—Giesbrecht 1892, p. 373, pl. 19, fig. 4; pl. 20, figs. 29, 30.

Heterorhabdus abyssalis Giesbrecht 1898, p. 116, fig. 26.

Description of female. Prosome length, 2.00-2.40 mm; body length, 2.88-3.36 mm. Similar in habitus (Fig.106-a) to *H. austrinus*. Prosome 71% length of body and approximately 2.3 times length of urosome. Laterally, prosome typically with a length-depth ratio of 100:36. Cephalosome 43% length of prosome.

Dorsally, genital somite (Fig.106-c) widest at middle, typically with a length-width ratio of 100:78, somewhat asymmetrical with a broadly curved right margin and more or less bilobed left margin. Laterally, genital somite (Figs.106-b, d) typically with a length-depth ratio of 100:87. Dorsal margin strongly sigmoid with a large rounded hump followed by a smoothly curved hollow. Ventral margin smoothly arched, with genital operculum extending to a level 7% length of somite from posterior end. Genital flange broadly arched, with posterior edge meeting ventral wall of somite in a notch. When genital operculum is open, left genital flange (Fig.106-f) extending into a triangular form, while right genital flange becomes depressed (Fig.106-e). Ventrally, genital operculum (Fig.106-g) longer than wide, typically with a length-width ratio of 100:79, and 73% as wide as somite itself. Right side of genital field with about 8 pores, but no pores were found on left side.

Urosome (Figs.106-b, c) 30% length of body. Length ratios of 4 urosomal somites and left caudal ramus, 38.1 : 18.2 : 13.3 : 7.7 : 22.7 = 100. Left caudal ramus extending beyond posterior end of right by about 1/10 its length as measured along medial margins. Dorsal appendicular seta of left ramus approximately 1.5 times length of that of right ramus. Anteriorly, each caudal ramus with 2 pores.

Antennule extending beyond posterior end of caudal ramus by its last 2 segments. All cephalosomal appendages and swimming legs similar to those of *H. austrinus* except for the following features: In Mx2, posterior subterminal spine of 4th lobe (Fig.106-h) 43% as long as 2nd saberlike spine; in P1 exopod (Fig.106-i), outer spines of 2nd and 3rd segments relatively small; outer spine of 2nd segment similar in size to 1st outer spine of 3rd segment.

Description of male. Prosome length, 1.96-2.28 mm; body length, 2.72-3.16 mm. Prosome 73% length of body and 2.6 times length of urosome. Urosome 28% length of body. Caudal rami and their setation as in female. Left, geniculate A1 similar in morphology to that of *H. spinifrons*. All other cephalosomal appendages and first 4 pairs of legs as in female.

In 5th pair of legs, right basal inner lobe (Figs.107-c-e) initially extending and tapering mediad for a distance nearly equal to length of segment and then widely recurved with lateral margin deeply folded, furnished with a single band of setules along whole medial margin. In right exopod, 2nd segment (Figs.107-g, h) approximately 1.5 times length of 1st, with a rounded tubercular process preceding the medial projection. Proximal margin of medial projection sloping at an angle of about 55° with reference to longitudinal axis of segment; distal end more or less truncated. Outer spine of 2nd segment well-developed, about 3/4 length of segment from proximal end. Third segment (Fig.107-a) rather strongly curved, about twice the length of the 2nd, with outer spine 37% length of segment from proximal end, terminal spine 63% as long as segment, terminal lobe (Fig.107-b)11% as long as terminal spine.

In left P5 (Figs.107-a, f), basis with strongly arched medial margin, of which the distal half fringed with a single band of setules. In left exopod, 2nd segment approximately 1.5 times length of 1st, without lateral conical process, with outer spine only slightly longer than that of 1st segment. Third segment tapering distally into a long spiniform process, with outer spine about 1/2 length of inner spine and a little longer than outer spine of 2nd segment. Together with distal spiniform process, 3rd segment 69% as long as whole exopod.

Distribution. The type locality of this species is 14°N, 132°W in the eastern tropical Pacific. In the present study, 301 specimens of this species including 201 females and 100 males were found in

28 samples, all taken in the eastern Pacific between 35°10'S and 32°35'N and from the west coast of the Americas westward to 118°05'W. These findings constitute the first valid records since the original description.

Remarks. The early descriptions by Giesbrecht (1889, 1892, 1898) were based solely on the type material (a male 2.75 mm long). Subsequently, the species was reported by Sewell (1947), Vervoort (1957), Tanaka (1964), Park (1968), and Bradford (1971), but none of the material of these authors is identical to *H. abyssalis* as described by Giesbrecht (1892). In the present study, a species fully in agreement in details of the male 5th pair of legs with *H. abyssalis* Giesbrecht was found in its type locality.

Heterorhabdus abyssalis is very close in habitus and details of the appendages to *H. austrinus* but can be distinguished from it in the female by the genital somite, of which the posterior edge of the left genital flange meets the ventral wall of the somite in a characteristic notch and in the male by the 5th pair of legs, of which the right 3rd exopodal segment has a long terminal spine and the outer spine of the left 2nd exopodal segment is not borne on a conical process.

Heterorhabdus pacificus Brodsky 1950
Figures 108 and 109

Heterorhabdus pacificus Brodsky 1950, p. 355, fig. 250.—Tanaka 1964, p. 4, fig. 176.

Description of female. Prosome length, 2.08-2.80 mm; body length, 2.88-3.84 mm. Body (Fig.108-a) relatively strongly built. Prosome 72% length of body and approximately 2.4 times length of urosome. Laterally, prosome typically with a length-depth ratio of 100:40. Cephalosome 47% length of prosome. Laterally, forehead broadly rounded, produced ventrad into a conical rostrum bearing 2 slender filaments. Dorsally, forehead (Fig.108-d) somewhat truncate, with a relatively low, more or less rectangular midanterior tubercular process bearing suprafrontal sensillae. Posterolateral corner of prosome (Fig.108-b) broadly rounded, overlapping anterior 1/10 length of genital somite.

Dorsally, genital somite (Fig.108-b) widest at middle, typically with a length-width ratio of 100:77, asymmetrical with broadly bulging left margin and wavy right margin. Posterior 1/4 of right margin bulging into a low, rounded lateral swelling. Laterally, genital somite (Fig.108-f) typically with a length-depth ratio of 100:84. Dorsal wall with a large rounded hump followed by a straight margin. Genital prominence high, with a rounded outline. Genital flange broadly rounded, its posterior edge meeting ventral wall of somite 7% length of somite from posterior end. Genital operculum extending beyond peak of genital flange and to level 23% length of somite from posterior end. Ventrally, genital operculum (Fig.108-e) wider than long, typically with a length-width ratio of 89:100 and 75% as wide as somite itself. Genital field with about 10 pores on each side.

Urosome (Fig.108-b) 30% length of body. Length ratios of 4 urosomal somites and left caudal ramus, 37.5 : 17.3 : 14.3 : 8.9 : 22.0 = 100. Left caudal ramus extending beyond posterior end of right by 1/8 its length as measured along medial margin. Dorsal appendicular seta of left caudal ramus almost twice as long as that of right ramus.

Antennule extending beyond posterior end of caudal ramus by its last 3 segments. All cephalosomal appendages similar to those of *H. habrosomus* described above except for the following features of Mx2 (Fig.108-g): Second seta of first lobe less than 1/2 length of 1st; spine of 2nd lobe relatively thin; posterior subterminal spine of 4th lobe 40% as long as 2nd saberlike spine; falcate spine of 5th lobe approximately 1.2 times length of saberlike spine, with proximal 1/2 of medial margin serrated with spinules; falcate spine of 6th lobe with proximal 68% of medial margin serrated with spinules.

Swimming legs also similar to those of *H. habrosomus* except for the following minor features of P5 (Figs.108-h, i): Outer spine of 2nd exopodal segment relatively small; in 3rd exopodal segment, 1st outer spine slightly posterior to middle of segment, terminal spine approximately 3/5 length of segment.

Description of male. Prosome length, 2.08-2.76 mm; body length, 2.92-3.96 mm. Prosome 73% length of body and approximately 2.7 times length of urosome. Urosome 27% length of body. Caudal rami and their setation as in female. Left, geniculate A1 similar in morphology to that of *H. spinifrons.* All other cephalosomal appendages and first 4 pairs of legs as in female.

Basal inner lobe of right P5 (Figs.109-b-d) arising from middle of segment, extending and tapering mediad to a distance equal to 2/3 length of segment and then recurved. Anteriorly and when tilted clockwise (Figs.109-b, d), lobe turning backward with lateral margin forming a relatively wide angle and without folding. Whole medial margin armed with a band of setules. In right exopod, 2nd segment (Figs.109-g, h) approximately 1.6 times length of 1st, with a low lobe proximally on medial margin; its medial projection roughly triangular, with proximal margin almost perpendicular to longitudinal axis of segment; outer spine about 3/4 length of segment from proximal end. Third segment (Fig.109-a) smoothly curved, with outer spine 30% length of segment from proximal end, terminal spine 70% as long as segment, terminal lobe (Fig.109-e) 8% as long as terminal spine.

Basis of left P5 (Fig.109-a) with strongly bulging medial margin armed with a band of setules along its distal half. Second exopodal segment (Fig.109-f) approximately 1.5 times length of 1st; its outer spine approximately 1/5 length of segment and not borne on a conical process. Third exopodal segment including distal spiniform process 71% as long as whole exopod.

Distribution. Since its original description from the northwestern Pacific, this species has been recorded only from off Japan (Tanaka 1964). In the present study, it is represented by a total of 473 specimens including 300 females and 173 males found in 49 samples—6 taken in the Atlantic between 33°47'S and 38°53'N; 38 throughout the Pacific between 47°08'S and 40°35'N and from the west coast of the Americas westward to the Malay Archipelago; 5 taken in the Indian Ocean between 35°00'S and 02°28'S. These findings establish *H. pacificus* as a circumglobal species occurring widely in all three major oceans.

Remarks. This species was originally described from both females 3.5 mm long and males 3.2 mm long collected from 4000-1000 m in the northwestern Pacific. Tanaka (1964) found the species in samples taken from 1000-0 m in the Izu region of Japan; his females were 3.40-3.73 mm long and males 3.07-3.35 mm.

The female can be distinguished from those of the other members of the species group by the genital somite, which in dorsal view has a low, rounded swelling posteriorly on the right margin and in lateral view has a genital flange reaching close to the posterior end of the somite and a genital operculum, which falls far short of the posterior edge of the genital flange. The male can be identified by the relatively smoothly curved basal inner lobe of the right P5, the long terminal spine and short terminal lobe of the right exopod, and the small outer spine of the left 2nd exopodal segment, which is not borne on a conical process.

Heterorhabdus confusibilis, new species
Figures 110 and 111

Type material. Holotype female (USNM 287083), 37 paratype females (USNM 287085), allotype male (USNM 287084), 23 paratype males (USNM 287086) found in IKMT sample from 4000-0 mwo, SIO Lusiad VII, station 79, 01°10'N, 11°36'W in the eastern tropical Atlantic, July 6, 1963.

Description of female. Prosome length, 1.80-2.44 mm; body length, 2.52-3.44 mm. Body (Fig.110-a) relatively slender. Prosome 73% length of body and approximately 2.5 times length of urosome. Laterally, prosome typically with a length-depth ratio of 100:38. Cephalosome 44% length of prosome. Laterally, forehead (Fig.110-c) broadly rounded, produced ventrad into a relatively large conical rostrum bearing 2 slender filaments. Dorsally, forehead (Fig.110-d) somewhat truncate, with a low, rectangular midanterior tubercular process bearing suprafrontal sensillae. Posterolateral corner of prosome (Fig.110-b) rounded, overlapping about anterior 1/10 length of genital somite.

Dorsally, genital somite (Fig.110-b) widest at middle, typically with a length-width ratio of 100:72, almost symmetrical with weakly bilobed lateral swellings extending to 87% length of somite from anterior end. Laterally, genital somite (Fig.11-e) typically with a length-depth ratio of 100:85. Dorsal wall more or less sigmoid, with a broadly bulging hump followed by a slight hollow. Genital prominence high, with rounded outline. Genital flange broadly rounded, slightly depressed when genital operculum open (Fig.110-f), its posterior edge meeting ventral wall of somite 6% length of somite from posterior end. Genital operculum extending beyond peak of genital flange and to level 18% length of somite from posterior end. Ventrally, genital operculum (Fig.110-g) about as long as wide and 74% as wide as somite. Genital field with about 12 pores on each side.

Urosome (Fig.110-b) 28% length of body. Length ratios of 4 urosomal somites and left caudal ramus, 38.7 : 17.4 : 13.6 : 8.4 : 21.9 = 100. Left caudal ramus extending beyond posterior end of right by about 1/6 its length as measured along medial margin. Dorsal appendicular seta of left caudal ramus approximately 1.5 times longer than the same seta of right ramus.

Antennule extending beyond posterior end of caudal ramus by its last 2 segments. All cephalosomal appendages similar to those of *H. habrosomus* described above except for the following features of Mx2 (Fig.110-h): Posterior subterminal spine of 4th lobe 38% as long as 2nd saberlike spine, which is about as long as 1st saberlike spine; falcate spine of 5th lobe approximately 1.4 times length of saberlike spine, with proximal 54% of medial margin serrated with spinules; falcate spine of 6th lobe with proximal 67% of medial margin serrated with spinules.

Swimming legs also similar to those of *H. habrosomus* except for the following features of P5 exopod (Fig.110-i): Inner spine of 2nd segment a little shorter than 3rd segment, 1st outer spine of 3rd segment located 59% length of segment from proximal end, terminal spine 56% as long as segment.

Description of male. Prosome length, 1.88-2.24 mm; body length, 2.56-3.08 mm. Prosome 72% length of body and approximately 2.6 times length of urosome. Urosome 28% length of body. Caudal rami and their setation as in female. Left, geniculate A1 similar in morphology to that of *H. spinifrons.* All other cephalosomal appendages and first 4 pairs of legs as in female.

In right P5, basal inner lobe (Figs.111-f-h) arising from middle of segment, extending and tapering mediad to a distance equal to 1/2 length of segment and recurved. Anteriorly and when tilted clockwise (Figs.111-f, g), recurved section with lateral margin flexed at a relatively sharp angle and medial margin curving in a smooth arch. Angle formed by flexed lateral margin somewhat variable (about from 50° to 60°). Whole length of medial margin fringed with a band of setules. Second exopodal segment (Figs.111-i, j) approximately 1.5 times length of 1st, with small tubercular outgrowth proximally on medial margin, followed by a roughly triangular medial projection. Proximal margin of medial projection about perpendicular to longitudinal axis of segment. Outer spine well developed and located about 3/4 length of segment from proximal end. Third exopodal segment (Figs.111-a, d) smoothly curved, with outer spine 30% length of segment from proximal end, terminal spine 78% as long as segment, terminal lobe (Fig.111-e) 9% as long as terminal spine.

In left P5 (Fig.111-b), basis with strongly bulging medial margin armed with a band of setules along distal 3/5. Second exopodal segment (Fig.111-c) approximately 1.5 times length of 1st, with a relatively long outer spine, which is about 1/3 length of segment and borne on a base only slightly raised from lateral margin of segment. Third exopodal segment tapering distally into a long spiniform process and, together with distal process, making up 68% length of whole exopod.

Etymology. The specific name *confusibilis,* formed with the Latin suffix *-bilis* meaning tending to be, alludes to the difficulty of distinguishing the species from its allies.

Distribution. This species is represented by a total of 162 specimens including 106 females and 56 males from 33 samples—3 taken in the southeastern Atlantic between 31°11'S and 01°25'N; 29 throughout the Pacific between 30°46'S and 40°35'N and from the west coast of the Americas westward to the Malay Archipelago and East and South China Seas; 1 in the Indian Ocean at 02°28'S, 86°23'E. The range encompasses all three major oceans.

Remarks. The genital somite of the female is similar in shape to those of *H. habrosomus* and *H. cohibilis,* but can be distinguished from them by the genital flange, which in lateral view extends close to the posterior end of the somite, instead of ending some distance before the posterior end of the somite. The male resembles closely that of *H. pacificus* but can be distinguished from it by the basal inner lobe of the right P5, of which the lateral margin is bent in a sharp angle at the turn, and the relatively long outer spine of the 2nd left exopodal segment.

Heterorhabdus tanneri (Giesbrecht 1895)
Figures 112 and 113

Heterochaeta tanneri Giesbrecht 1895, p. 259, pl. 4, figs. 5, 6.

Heterorhabdus tanneri; Giesbrecht 1898, p. 115.—Brodsky 1950, p. 357, fig. 251.—Tanaka 1964, p. 11, fig. 178.

Heterorhabdus proximus Davis 1949, p. 57, pl. 10, figs. 119-125.

Description of female. Prosome length, 2.12-3.08 mm; body length, 3.08-4.41 mm. Body (Fig.112-a) relatively elongate. Prosome 68% length of body and approximately 2.1 times length of urosome. Laterally, prosome typically with a length-depth ratio of 100:39. Cephalosome 44% length of prosome. Forehead rounded, produced ventrad into a conical rostrum bearing 2 slender filaments. Posterolateral corner of prosome (Fig.112-b) rounded, overlapping only anterior 5% length of genital somite.

Dorsally, genital somite (Fig.112-b) widest at middle, typically with a length-width ratio of 100:63, asymmetrical, with a wavy but relatively straight left margin and a low triangular right margin. Laterally, genital somite (Fig.112-d) typically with a length-depth ratio of 100:73. Dorsal wall with a low hump, 34% length of somite from anterior end, followed by a notch, which is in turn followed by a straight margin. Genital prominence relatively low and more or less rounded. Genital flange low, smoothly curved, followed by a bilobed ventral wall of somite; the posterior lobe larger, roughly triangular, with its posterior edge reaching posterior margin of somite. Genital operculum reaching between 2 posterior ventral lobes of somite. Genital flange becoming straight as genital operculum opens (Fig.112-e). Ventrally, genital operculum (Fig.112-c) longer than wide, typically with a length-width ratio of 100:74, somewhat tapering posteriorly and slanted. Genital field with 17-21 pores on each side.

Urosome (Fig.112-b) 32% length of body. Length ratios of 4 urosomal somites and left caudal ramus, 40.7 : 18.1 : 13.5 : 7.7 : 20.0 = 100. Left caudal ramus extending beyond posterior end of right by 1/7 its length as measured along medial margin. Dorsal appendicular seta of left caudal ramus approximately 1.2 times length of that of right ramus.

Antennule extending beyond posterior end of caudal ramus by its last 2 segments. Cephalosomal appendages similar to those of *H. habrosomus* described above except for the following features of Mx2 (Fig.112-f): Posterior subterminal spine of 4th lobe 58% as long as 2nd saberlike spine, which is a little longer than 1st saberlike spine; falcate spine of 5th lobe approximately 1.3 times length of saberlike spine, with proximal 52% of medial margin serrated with spinules; falcate spine of 6th lobe with proximal 68% of medial margin serrated with spinules.

Swimming legs also similar to those of *H. habrosomus* except for the following features of P5 (Fig.112-g): Endopod about as long as basipod; third exopodal segment 38% as long as whole exopod, while it is 44% length of whole exopod in *H. habrosomus*; exopod with first 2 outer spines of similar size; inner spine of 2nd exopodal segment 89% as long as 3rd exopodal segment; terminal spine of 3rd exopodal segment 57% as long as segment.

Description of male. Prosome length, 2.12-2.80 mm; body length, 3.04-4.12 mm. Prosome 71% length of body and approximately 2.4 times length of urosome. Urosome 30% length of body. Caudal rami and their setation as in female. Left, geniculate A1 similar in morphology to that of *H. spinifrons.*

Right P5 (Fig.113-a) very characteristic in having basal inner lobe arising from anterior surface close to medial margin. Anteriorly, basis (Fig.113-e) somewhat rectangular. Basal inner lobe relatively narrow, extending mediad for a short distance and then recurved making a relatively smooth curve, without folding of lateral margin; medial margin armed with 2 separate rows of setules. In right exopod, 1st segment (Fig.113-d) with a large pore at distomedial corner; 2nd segment with a relatively low medial projection 40% length of segment from proximal end, a conspicuous, rounded lobe on anterior surface (Fig.113-e) close to proximal end, outer spine 62% length of segment from proximal end; 3rd segment (Fig.113-a) smoothly curved, with outer spine 26% length of segment from proximal end, terminal spine 62% as long as segment, terminal lobe (Fig.113-b) large, approximately 1/3 length of terminal spine.

In left P5 (Fig.113-a), basis with moderately bulging medial margin fringed with 2 separate rows of setules; 2nd exopodal segment 1.7 times length of 1st, with a relatively short lateral process terminating with outer spine (Fig.113-f); 3rd segment including long distal spiniform process making up 59% length of whole exopod.

Distribution. This species has been recorded from off California (Giesbrecht 1895), the northeastern Pacific (Davis 1949 as *H. proximus*), the northwestern Pacific (Brodsky 1950), and the Izu region of Japan (Tanaka 1964). In the present study, the species is represented by a total of 499 specimens including 264 females and 235 males from 12 samples—10 taken off California between 27°17'N and 40°35'N; 1 at 52°20'N, 155°16'W in the Gulf of Alaska; 1 at 34°33'N, 141°55'E off the east coast of central Japan. According to these findings and the previous records, the range of this species is limited to the coastal waters of the northern Pacific.

Remarks. The original description by Giesbrecht (1895) was based on a male specimen 3.6 mm long collected from about 540-0 m at 35°N, 125°W off California. The species has been redescribed by Brodsky (1950) from both females (4.0-4.1 mm long) and males (3.6-3.9 mm long) obtained in the northwestern Pacific including the Sea of Okhotsk and the Bering Sea and by Tanaka (1964) from females 4.01 mm long and males 3.75-3.85 mm long taken in the Izu region of Japan. *Heterorhabdus proximus* Davis 1949 based on females 3.2-3.7 mm long and males 3.2-3.4 mm long collected from the northeastern Pacific agrees well in both morphology and distribution with this species.

The female is readily distinguishable from the other members of the species group by its characteristic genital somite, which is relatively long and in lateral view has a large lobular process posteriorly on the ventral wall. The male can be identified by the right P5, of which the basal inner lobe arises from the anterior surface of the segment and the 2nd exopodal segment has a low medial projection.

Heterorhabdus norvegicus (Boeck 1872)
Figures 114 and 115

Heterochaeta norvegicus Boeck 1872, p. 40.—Sars 1900, p. 79, pl. 23.
Heterorhabdus norvegicus; Giesbrecht 1898, p. 115.—Sars 1902, p. 118, pls. 80, 81.—Sars 1925, p. 226.—Wilson 1932, p. 132, fig. 89.

Description of female. Prosome length, 2.36-3.00 mm; body length, 3.40-4.24 mm. Body (Fig.114-a) relatively elongate. Prosome 72% length of body and approximately 2.4 times length of urosome. Laterally, prosome typically with a length-depth ratio of 100:36. Cephalosome 43% length of prosome. Laterally, forehead rounded. Dorsally, forehead with a rectangular midanterior tubercular process. Posterolateral corner of prosome (Fig.114-b) broadly rounded, overlapping anterior 1/10 length of genital somite.

Dorsally, genital somite (Fig.114-b) widest about 1/3 length of somite from anterior end, typically with a length-width ratio of 100:69, symmetrical with smoothly curved lateral swellings and gradually narrowing toward posterior end. Laterally, genital somite (Figs.114-f, g) typically with a length-depth ratio of 100:78. Dorsal wall sigmoid with a broadly rounded hump followed by a smooth shallow concavity. Genital prominence rounded. Genital flange relatively high, smoothly rounded, followed by a small lobular outgrowth, which is in turn followed by several concentric ridges. Posterior slope of lobular outgrowth extending linearly to posterior end of somite. Genital operculum reaching anterior edge of lobular outgrowth. Shape of genital flange remaining unchanged when genital operculum is open (Figs.114-d, e). Ventrally, genital operculum (Fig.114-h) asymmetrical, tapering posteriorly, typically with a length-width ratio of 100:79 and 70% as wide as somite itself. Genital field with a cluster of about 40-50 small pores on each side.

Urosome (Figs.114-b, c) 30% length of body. Length ratios of 4 urosomal somites and left caudal ramus, 38.9 : 18.0 : 13.2 : 8.4 : 21.5 = 100. Left caudal ramus extending beyond posterior end of right by about 1/7 its length as measured along medial margin. Dorsal appendicular seta of left ramus approximately 1.5 times length of that of right ramus. Anteriorly, each caudal ramus with 5 pores.

Antennule extending beyond posterior end of caudal ramus by its last 2 segments. All cephalosomal appendages similar to those of *H. habrosomus* described above except for the following features of Mx2 (Fig.115-a): Posterior subterminal spine of 4th lobe 34% as long as 2nd saberlike spine; falcate spine of 5th lobe approximately 1.3 times length of saberlike spine, with proximal 58% of medial margin serrated with spinules; falcate spine of 6th lobe with proximal 67% of medial margin serrated with spinules.

Swimming legs also similar to those of *H. habrosomus* except for the following minor features of P5 exopod (Fig.115-b): Outer spine of 2nd segment relatively large, inner spine 88% length of 3rd segment, terminal spine of 3rd segment 49% as long as segment.

Description of male. Prosome length, 2.20-2.32 mm; body length, 3.08-3.28 mm. Prosome 71% length of body and approximately 2.4 times length of urosome. Urosome 30% length of body. Caudal rami and their setation as in female. Left, geniculate A1 similar in morphology to that of *H. spinifrons.* All other cephalosomal appendages and first 4 pairs of legs as in female.

In right P5, basal inner lobe (Figs.115-g, h) relatively narrow, arising from middle of segment, extending and tapering mediad for a distance equal to about 2/3 length of segment and then recurved with lateral margin folding only slightly; whole length of medial margin fringed with a band of setules. Second exopodal segment (Figs.115-e, f) approximately 1.5 times length of 1st, with a rounded tubercular process proximally on medial margin followed by roughly triangular medial projection, outer spine about 3/4 length of segment from proximal end. Third segment only slightly curved, with outer spine about 1/3 length of segment from proximal end, terminal spine 24% as long as segment, terminal lobe (Fig.115-i) 70% length of terminal spine.

In left P5 (Fig.115-c), basal inner lobe low but wide, armed with a band of setules. Second exopodal segment (Fig.115-d) approximately 1.5 times length of 1st, with a conical lateral process; as measured along medial margin, lateral process approximately 1/8 length of segment and about 1/3 length of outer spine. Third exopodal segment including long distal spiniform process making up 61% length of whole exopod.

Distribution. Since its original discovery from off the west coast of Norway, *H. norvegicus* has been found commonly from the Arctic Ocean (Sars 1900) to the northern Atlantic (Sars 1902). In the northeastern Atlantic, it has been recorded as far south as 28°04'N (Sars 1925) and in the northwestern Atlantic, it has been recorded as far south as the vicinity of Woods Hole (Wilson 1932). In the present study the species is represented by 13 females and 4 males found in 3 samples—2 taken in the Norwegian Sea at 63°25'N, 02°40'W and 63°24'N, 12°38'W, respectively; 1 at 38°53'N, 72°22'W off Delaware.

Remarks. Originally this species was briefly described from the west coast of Norway and subsequently fully redescribed by Sars (1900, 1902). The female can be distinguished from the other members of the species group by the characteristic genital somite, which in dorsal view is symmetrical, widest close to the anterior end and gradually narrows posteriorly and in ventral view has a small lobe posterior to the genital operculum followed by concentric ridges. The male can be identified by the 5th pair of legs, of which the right exopod has a very small terminal spine, the left basis has a low but broad inner lobe, and the outer spine of the left 2nd exopodal segment is borne on a conical process.

PHYLOGENY

As members of the superfamily Arietelloidea, the Heterorhabdidae share the general morphology of the body and appendages with the other members of the superfamily, in particular with the Lucicutiidae and Augaptilidae. The Heterorhabdidae can readily be distinguished from them and any other calanoid families by the following synapomorphies: 1) Left caudal ramus longer than right and normally fused with anal segment, 2) fourth marginal seta of left caudal ramus unarmed and greatly elongated, 3) basis of male right P5 with a large plumose inner lobe. Based on these synapomorphies, only the following 7 genera are referable to the family: *Disseta, Mesorhabdus, Heterostylites, Hemirhabdus, Neorhabdus, Paraheterorhabdus,* and *Heterorhabdus.* As discussed in detail under the heading of "List of Genera of the Family Heterorhabdidae," *Microdisseta* and *Alrhabdus* have been removed from the family because they do not possess the above character states.

The evolutionary changes among these genera involve the feeding appendages, which show a transformation from those of typical particle feeders to those of highly specialized carnivores. The most plesiomorphic conditions in such feeding appendages as the mandible, maxilla, and maxilliped are found in *Disseta* and the most specialized conditions in *Heterorhabdus.* The appendages of *Disseta* are practically the same as those of the Lucicutiidae, while some appendages in *Heterorhabdus* are so specialized as to be unique. Cladograms based on 20 pairs of character states are shown in Figure 116. The clade separated from *Disseta* is characterized by the following synapomorphies: 1) Endopod of Mx1 small, 1-segmented, 2) exopod of Mx1 elongated, extending beyond distal end of endopod, 3) setae of 5th and 6th lobes of Mx2 reduced, and 4) basis of P1 lost outer seta. *Mesorhabdus* is the second offshoot and shares a number of plesiomorphic conditions with *Disseta,* such as the group of contiguous teeth next to the basal spine of Md, 9 setae on the outer lobe of Mx1, more than 2 groups of setae on the Mxp coxa, and 8 setae on the 3rd endopodal segment of P2. However, *Mesorhabdus* has a large spiniform seta on the 1st inner lobe of Mx1, by which it can easily be distinguished not only from *Disseta* but also all other genera of the family. The stock diverged from *Mesorhabdus* can be regarded as monophyletic by the following synapomorphies: 1) Masticatory edge of Md lost the group of contiguous teeth next to basal spine, 2) coxa of Mxp with only 1 inner marginal seta in addition to a group of distal setae, 3) endopod of P2 with only 7 setae on 3rd segment, 4) outer lobe of Mx1 without 3 small proximal setae. It is therefore easy to recognize *Disseta* and *Mesorhabdus* as the first and second branches of the heterorhabdid evolution.

It is equally easy to recognize three terminal clades in the family: the *Heterostylites, Hemirhabdus-Neorhabdus,* and *Paraheterorhabdus-Heterorhabdus* clades. The *Heterostylites* clade is distinguished by a serrated lappet on the 2nd exopodal segment of P5. The *Hemirhabdus-Neorhabdus* clade is distinguished by Mx2, in which the endopod has only 2 setae and spines of the 5th and 6th lobes are conspicuously serrated with large teeth. The *Paraheterorhabdus-Heterorhabdus* clade is distinguished by a highly developed middle seta of the Mxp coxa. However, the phylogenetic relationships among these three terminal clades remain unclear. According to the morphology of the masticatory edge of Md, of which the ventralmost tooth is peculiarly ribbed and separated from the rest of teeth by a wide gap, *Hemirhabdus, Neorhabdus, Paraheterorhabdus,* and *Heterorhabdus* seem to form a separate clade, which diverged from *Heterostylites.* It seems to be also possible that *Heterostylites, Paraheterorhabdus,* and *Heterorhabdus* form a single monophyletic line separate from the *Hemirhabdus-Neorhabdus* clade. The following character states support this possibility: 1) Basal spine of right mandibular cutting edge followed by 3 dorsal teeth, 2) exopod of Mx1 modified into an elongate rectangle and attached to terminal end of appendage, and 3) first segment of Mx2 elongated. However, it is not possible to determine which of these is the more likely. The choice hinges on knowing which character state is more likely to have come from the common ancestor, and which of the two evolved independently of the other.

On the other hand, derived characters common to *Heterostylites* and the *Hemirhabdus-Neorhabdus* clade (which could be regarded as evidence for their close relationships) have not been found in the present study. Ohtsuka et al. (1997) proposed a phylogenetic relationship among the heterorhabdid genera showing a single clade comprising *Heterostylites*, *Hemirhabdus*, and *Neorhabdus*. Their arguments are based on the presence or absence of aesthetes on certain segments of A1 and the number of setae on the maxillular exopod. These character states in phylogenetic analyses seem to be poor choices. In the present study, the number of aesthetes on certain segments of A1 was closely examined for the species of *Neorhabdus* and found to be quite variable not only among species but also among individuals of the same species. The number of setae on the Mx1 exopod was also found to be quite variable among individuals of the same species.

DISTRIBUTION

All of the genera, subgenera, and species groups of the Heterorhabdidae defined in this study are circumglobal in distribution, occurring widely in all three major oceans. *Disseta* is a bathypelagic genus comprising 3 species, of which *D. palumbii* is a common circumglobal species occurring widely in the low and midlatitudes of all three major oceans. *Disseta scopularis* was found only in the low and midlatitudes of the Pacific from the west coast of the Americas to the Malay Archipelago. *Disseta magna* was found in the low and midlatitudes of both the Pacific and Indian Oceans. *Disseta palumbii* was the most common, comprising 63% of the 1036 specimens of *Disseta* found in the study. *Disseta scopularis* and *D. magna* accounted for 31% and 6%, respectively.

Mesorhabdus is a meso-bathypelagic genus occurring widely in low and midlatitudes of the world's oceans but it is relatively rare. Only 139 specimens were found in the study. Of the 5 species of the genus, *M. brevicaudatus* occurred in all three major oceans and was the most common, representing 44% of the total found for the genus. *Mesorhabdus gracilis* and *M. paragracilis* also occurred in all three major oceans but accounted for only 8% and 14% of specimens, respectively. *Mesorhabdus angustus* and *M. poriphorus* were found only in Atlantic and Pacific Ocean samples, representing 29% and 6%, respectively. *Mesorhabdus angustus*, however, has been recorded from the Indian Ocean (Sewell 1932, 1947).

Heterostylites is also a meso-bathypelagic genus occurring widely in the world's oceans. Of the 6 species found in the study, *H. major* and *H. longicornis* have the widest ranges, occurring throughout the Pacific, Atlantic, and Indian Oceans except for Arctic and Antarctic waters. *Heterostylites submajor* also occurred in all three major oceans but was confined mainly to coastal waters. *Heterostylites longioperculis* and *H. echinatus* are common Pacific species, the former occurring widely in the low and midlatitudes of the whole Pacific and the latter confined to coastal waters of Central and South America. *Heterostylites nigrotinctus* is an Antarctic species. The most common of the six species found for the genus were the two Pacific species, *H. longioperculis* and *H. echinatus*, each representing 30% of the 857 specimens found for the genus, followed by *H. longicornis* and *H. major* accounting for 20% and 16%, respectively.

Hemirhabdus is a bathypelagic genus comprising two species. *Hemirhabdus grimaldii* is a relatively common circumglobal species occurring throughout the low and midlatitudes of the three major oceans, and it accounted for 89% of the 124 specimens found for the genus. *Hemirhabdus amplus* was relatively rare and occurred exclusively in the Pacific and Indian Oceans, mainly in the low latitudes.

Neorhabdus is also a bathypelagic genus occurring mainly in the low latitudes. Of the 5 species of *Neorhabdus* found in the study, *N. latus* and *N. brevicornis* were commonly found throughout the low latitudes of the world's oceans and represented 61% and 20%, respectively, of the 133 specimens found for the genus. *Neorhabdus falciformis* and *N. capitaneus* were also found in all three major oceans but they were very rare, each representing only 8% of the total found for the genus. *Neorhabdus subcapitaneus* was represented by only 3 specimens found in the northwestern Atlantic and southeastern Pacific.

The subgenus *Paraheterorhabdus* comprising 6 species seems to be mesopelagic, occurring widely in the world's oceans. *Paraheterorhabdus robustus, P. illgi, P. medianus,* and *P. vipera* all have a similar range of distribution, occurring widely in the low and midlatitudes of all three major oceans. *Paraheterorhabdus robustus* was the most common, representing 26% of the 369 specimens found for the subgenus, followed in order by *P. illgi, P. medianus,* and *P. vipera* accounting for 14%, 10% and 8%, respectively. *Paraheterorhabdus longispinus* is endemic to the northern Pacific, where it seems to be quite common. A total of 58 specimens belonging to this species were found in 5 samples taken in the northeastern Pacific, and one of the samples yielded as many as 40 specimens. *Paraheterorhabdus farrani* is distributed exclusively in the Southern Ocean. Six samples from the Southern Ocean that were examined in the study yielded 95 specimens of this species, accounting for

26% of the total found for the subgenus, and one of the samples, taken in the Drake Passage, contained as many as 61 specimens.

The monotypic subgenus *Antirhabdus* is bathypelagic and circumglobal, occurring commonly throughout the low and midlatitudes of the world's oceans. The species was represented in the study by a total of 266 specimens.

The *spinifrons* species group of *Heterorhabdus* comprises 7 species, of which *H. caribbeanensis* and *H. insukae* are epipelagic and the rest are epi-mesopelagic. *Heterorhabdus spinifrons, H. subspinifrons, H. caribbeanensis,* and *H. insukae* are all common circumglobal species occurring widely in the low and midlatitudes of the world's oceans. The most common of these 4 circumglobal species was *H. spinifrons* followed by *H. insukae.* They accounted for 28% and 26%, respectively, of the 1217 specimens found for the species group, and *H. subspinifrons* and *H. caribbeanensis* for 11% and 6%, respectively. *Heterorhabdus ankylocolus* representing only 6% of the species group, is an Indo-Pacific equatorial species. *Heterorhabdus heterolobus,* representing 7% of the species group, was found exclusively in the equatorial waters of the whole Pacific. *Heterorhabdus quadrilobus* is endemic to the eastern Pacific, where it is distributed mainly along the coasts and extends westward for some distance in equatorial waters. A total of 200 specimens were found for the species, which account for 16% of the species group.

The *papilliger* species group, including 5 species, seems to be epi-mesopelagic. *Heterorhabdus papilliger* and *H. spinifer* are circumglobal species occurring throughout the low and midlatitudes of the world's oceans. *Heterorhabdus prolatus* is an Indo-Pacific equatorial species, and *H. guineanensis* was found exclusively in warm waters of the eastern Atlantic. *Heterorhabdus lobatus* has a rather unusual distribution, occurring along the southern coast of South Africa, around New Zealand, and along the west coast of the Americas from 40°S to 40°N. The most common species was *H. papilliger*, representing 51% of 1482 specimens found for the species group, followed by *H. lobatus* accounting for 29%.

The *fistulosus* species group comprises only two species, of which *H. egregius* is a circumglobal species occurring widely in the low and midlatitudes of the world's oceans and *H. fistulosus* seems to be endemic to the northern Pacific. *Heterorhabdus egregius* was represented in the study by 111 specimens. *Heterorhabdus fistulosus* has previously been recorded only from the northwestern Pacific, and in the present study was represented by 40 specimens in a single sample which was taken at a northeastern locality.

The *abyssalis* species group, comprising 17 species, seems to be meso-bathypelagic. Four species—*H. habrosomus, H. oikoumenikis, H. pacificus,* and *H. confusibilis*—are circumglobal species occurring throughout the low and midlatitudes of the world's oceans. Of these circumglobal species, *H. pacificus* was the most common, accounting for 14% of the 3422 specimens found for the group, followed by *H. oikoumenikis* accounting for 10%. *Heterorhabdus longisegmentus* is also circumglobal but occurred only in the low and midlatitudes of the Southern Hemisphere and it was represented in the study by 161 specimens or 5% of the group.

Only one species, *H. tuberculus,* occurred widely in the low and midlatitudes of the whole Indo-Pacific and it was represented by 370 specimens or 11% of the group. Six species—*H. cohibilis, H. clausii, H. tanneri, H. abyssalis, H. americanus* and *H. prolixus*—were found only in the Pacific. *Heterorhabdus cohibilis* and *H. clausii* are Pacific equatorial species, the former occurring in the whole equatorial waters and the latter only in the central and eastern equatorial waters. *Heterorhabdus tanneri* is endemic to the coastal waters of the northern Pacific. *Heterorhabdus abyssalis, H. americanus,* and *H. prolixus* are confined to the eastern Pacific, where they occur mainly in the coastal waters of the low latitudes. The most common of the Pacific species was *H. tanneri,* represented by 499 specimens or 15% of the group, followed in order by *H. abyssalis* and *H. prolixus,* which were represented by 301 and 265 specimens, respectively.

Heterorhabdus norvegicus is known to be distributed in the northern Atlantic and Arctic Ocean (Sars 1902). In the present study the species was found in the Norwegian Sea and northwestern Atlantic down to 38°N.

Heterorhabdus austrinus, H. pustulifer, H. spinosus, and *H. paraspinosus* are confined to the Southern Ocean. *H. austrinus* and *H. pustulifer* are distributed mainly in the Antarctic waters, while *H. spinosus* and *H. paraspinosus* seem to be subantarctic species, which occurred in the study around South Africa, the southern end of South America, and New Zealand..

Only two species of the family—*Heterorhabdus quineanensis* and *H. norvegicus*—are endemic to the Atlantic. The following 13 species were found exclusively in the Pacific: *Disseta scopularis, Heterostylites longioperculis, H. echinatus, Paraheterorhabdus longispinus, Heterorhabdus heterolobus, H. quadrilobus, H. fistulosus, H. prolixus, H. clausii, H. americanus, H. cohibilis, H. abyssalis,* and *H. tanneri.* Of these 13 Pacific species, *Paraheterorhabdus longispinus, Heterorhabdus fistulosus,* and *H. tanneri* are endemic to the northern Pacific and the following five species are endemic to the eastern Pacific: *Heterostylites echinatus, Heterorhabdus quadrilobus, H. prolixus, H. americanus,* and *H. abyssalis.* The following six species were found exclusively in the Southern Ocean: *Heterostylites nigrotinctus, Paraheterorhabdus farrani, Heterorhabdus pustulifer, H. spinosus, H. paraspinosus,* and *H. austrinus.* No endemic heterorhabdid species were found in the Indian Ocean. In short, of the four oceans including the Southern Ocean, the Pacific has the highest number of endemic species (13), followed by the Southern Ocean (6). In the Pacific, the eastern Pacific has the highest number of endemic species (5), followed by the northern Pacific (3).

The following five species are Indo-Pacific in distribution: *Disseta magna, Hemirhabdus amplus, Heterorhabdus ankylocolus, H. prolatus,* and *H. tuberculus* and the following 31 species are circumglobal in distribution: *Disseta palumbii,* 4 of the 5 species of *Mesorhabdus* (except *M. poriphorus*), *Heterostylites major, H. submajor, H. longicornis, Hemirhabdus grimaldii,* 4 of the 5 species of *Neorhabdus* (except *N. subcapitaneus*), *Paraheterorhabdus robustus, P. illgi, P. medianus, P. vipera, P. compactus, Heterorhabdus spinifrons, H. subspinifrons, H. caribbeanensis, H. insukae, H. lobatus, H. papilliger, H. spinifer, H. egregius, H. habrosomus, H. oikoumenikis, H. longisegmentus, H. pacificus,* and *H. confusibilis.* It is interesting that 53% of the 59 species found in the study have a circumglobal distribution, while only 5 (29%) of the 17 species of the *H. abyssalis* species group are circumglobal in distribution. In *Mesorhabdus* and *Neorhabdus,* 80% of the species found were circumglobal.

The distribution pattern of the Heterorhabdidae described above is very similar to that of the bathypelagic genus *Paraeuchaeta* and thus supports the hypothesis proposed by Park (1994), that endemic species with a limited geographic range are found exclusively in highly productive waters, where they are usually abundant. They are therefore believed to be eutrophic. Species with a wide range of distribution are generally rare and are believed to be oligotrophic, although the total number of specimens found in the study for some of these rare species was actually higher than the eutrophic species, because they occurred in more samples. The typical endemic species in each eutrophic area are as follows: *Heterorhabdus norvegicus* in the northern Atlantic; *Heterorhabdus guineanensis* in the eastern tropical Atlantic; *Paraheterorhabdus longispinus, Heterorhabdus fistulosus,* and *H. tanneri* in the northern Pacific; *Heterostylites echinatus, Heterorhabdus quadrilobus, H. prolixus, H. americanus,* and *H. abyssalis* in coastal waters of the eastern Pacific; *Heterorhabdus farrani, H. pustulifer, H. spinosus, H. paraspinosus,* and *H. austrinus* in the Southern Ocean.

LITERATURE CITED

Boeck, Axel. 1872. Nye Slaegter og Arter af Saltvands-Copepoder. Christiania Videnskabers Selskab Forhandlinger, 1872, Christiania, Norway pp. 35-60.

Bradford, Janet M. 1971. New and little-known species of Heterorhabdidae (Copepoda: Calanoida) from the southwest Pacific. New Zealand Journal of Marine and Freshwater Research 5(1):120-140, figs. 1-11, table 1.

Brady, George Stewardson. 1883. Report on the Copepoda collected by H.M.S. *Challenger* during the years 1873-76. In: Report on the Scientific Results of the Voyage of H.M.S. *Challenger* during the Years 1873-76, edited by C. W. Thomson and J. Murray. Zoology, 8(23):1-142, pls. 1-55.

Brady, George Stewardson. 1918. Copepoda. Australasian Antarctic Expedition 1911-14 under the leadership of Sir Douglas Mawson, Scientific Reports. Series C—Zoology and Botany, 5(3):48p., pls. 1-15.

Brodsky, Konstantin A. 1950. Calanoida of the far eastern seas and polar basin of the USSR (in Russian). Opredeliteli Po Faune SSSR no.35:1-442, figs. 1-306.

Claus, Carl. 1863. Die frei lebenden Copepoden mit besonderer Berücksichtigung der Fauna Deutschlands, der Nordsee, und des Mittelmeeres. Verlag von Wilhelm Engelmann, Leipzig. 230p., pls. 1-37.

Dahl, F. 1894. Über die horizontale und verticale Verbreitung der Copepoden im Ocean. Verhandlungen der Deutschen Zoologischen Gesellschaft 4:61-80, figs. 1-4.

Davis, Charles C. 1949. The pelagic Copepoda of the northeastern Pacific Ocean. University of Washington Publications in Biology 14:1-117, pls. 1-15.

Esterly, Calvin Olin. 1905. The pelagic Copepoda of the San Diego region. University of California Publications in Zoology 2(4):113-233, figs. 1-62.

Esterly, Calvin Olin. 1906. Additions to the copepod fauna of the San Diego region. University of California Publications in Zoology 3(5):53-92, pls. 9-14.

Esterly, Calvin Olin. 1911. Third report on the Copepoda of the San Diego region. University of California Publications in Zoology 6(14):313-352, pls. 26-32.

Farran, G. P. 1908. Second report on the Copepoda of the Irish Atlantic slope. Department of Agriculture and Technical Instruction for Ireland. Fisheries Branch Scientific Investigations, 1906 no.2:1-104, pls. 1-11.

Farran, G. P. 1926. Biscayan plankton collected during a cruise of H.M.S. *Research*, 1900. Part 14. The Copepoda. The Journal of the Linnean Society of London. Zoology 36:219-310, text figs. 1, 2, pls. 5-10.

Farran, G. P. 1929. Crustacea. 10. Copepoda. Natural History Reports, British Antarctica "Terra Nova" Expedition, 1910. Zoology 8(3):203-306, figs. 1-37, maps 1-4.

Farran, G. P. 1936. Copepoda. Great Barrier Reef Expedition Scientific Reports, 1928-1929 5(3):73-142, text figs. 1-30.

Ferrari, Frank D. and Deborah K. Steinberg. 1993. *Scopalatum vorax* (Esterly 1911) and *Scolecithricella lobophora* Park 1970, calanoid copepods (Scolecithrichidae) associated with a pelagic tunicate in Monterey Bay. Proceedings of the Biological Society of Washington 106(3):467-489, figs. 1-12.

Ferrari, Frank D. and Jennifer Saltzman. 1998. *Pleuromamma johnsoni*, a new looking-glass copepod from the Pacific Ocean with redescriptions of *P. robusta* (Dahl 1893), *P. antarctica* Steuer, 1931 new rank, and *P. scutullata* Brodsky, 1950 (Crustacea: Calanoida: Metridinidae). Plankton Biology and Ecology 45(2):203-223, figs. 1-9.

Giesbrecht, Wilhelm. 1889. Zoologia - Elenco dei Copepodi pelagici raccolti dal tenente di vascello Gaetano Chierchia durante il viaggio della R. Corvetta "Vettor Pisani" negli anni 1882-

1885, e dal tenente di vascello Francesco Orsini nel Mar Rosso, nel 1884. Rendiconti Accademia dei Lincei, Roma, ser. 4, 5(1):811-815.

Giesbrecht, Wilhelm.1892. Systematik und Faunistik der pelagischen Copepoden des Golfes von Neapel und der angrenzenden Meeres-abschnitte. Neapel Zoologischen Station, Fauna and Flora no.19:1-831, pls. 1-54.

Giesbrecht, Wilhelm. 1895. Reports on the dredging operations off the west coast of Central America to the Galapagos, to the west coast of Mexico, and in the Gulf of California, in charge of ALEXANDER AGASSIZ, carried on by the U.S. Fish Commission steamer *Albatross*, during 1891, Lieut.-Commander Z. L. Tanner, U.S.N., commanding. 16. Die pelagischen Copepoden. Bulletin of the Museum of Comparative Zoölogy at Harvard College 25(12):243-263, pls. 1-4.

Giesbrecht, Wilhelm. 1902. Copepoden. In Resultats du voyage du S.Y. *Belgica* en 1897-1898-1899. Rapports Scientifiques, Expédition Antarctique Belge, Zoologie, 49p., pls. 1-13.

Giesbrecht, Wilhelm and O. Schmeil. 1898. Copepoda: I. Gymnoplea. Das Tierreich, Lieferung 6, Crustacea. 169p., figs. 1-31.

Grice, George D. 1962. Calanoid copepods from equatorial waters of the Pacific Ocean. Fishery Bulletin of the Fish and Wildlife Service, U.S. 61(186):171-246, fig. 1, pls. 1-34, tables 1, 2.

Grice, George D. 1973. *Alrhabdus johrdeae*, a new genus and species of benthic calanoid copepods from the Bahamas. Bulletin of Marine Science 23(4):942-947, figs. 1-17.

Grice, George D. and Kuni Hulsemann. 1965. Abundance, vertical distribution and taxonomy of calanoid copepods at selected stations in the northeast Atlantic. Journal of Zoology 146:213-262, figs. 1-26.

Heptner, M. V. 1971. On the copepod fauna of the Kurile-Kamchatka Trench. The families Euchaetidae, Lucicutiidae, Heterorhabdidae (in Russian). Trudy Institua Okeanologii im P. P. Shirshova 92:73-161, figs. 1-39.

Heptner, M. V. 1972a. New species of deep-water genera *Disseta* and *Heterorhabdus* (Copepoda, Calanoida) from the Pacific. Zoologicheskii Zhurnal 51(2):1645-1650, figs. 1, 2.

Heptner, M. V. 1972b. A review of the generic structure of the family Heterorhabdidae (Copepoda, Calanoida). Biulleten Moskovskoe Obshchestvo Ispytatelei Prirody. Otdel Biologicheskii 77(6):54-64, figs. 1, 2.

Jespersen, P. 1940. Non-parasitic Copepoda. In: The Zoology of Iceland, edited by A. Frioriksson and S. L. Tuxen. Ejnar Munksgaard, Copenhagen and Reykjavík. Volume 3, part 33, 116p., 8 figs., 5 tables.

Mauchline, John. 1988. Taxonomic value of pore pattern in the integument of calanoid copepods (Crustacea). Journal of Zoology, London 214:697-749, figs. 1-27.

Mori, Takamochi. 1937. The pelagic Copepoda from the Neighbouring Waters of Japan. Yokendo, Tokyo. 150p., 80 pls.

Ohtsuka, Susumu, Ho Young Soh, and Shuhei Nishida. 1997. Evolutionary switching from suspension feeding to carnivory in the calanoid family Heterorhabdidae (Copepoda). Journal of Crustacean Biology 17(4):577-595, figs.1-9, tables 1-4.

Park, Taisoo. 1966. The biology of a calanoid copepod, *Epilabidocera amphitrites* McMurrich. La Cellule 66(2):129-251, text figs. 1-18, pls. 1-10.

Park, Taisoo. 1968. Calanoid copepods from the central North Pacific Ocean. Fishery Bulletin of the Fish and Wildlife Service, U.S. 66(3):527-572, pls. 1-13.

Park, Tai Soo [sic]. 1970. Calanoid copepods from the Caribbean Sea and Gulf of Mexico. 2. New species and new records from plankton samples. Bulletin of Marine Science 20(2):472-546, figs. 1-402, table 1.

Park, Taisoo. 1986. Phylogeny of calanoid copepods. Syllogeus, National Museums of Canada, no.58:191-196, figure 1.

Park, Taisoo. 1994. Geographic distribution of the bathypelagic genus *Paraeuchaeta* (Copepoda, Calanoida). In: Ecology and Morphology of Copepods, Proceedings of the 5th International

Conference on Copepoda, Baltimore, USA, June 6-13, 1993, edited by Frank D. Ferrari and Brian P. Bradley. Hydrobiologia 292/293:317-332, figs. 1-4, tables 1-6.

Richard, Jules. 1893. *Heterochaeta grimaldii*, n. sp., Calanide nouveau provenant de la troisième campagne scientifique du yacht *L'Hirondelle*. Bulletin de la Société Zoologique de France 18:151-152.

Rose, Maurice. 1929. Copépodes pélagiques particulièrement de surface provenant des campagnes scientifiques du Prince Albert Ier de Monaco. Résultats des Campagnes Scientifiques du Prince de Monaco 78:1-126, pls. 1-6.

Rose, Maurice. 1933. Copépodes pélagiques. Faune de France 26:1-374, figs. 1-456, pls. 1-19.

Sars, Georg Ossian. 1900. Crustacea. The Norwegian North Polar Expedition 1893-1896. Scientific Results edited by Fridtjof Nansen 1(5):1-141, pls. 1-36.

Sars, Georg Ossian. 1902. Copepoda Calanoida. An Account of the Crustacea of Norway with Short Descriptions and Figures of all the Species. Bergen Museum, Bergen, Norway 4(9,10):97-120, pls. 65-80.

Sars, Georg Ossian. 1905. Liste préliminaire des *Calanoïdés* recueillis pendant les campagnes de S. A. S. le Prince Albert de Monaco, avec diagnoses des genres et espèces nouvelles (2e partie) (I). Bulletin du Musée Océanographique de Monaco no.40:1-24.

Sars, Georg Ossian.1907. Notes supplémentaires sur les *Calanoïdés* de la *Princesse-Alice*. Bulletin du Musée Océanographique de Monaco no.101:1-27.

Sars, Georg Ossian.1920. Calanoidés recueillis pendant les campagnes de S. A. S. le Prince Albert de Monaco. (Nouveau Supplément) Bulletin du Musée Océanographique de Monaco no.377:1-20.

Sars, Georg Ossian. 1925. Copépodes particulièrement bathypélagiques provenant des campagnes scientifiques du Prince Albert Ier de Monaco. Resultats des Campagnes Scientifiques Prince Albert I 69:1-408, pls. 1-127.

Scott, Andrew. 1909. The Copepoda of the Siboga Expedition. 1. Free-swimming, littoral and semi-parasitic Copepoda. Siboga-Expeditie, no.29a:1-323, pls. 1-69.

Sewell, R. B. Seymour. 1932. The Copepoda of Indian seas. Calanoida. Memoirs of the Zoological Survey of India 10:223-407, text figs. 82-131, pls. 1-6.

Sewell, R. B. Seymour. 1947. The free-swimming planktonic Copepoda. Systematic account. John Murray Expedition (1933-34). Scientific Reports 8(1):1-303, text figs. 1-71.

Tanaka, Otohiko. 1964. The pelagic copepods of the Izu region, middle Japan. Systematic account 10. Family Heterorhabdidae. Publications of the Seto Marine Biological Laboratory 12(1):1-37, figs. 175-191.

Vervoort, W. 1951. Plankton copepods from the Atlantic sector of the Antarctic. Verhandelingen der Koninklijke Nederlandse Akademie van Wetenschappen, Afd. Natuurkunde, sect. 2, 47(4):1-156, figs. 1-82.

Vervoort, W. 1957. Copepods from Antarctic and sub-antarctic plankton samples. In: British Australian and New Zealand Antarctic Expedition Research (1929-1931). Reports. Series B. (Zoology and Botany) edited by T. Harvey Johnston. 3:1-160, figs. 1-138.

Vervoort, W. 1965. Pelagic Copepoda. Part II. Copepoda Calanoida of the families Phaennidae up to and including Acartiidae, containing the description of a new species of Aetideidae. Atlantide - Report no.8:9-216, figs. 1-41.

Von Vaupel-Klein, J. C. 1972. A new character with systematic value in *Euchirella* (Copepoda, Calanoida). Zoologische Mededelingen 47:497-512, figs. 1-5, pls. 1-6.

Wilson, Charles Branch. 1932. The copepods of the Woods Hole region, Massachusetts. Bulletin U.S. National Museum no.158:1-635, figs. 1-316, pls. 1-41.

Wilson, Charles Branch. 1950. Copepods gathered by the United States Fisheries Steamer "Albatross" from 1887 to 1909, chiefly in the Pacific Ocean. U.S. National Museum Bulletin 100, 14(4):141-441, pls. 2-36.

Wolfenden, R. Norris. 1902. The plankton of the Faröe Channel and Shetlands. Preliminary notes on some Radiolaria and Copepoda. Journal of the Marine Biological Association of the United Kingdom 6(3):344-372, pls. 1-4.

Wolfenden, R. Norris. 1904. Notes on the Copepoda of the North Atlantic Sea and the Faröe Channel. Journal of the Marine Biological Association of the United Kingdom 7:110-146, text fig. 1, pl. 1.

Wolfenden, R. Norris. 1905. Plankton studies: Preliminary notes upon new or interesting species. Part 1: Copepoda, pp. 1-24, pls. 1-7. Rebman Limited, London.

Wolfenden, R. Norris. 1911. Die marinen Copepoden der deutschen Südpolar-Expedition 1901-1903. 2. Die pelagischen Copepoden der Westwinddrift und des südlichen Eismeers. Deutsche Südpolar-Expedition, 1901-1903 12(2):181-380, text figs.

FIGURES

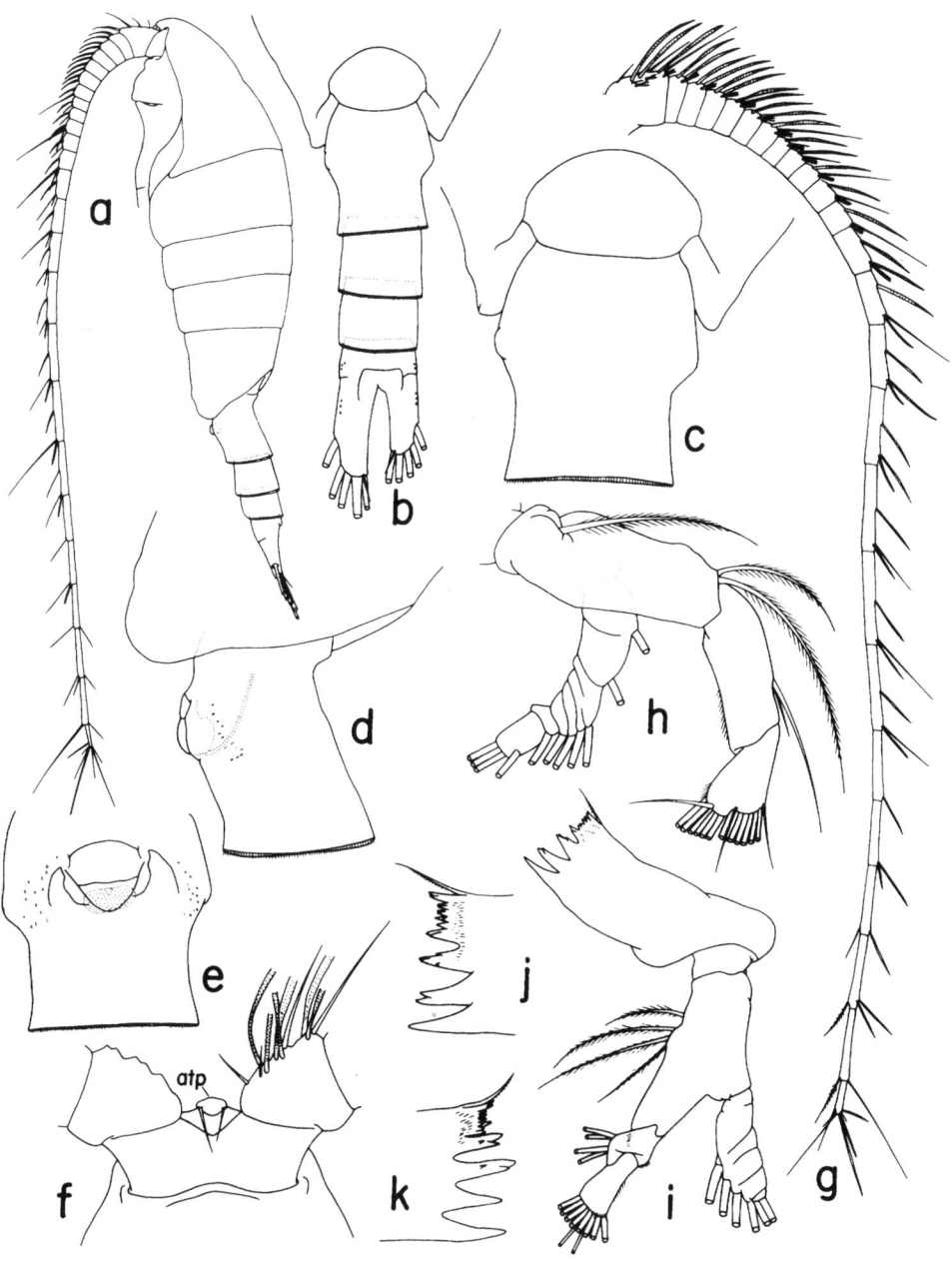

Figure 1. *Disseta palumbii* female: **a**, habitus, left; **b**, urosome, dorsal; **c**, genital somite, dorsal; **d**, do, left; **e**, do, ventral; **f**, forehead, ventral; **g**, left antennule, ventral; **h**, left antenna, posterior; **i**, left mandible, posterior; **j**, masticatory edge of left mandible; **k**, masticatory edge of right mandible. atp = midanterior tubercular process.

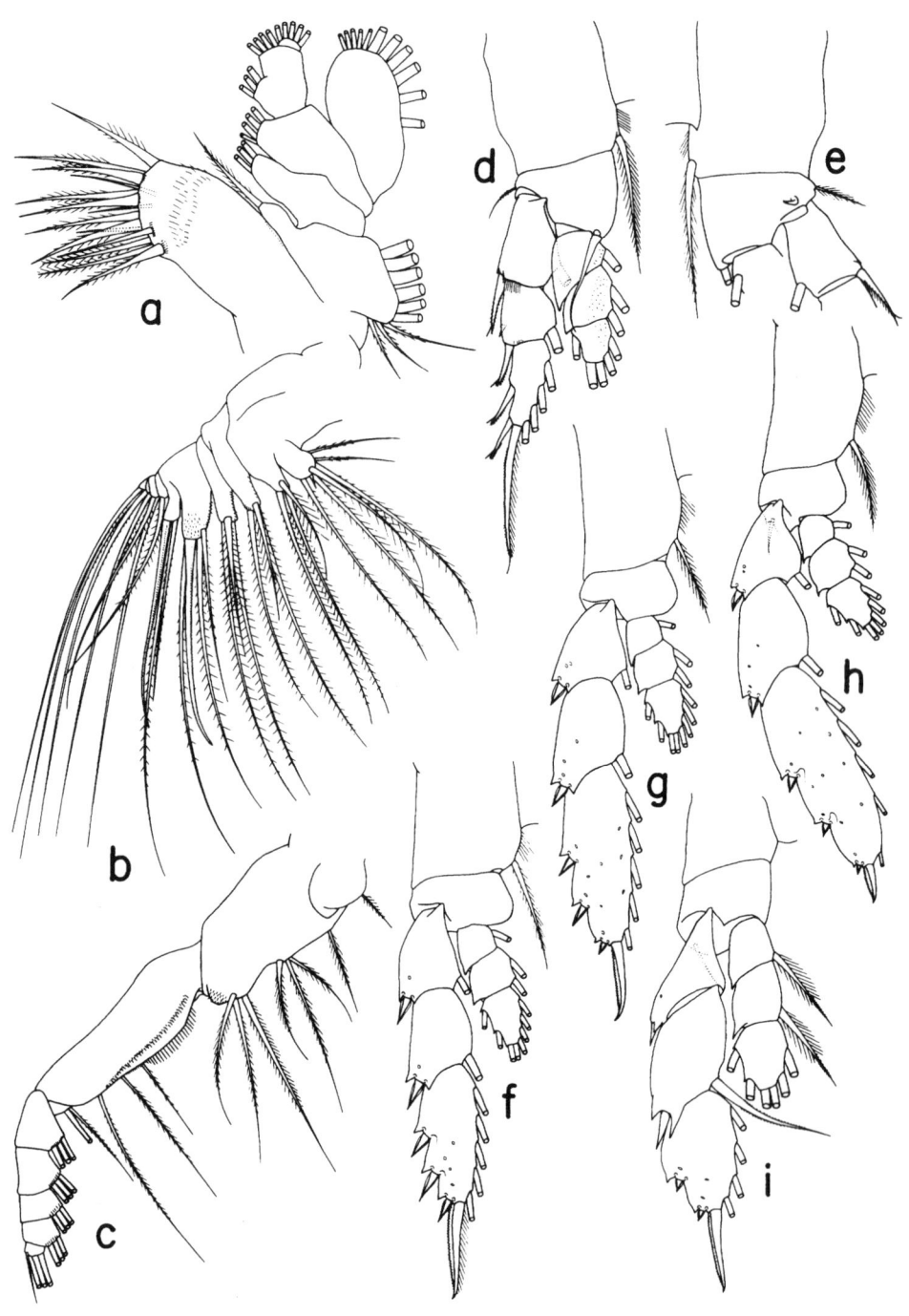

Figure 2. *Disseta palumbii* female: **a**, left maxillule, posterior; **b**, left maxilla, posterior; **c**, right maxilliped, anterior; **d**, first leg, anterior; **e**, first leg with distal part omitted, posterior; **f**, second leg, anterior; **g**, third leg, anterior; **h**, fourth leg, anterior; **i**, fifth leg, anterior.

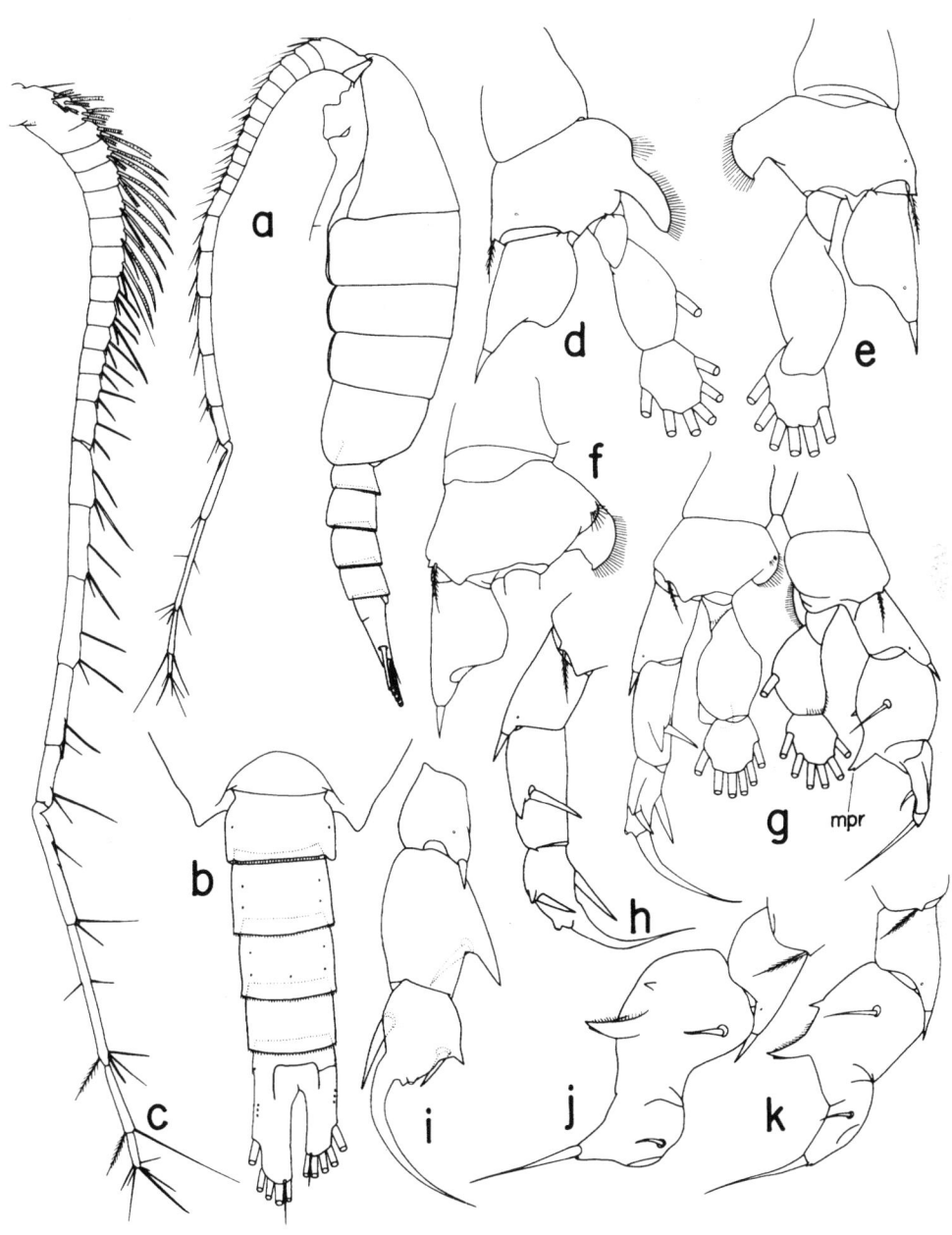

Figure 3. *Disseta palumbii* male: **a**, habitus, left; **b**, urosome, dorsal; **c**, left antennule, ventral; **d**, right 5th leg with 2 distal exopodal segments omitted, anterior; **e**, left 5th leg with 2 distal exopodal segments omitted, anterior; **f**, basipod and 1st exopodal segment of left 5th leg, posterior; **g**, 5th pair of legs, posterior; **h**, exopod of left 5th leg, posterior; **i**, do, anterior; **j**, exopod of right 5th leg, posterior, tilted clockwise; **k**, do, posterior. mpr = medial projection.

155

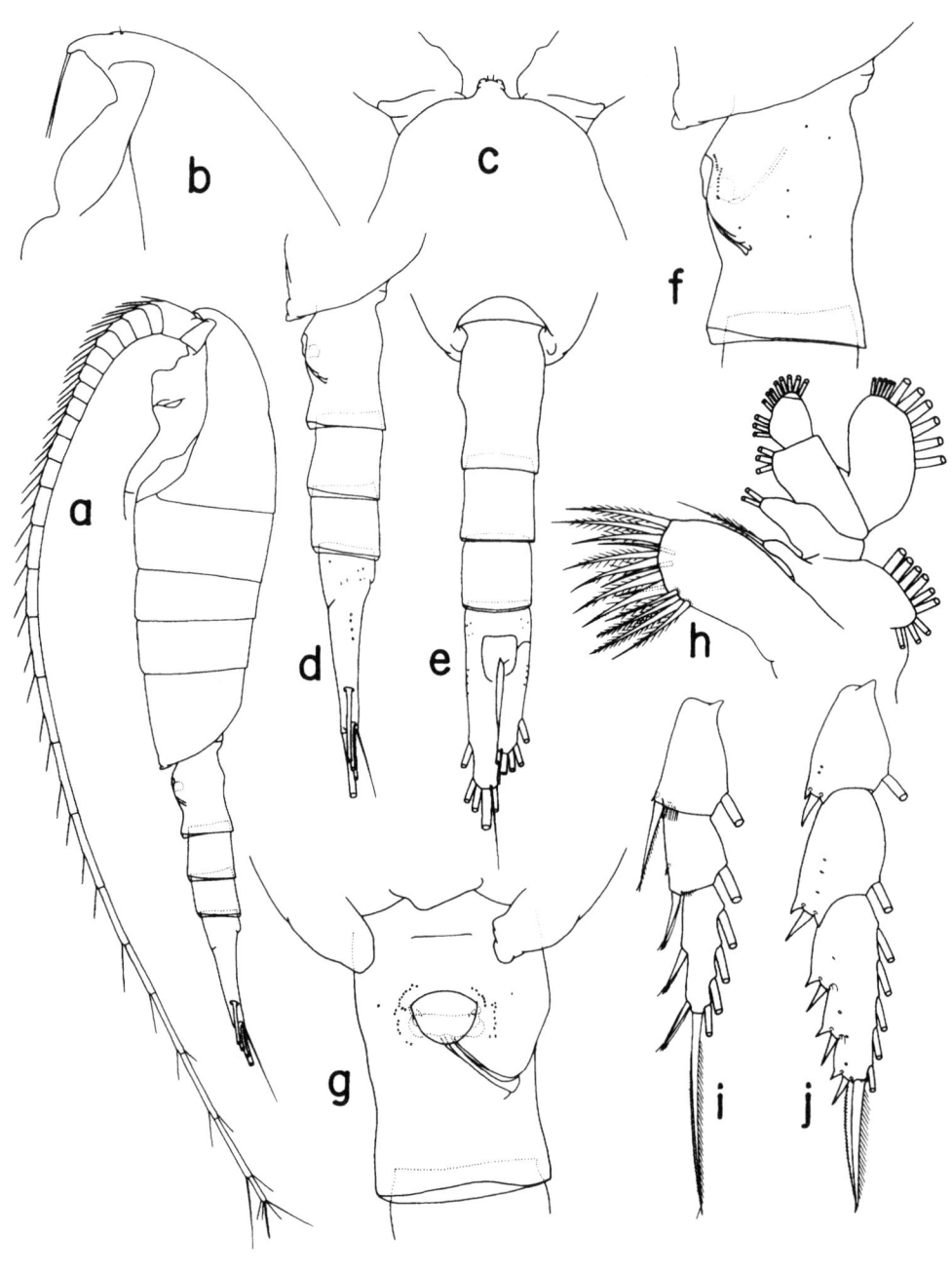

Figure 4. *Disseta scopularis* female: **a**, habitus, left; **b**, forehead, left; **c**, do, dorsal; **d**, urosome, left; **e**, do, dorsal; **f**, genital somite, left; **g**, do, ventral; **h**, left maxillule, posterior; **i**, exopod of 1st leg, anterior; **j**, exopod of 2nd leg, anterior.

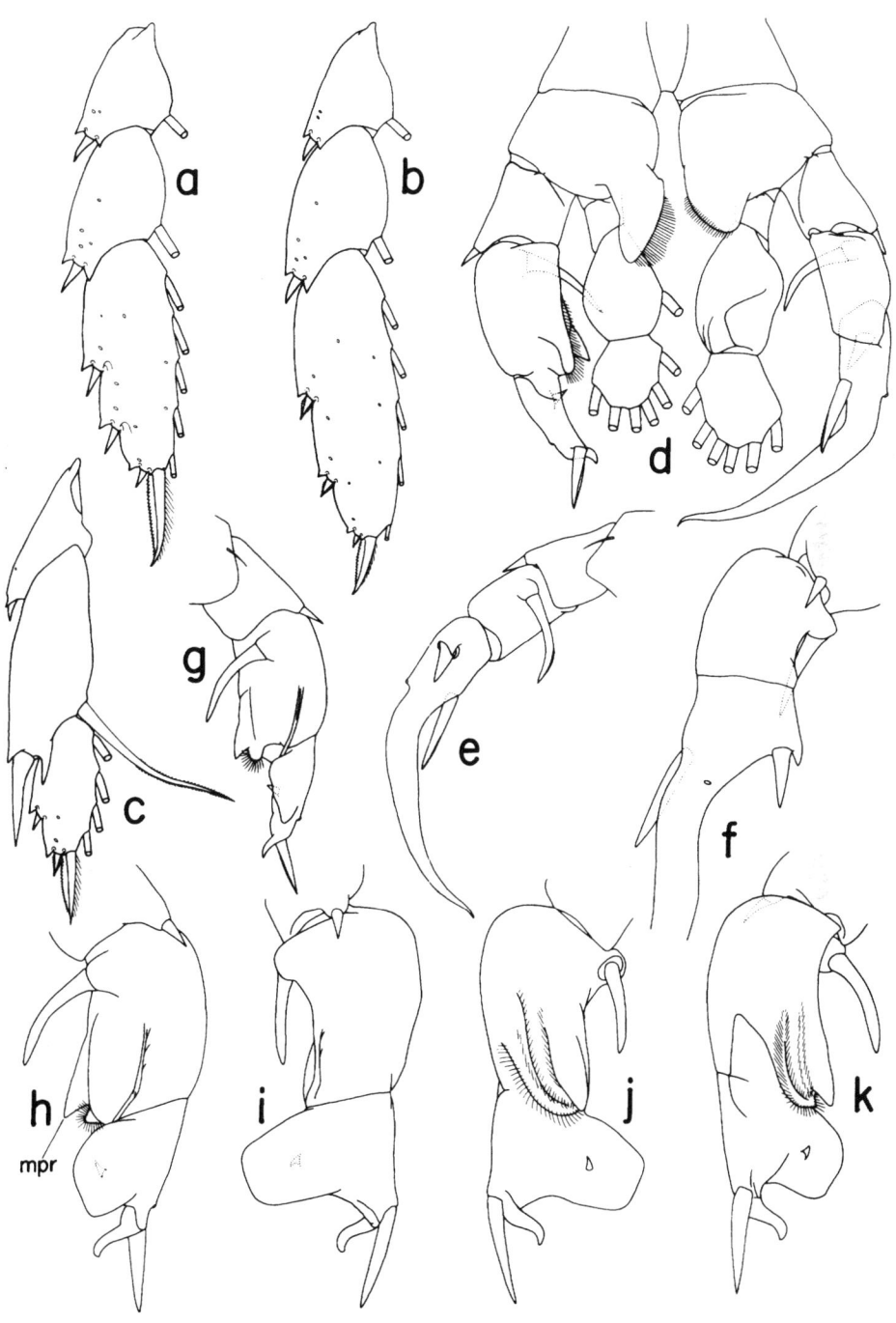

Figure 5. *Disseta scopularis* female: **a**, exopod of 3rd leg, anterior; **b**, exopod of 4th leg, anterior; **c**, exopod of 5th leg, anterior. Male: **d**, fifth pair of legs, anterior; **e**, exopod of left 5th leg, posterior; **f**, do, with proximal and distal parts omitted, anterior; **g**, exopod of right 5th leg, posterior; **h**, do, anterior; **i**, do, anterior, tilted counterclockwise; **j**, do, posterior; **k**, do, posterior, tilted clockwise. mpr = medial projection.

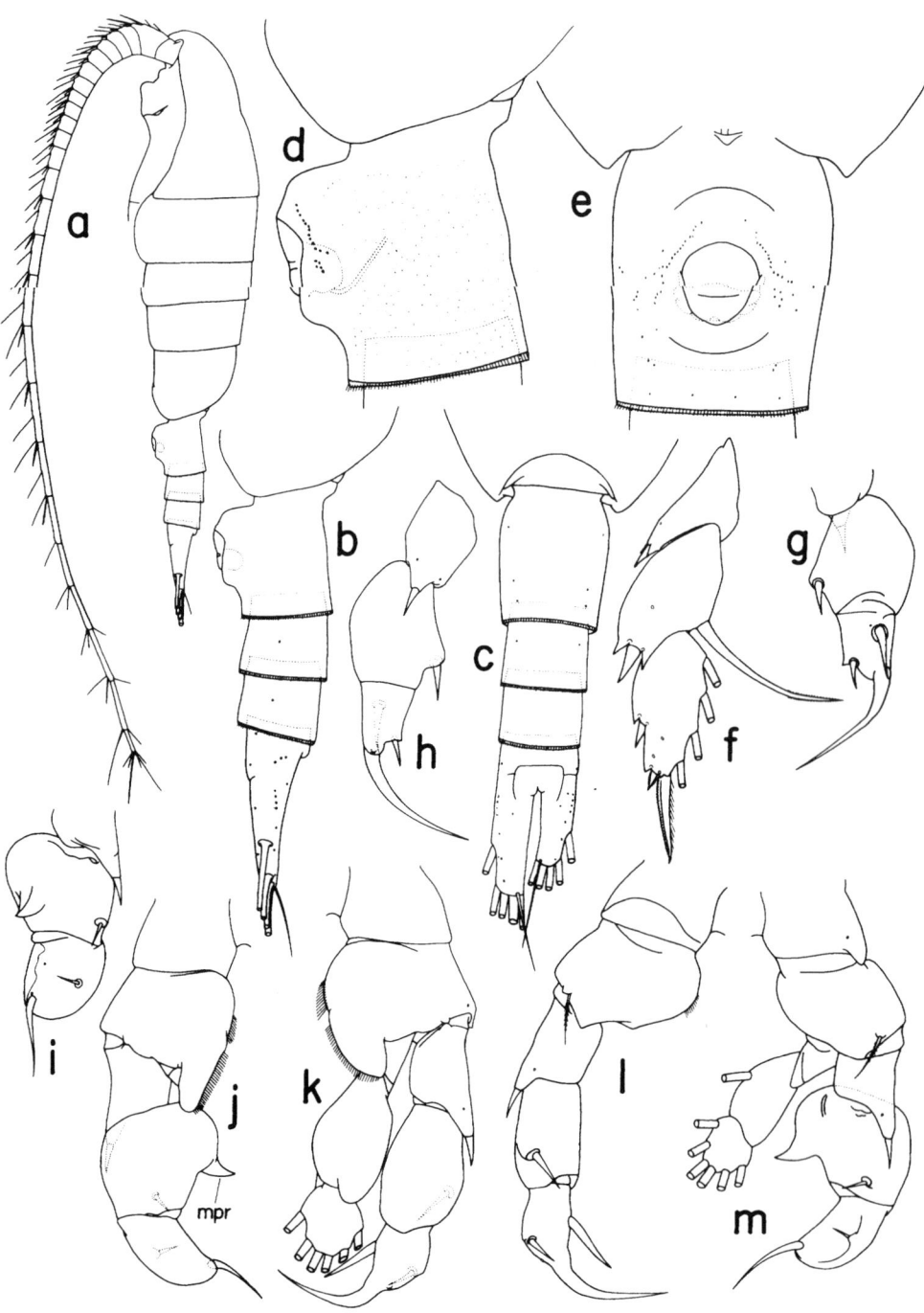

Figure 6. *Disseta magna* female: **a**, habitus, left; **b**, urosome, left; **c**, do, dorsal; **d**, genital somite, left; **e**, do, ventral; **f**, exopod of 5th leg, anterior. Male: **g**, exopod of left 5th leg, posterior; **h**, do, anterior; **i**, exopod of right 5th leg, posterior; **j**, right 5th leg with endopod omitted, anterior; **k**, left 5th leg, anterior; **l**, left 5th leg with endopod omitted, posterior; **m**, right 5th leg, posterior. mpr = medial projection.

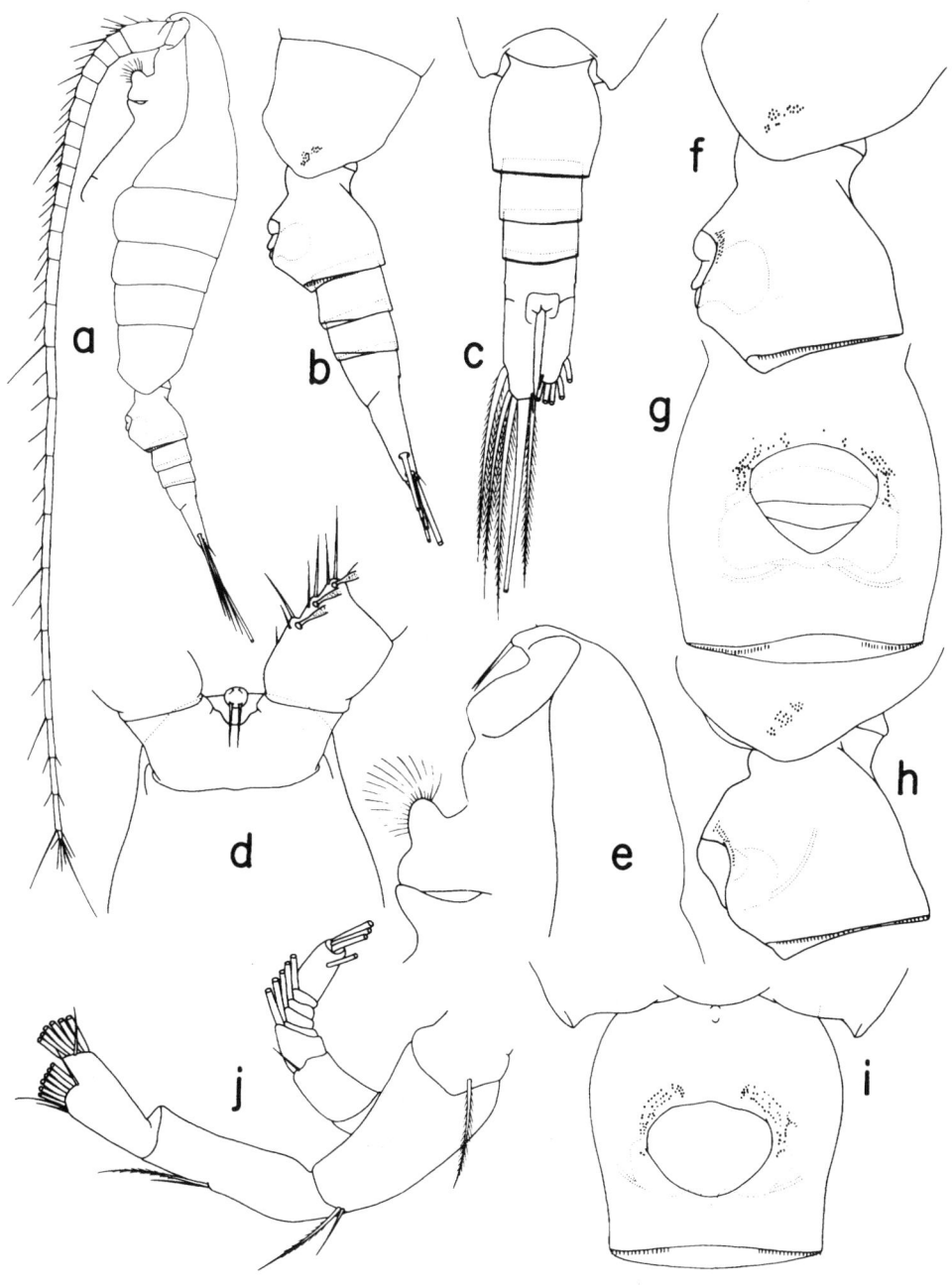

Figure 7. *Mesorhabdus angustus* female: **a**, habitus, left; **b**, urosome, left; **c**, do, dorsal; **d**, forehead, ventral; **e**, do, left; **f**, genital somite with operculum open, left; **g**, do, ventral; **h**, genital somite with operculum closed, left; **i**, do, ventral; **j**, left antenna, posterior.

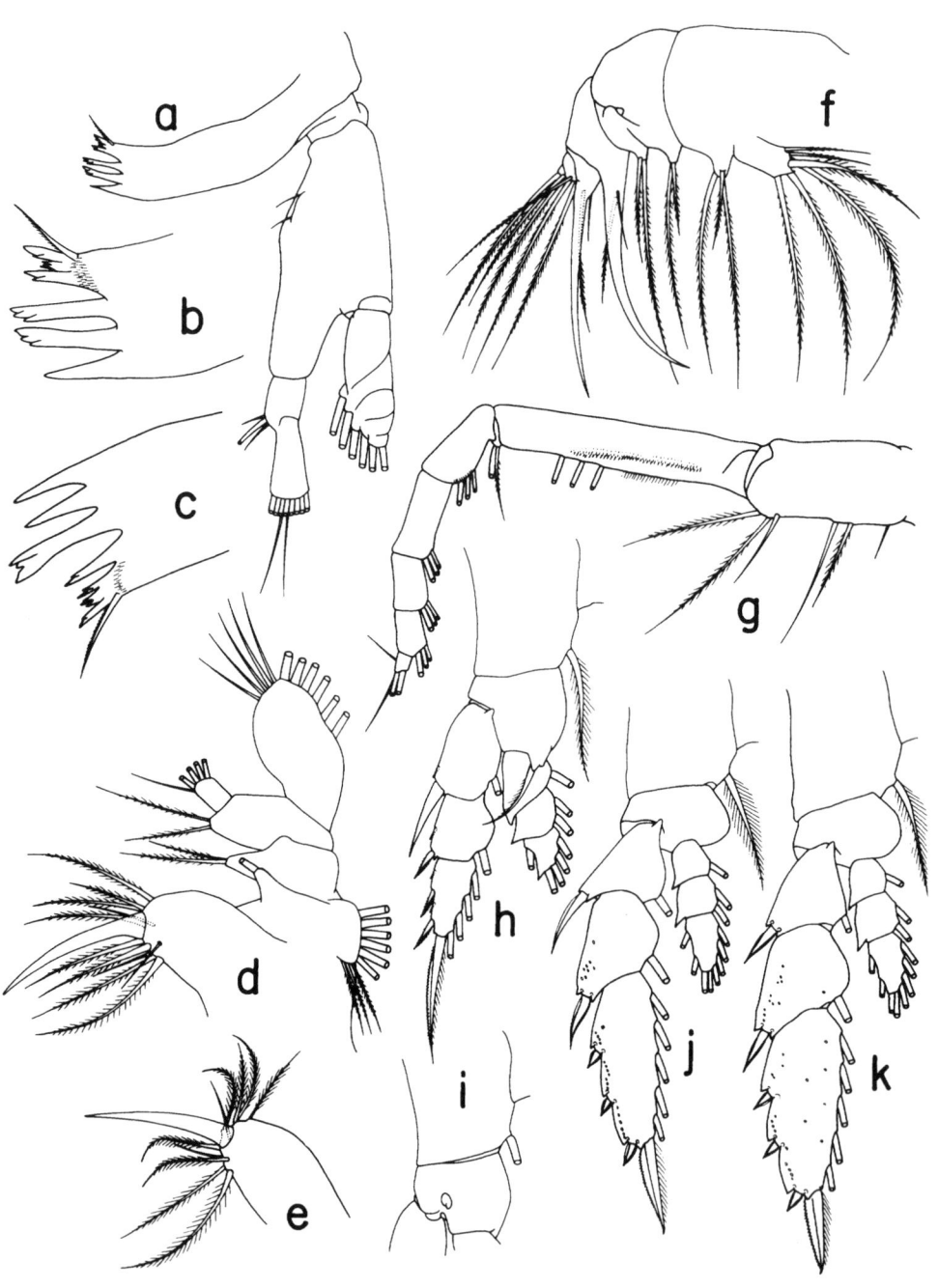

Figure 8. *Mesorhabdus angustus* female: **a**, left mandible, posterior; **b**, masticatory edge of left mandible, posterior; **c**, masticatory edge of right mandible, posterior; **d**, left maxillule, posterior; **e**, first inner lobe of left maxillule of a specimen from the central North Pacific, posterior; **f**, left maxilla, posterior; **g**, right maxilliped, anterior; **h**, first leg, anterior; **i**, basipod of 1st leg, posterior; **j**, second leg, anterior; **k**, third leg, anterior.

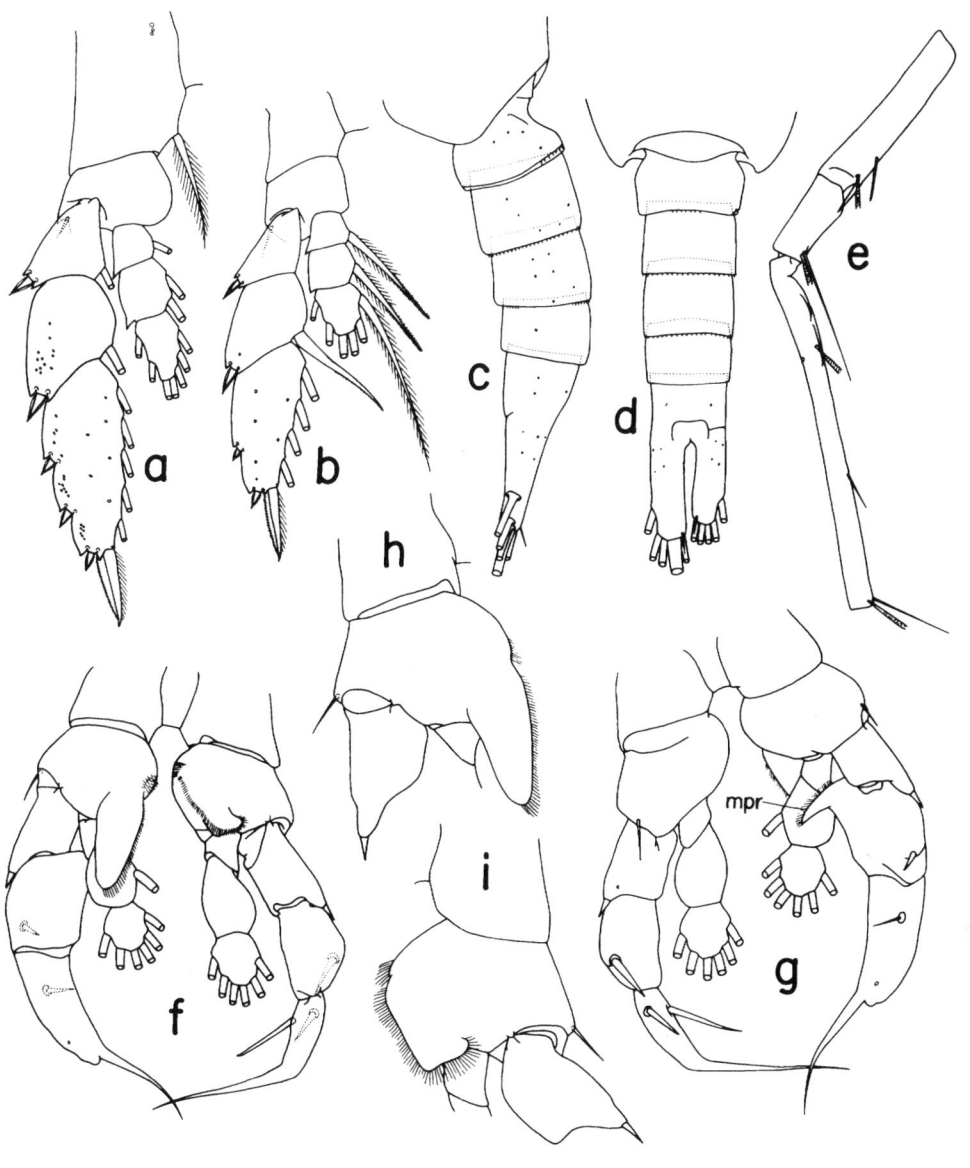

Figure 9. *Mesorhabdus angustus* female: **a**, fourth leg, anterior; **b**, fifth leg, anterior. Male: **c**, urosome, left; **d**, do, dorsal; **e**, segments 17-21 of left antennule, ventral; **f**, fifth pair of legs, anterior; **g**, do, posterior; **h**, basipod of right 5th leg, anterior; **i**, basipod of left 5th leg, anterior. mpr = medial projection.

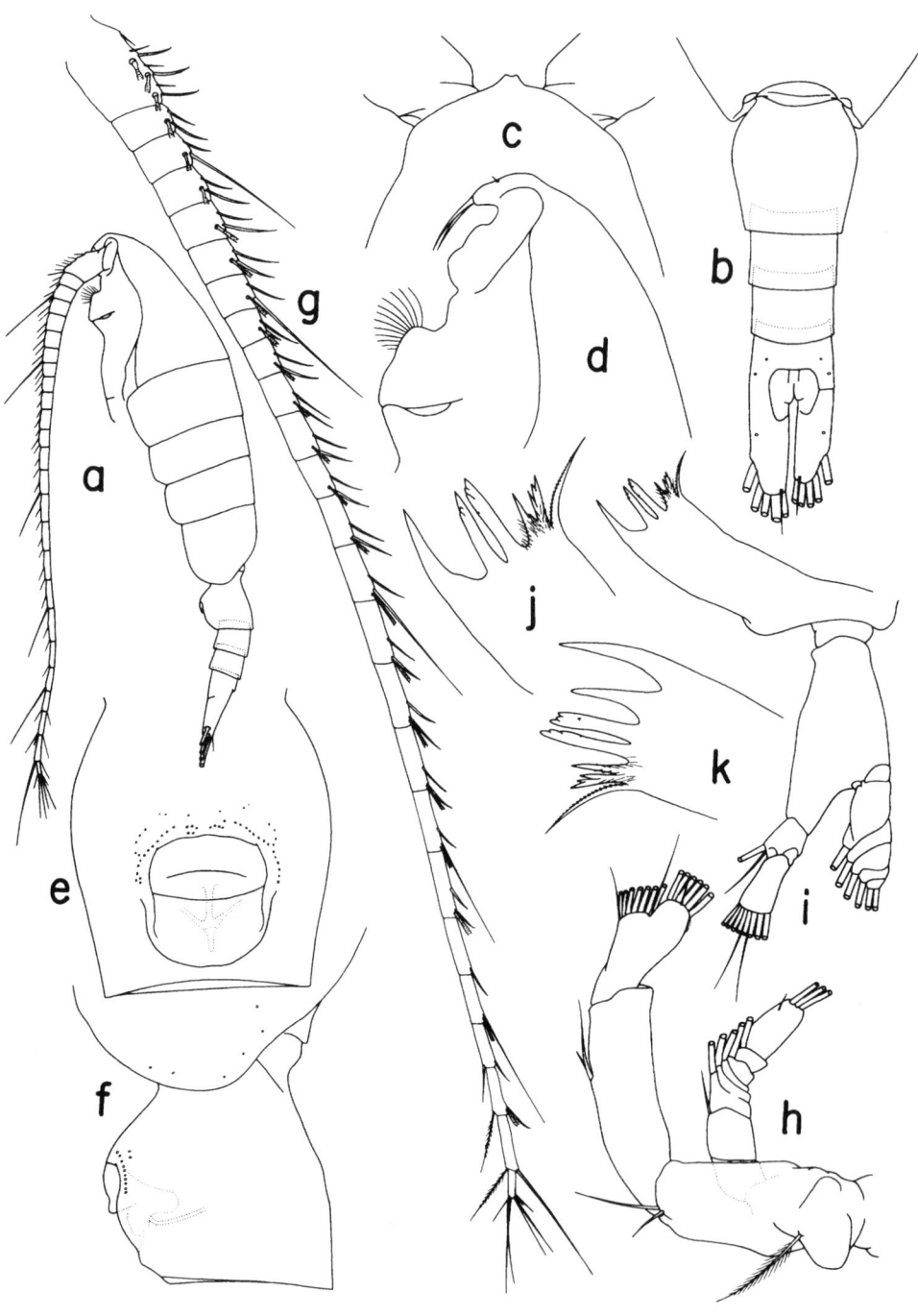

Figure 10. *Mesorhabdus brevicaudatus* female: **a**, habitus, left; **b**, urosome, dorsal; **c**, forehead, dorsal; **d**, do, left; **e**, genital somite, ventral; **f**, do, left; **g**, left antennule, ventral; **h**, left antenna, posterior; **i**, left mandible, posterior; **j**, masticatory edge of left mandible, posterior; **k**, masticatory edge of right mandible, posterior.

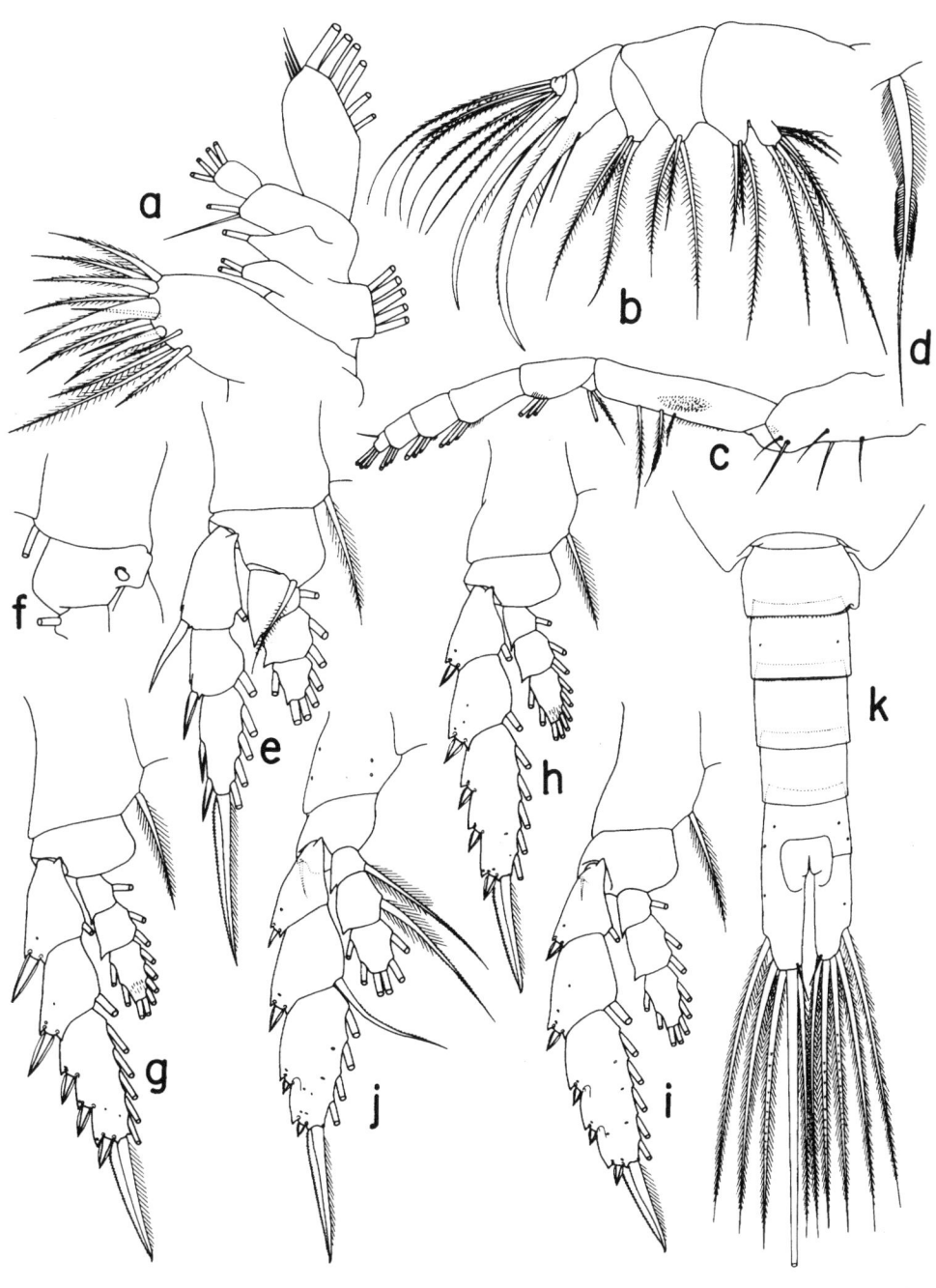

Figure 11. *Mesorhabdus brevicaudatus* female: **a**, left maxillule, posterior; **b**, left maxilla, posterior; **c**, right maxilliped, anterior; **d**, second middle marginal seta of maxilliped basis; **e**, first leg, anterior; f, basipod of 1st leg, posterior; **g**, second leg, anterior; **h**, third leg, anterior; **i**, fourth leg, anterior; **j**, fifth leg, anterior. Male: **k**, urosome, dorsal.

163

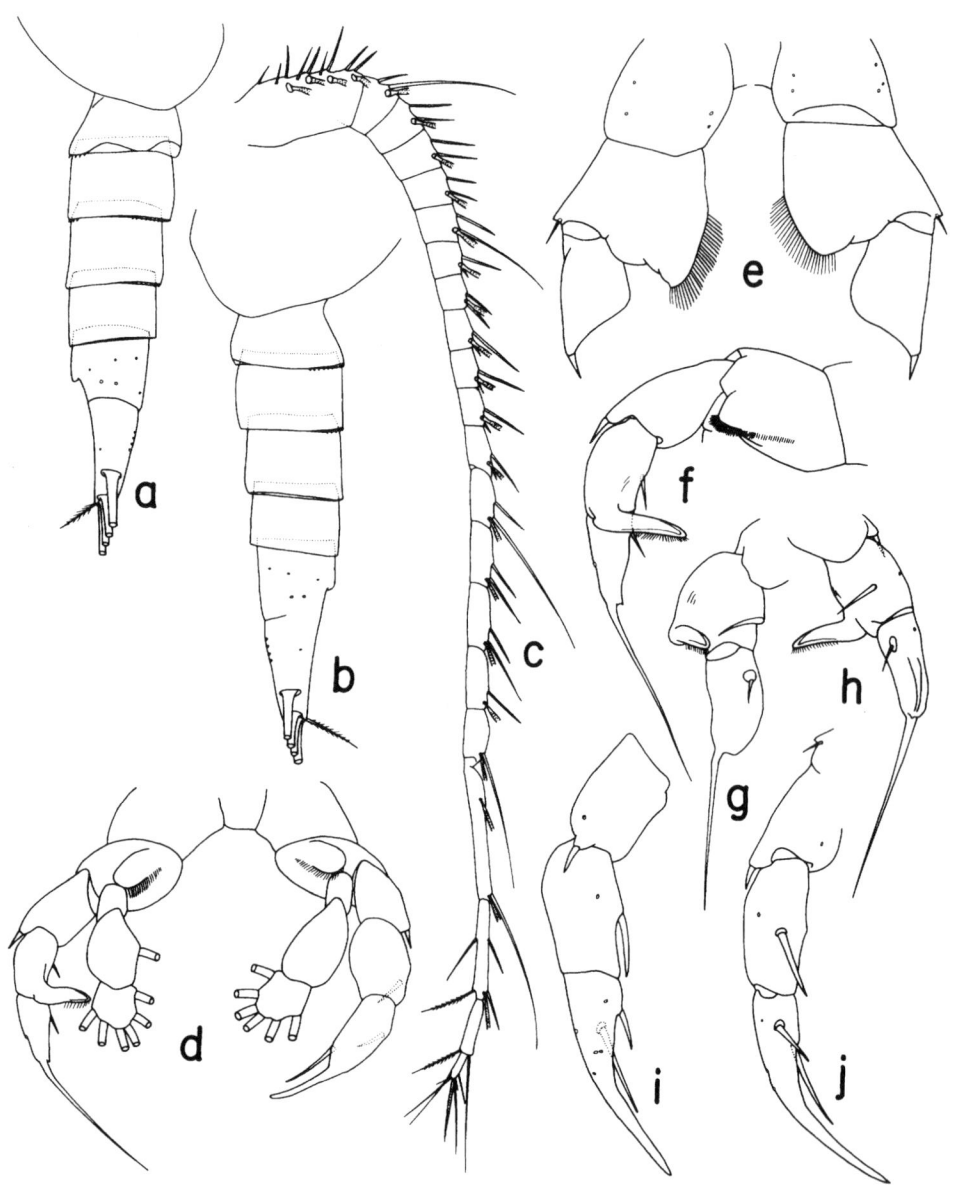

Figure 12. *Mesorhabdus brevicaudatus* male: **a**, urosome, right; **b**, do, left; **c**, left antennule, ventral; **d**, fifth pair of legs, anterior; **e**, fifth pair of legs with endopods and distal exopodal segments omitted, anterior; **f**, right 5th leg with endopod omitted, anterior; **g**, exopod of right 5th leg, posterior; **h**, do, tilted counterclockwise; **i**, exopod of left 5th leg, anterior; **j**, do, posterior.

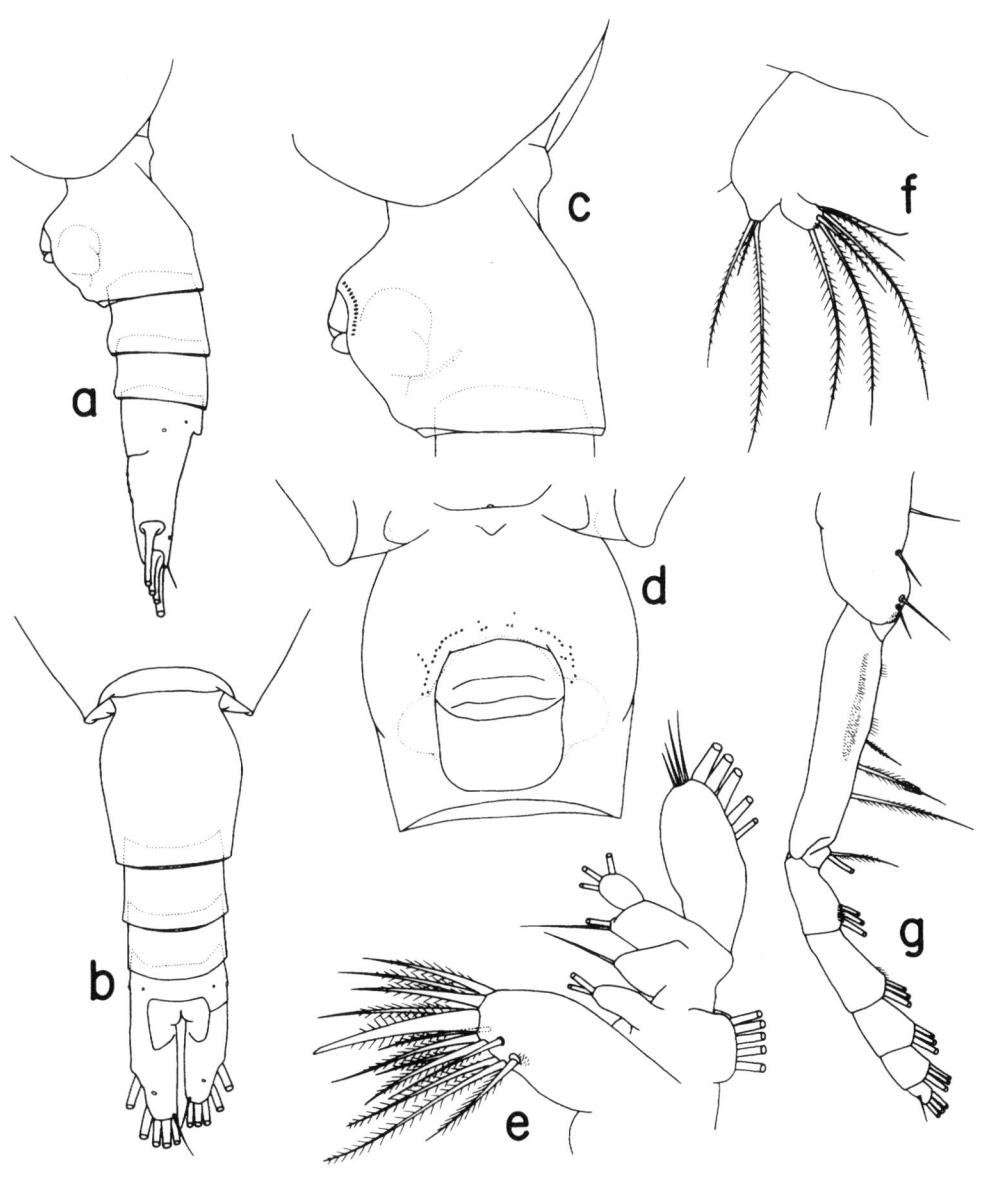

Figure 13. *Mesorhabdus gracilis* female: **a**, urosome, left; **b**, do, dorsal; **c**, genital somite, left; **d**, do, ventral; **e**, left maxillule, posterior; **f**, first segment of left maxilla, posterior; **g**, right maxilliped, anterior.

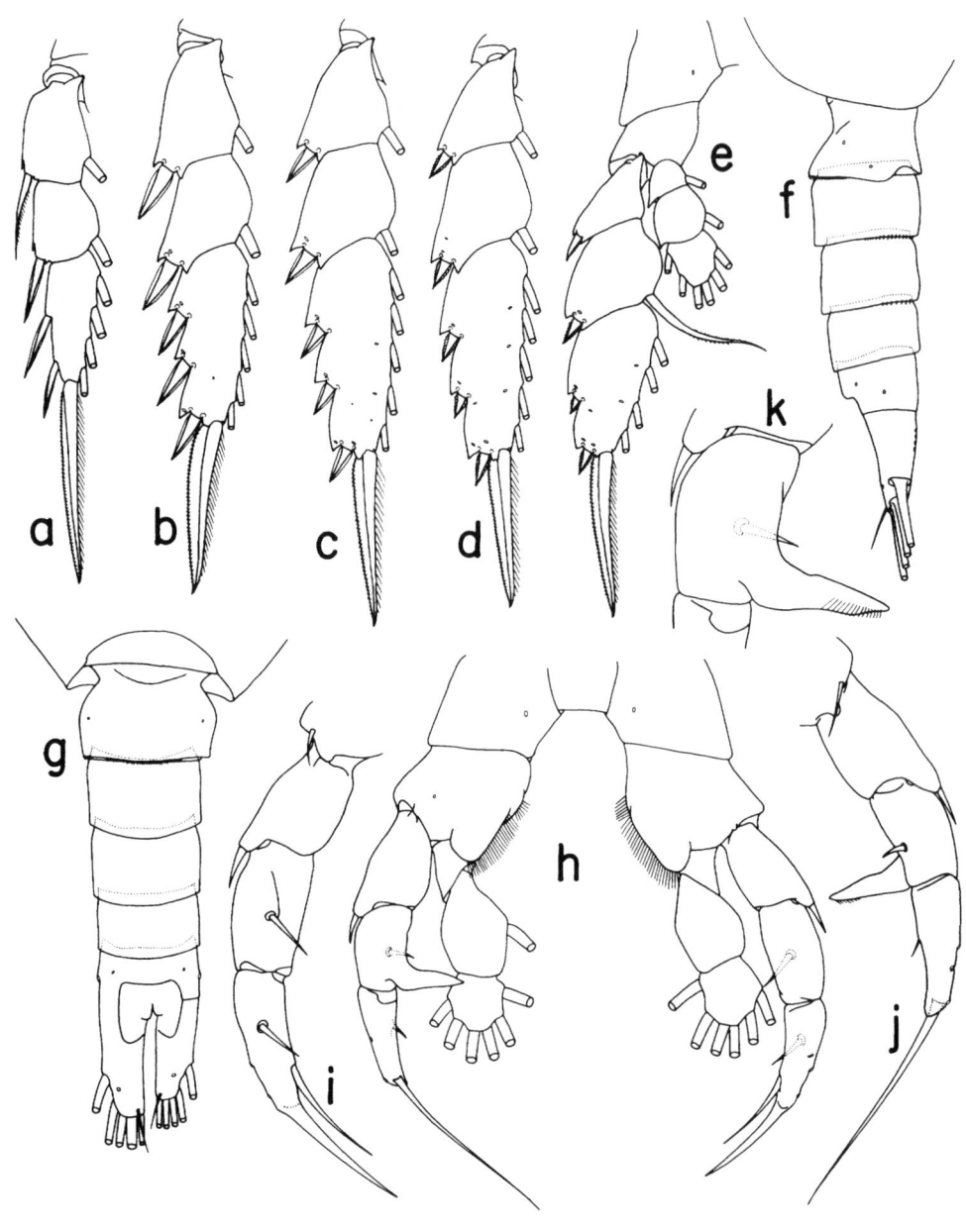

Figure 14. *Mesorhabdus gracilis* female: **a**, exopod of 1st leg, anterior; **b**, exopod of 2nd leg, anterior; **c**, exopod of 3rd leg, anterior; **d**, exopod of 4th leg, anterior; **e**, fifth leg, anterior. Male: **f**, urosome, right; **g**, do, dorsal; **h**, fifth pair of legs, anterior; **i**, exopod of left 5th leg, posterior; **j**, exopod of right 5th leg, posterior; **k**, second exopodal segment of right 5th leg, anterior.

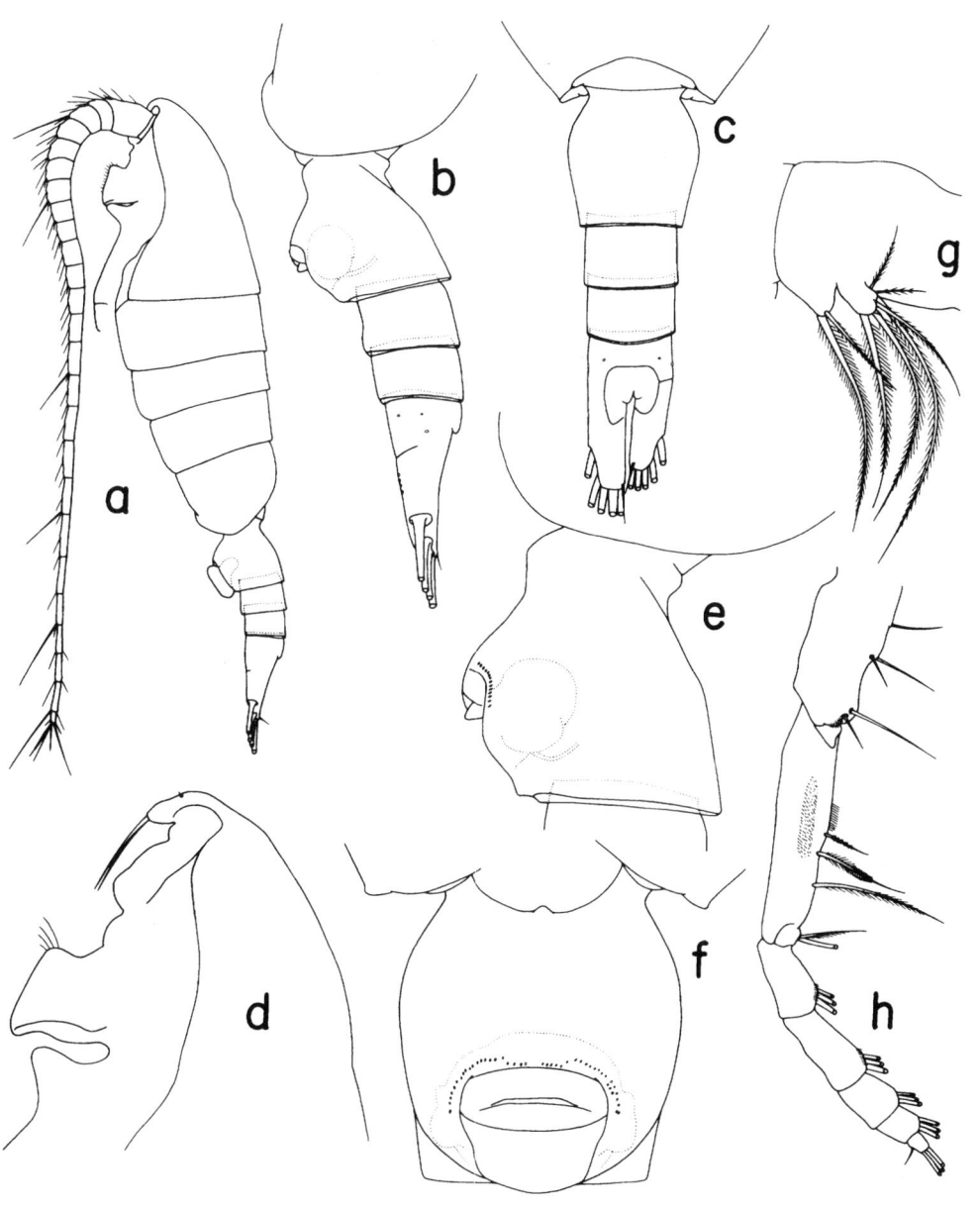

Figure 15. *Mesorhabdus paragracilis*, new species, female: **a**, habitus, left; **b**, urosome, left; **c**, do, dorsal; **d**, forehead, left; **e**, genital somite, left; **f**, do, ventral; **g**, first segment of left maxilla, posterior; **h**, right maxilliped, anterior.

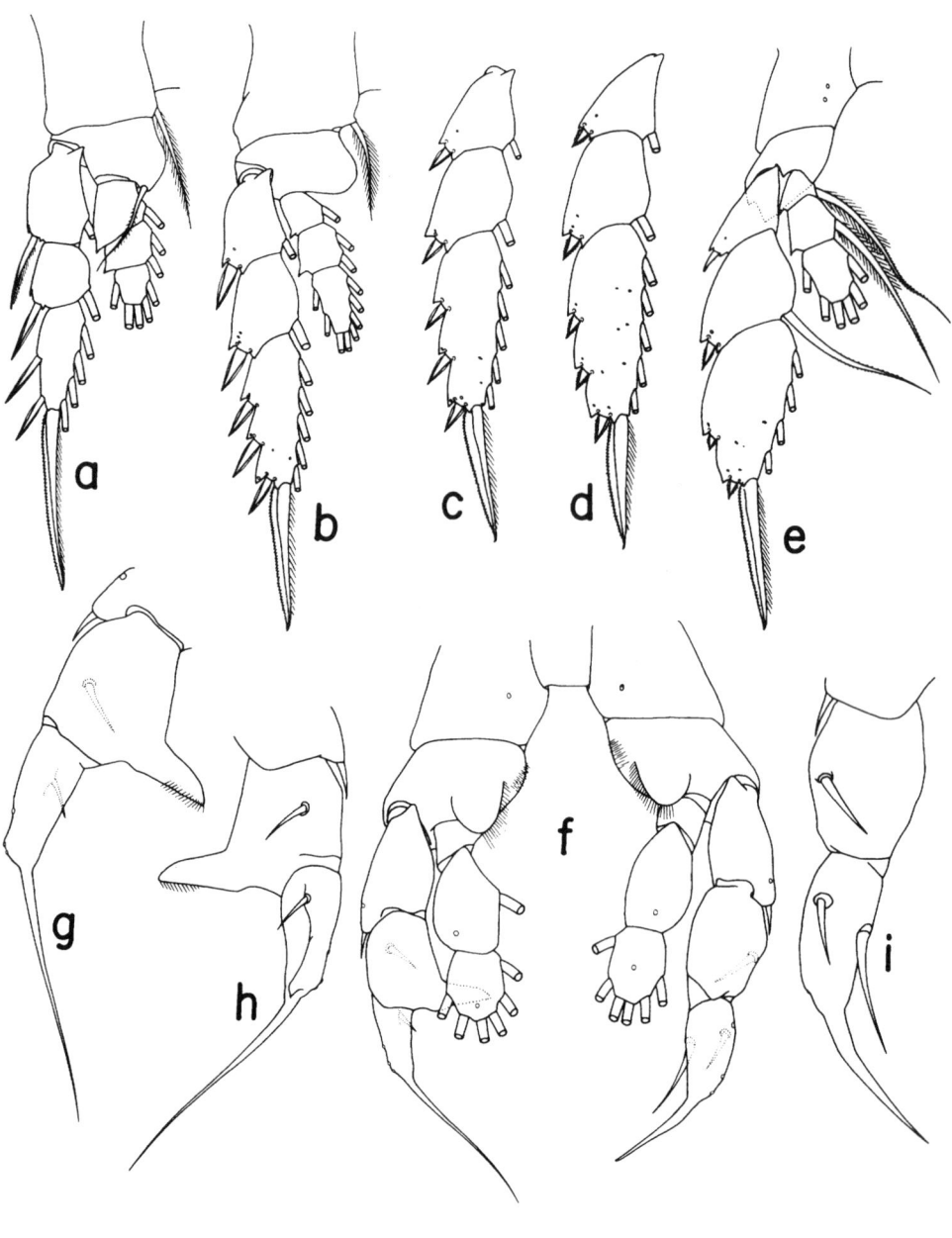

Figure 16. *Mesorhabdus paragracilis*, new species, female: **a**, first leg, anterior; **b**, second leg, anterior; **c**, exopod of 3rd leg, anterior; **d**, exopod of 4th leg, anterior; **e**, fifth leg, anterior. Male: **f**, fifth pair of legs, anterior; **g**, exopod of right 5th leg, anterior; **h**, do, posterior; **i**, exopod of left 5th leg, posterior.

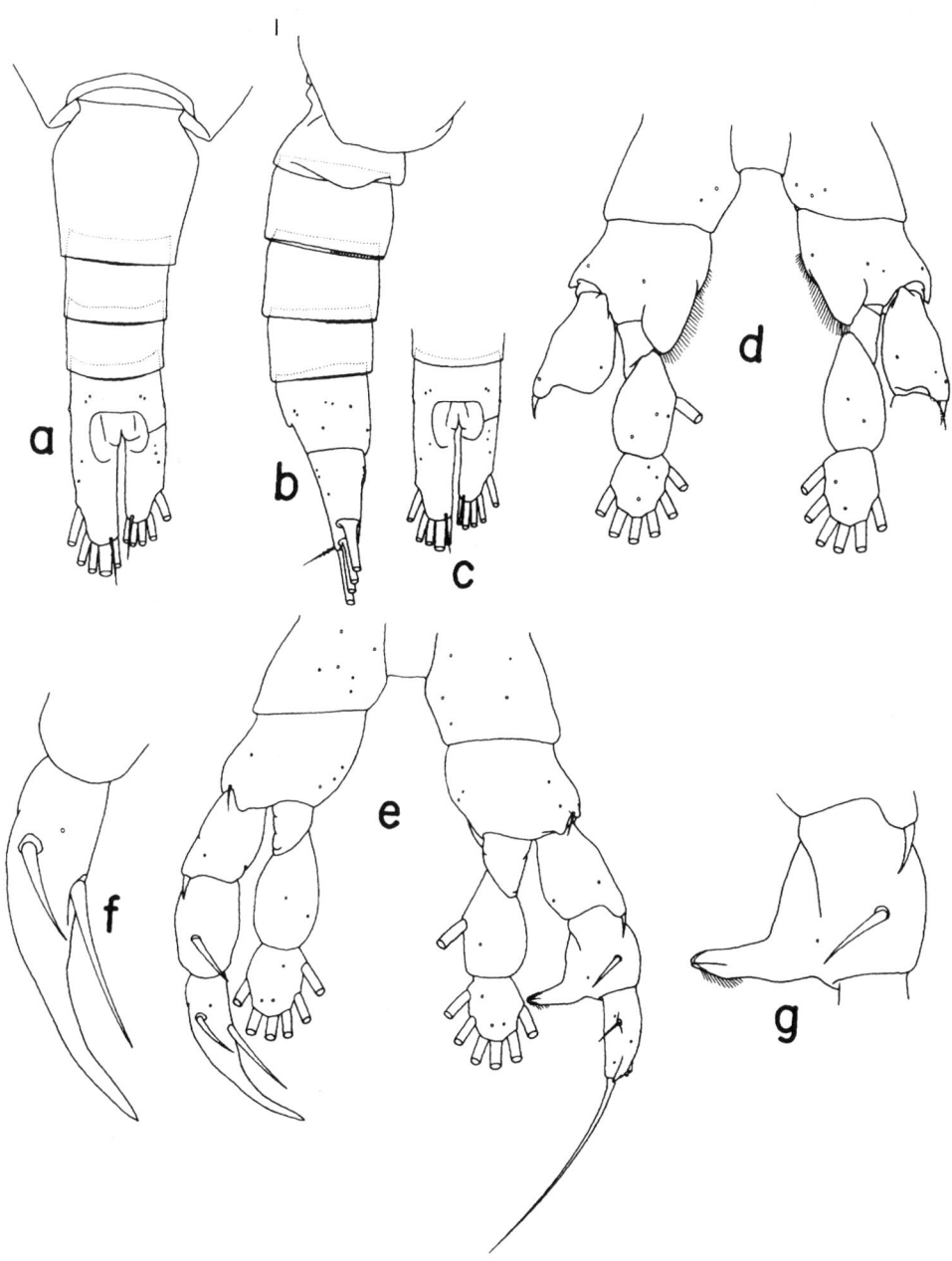

Figure 17. *Mesorhabdus poriphorus*, new species, female: **a**, urosome, dorsal. Male: **b**, urosome, right; **c**, posterior part of urosome, dorsal; **d**, fifth pair of legs with distal exopodal segments omitted, anterior; **e**, fifth pair of legs, posterior; **f**, distal exopodal segment of left 5th leg, posterior; **g**, second exopodal segment of right 5th leg, posterior.

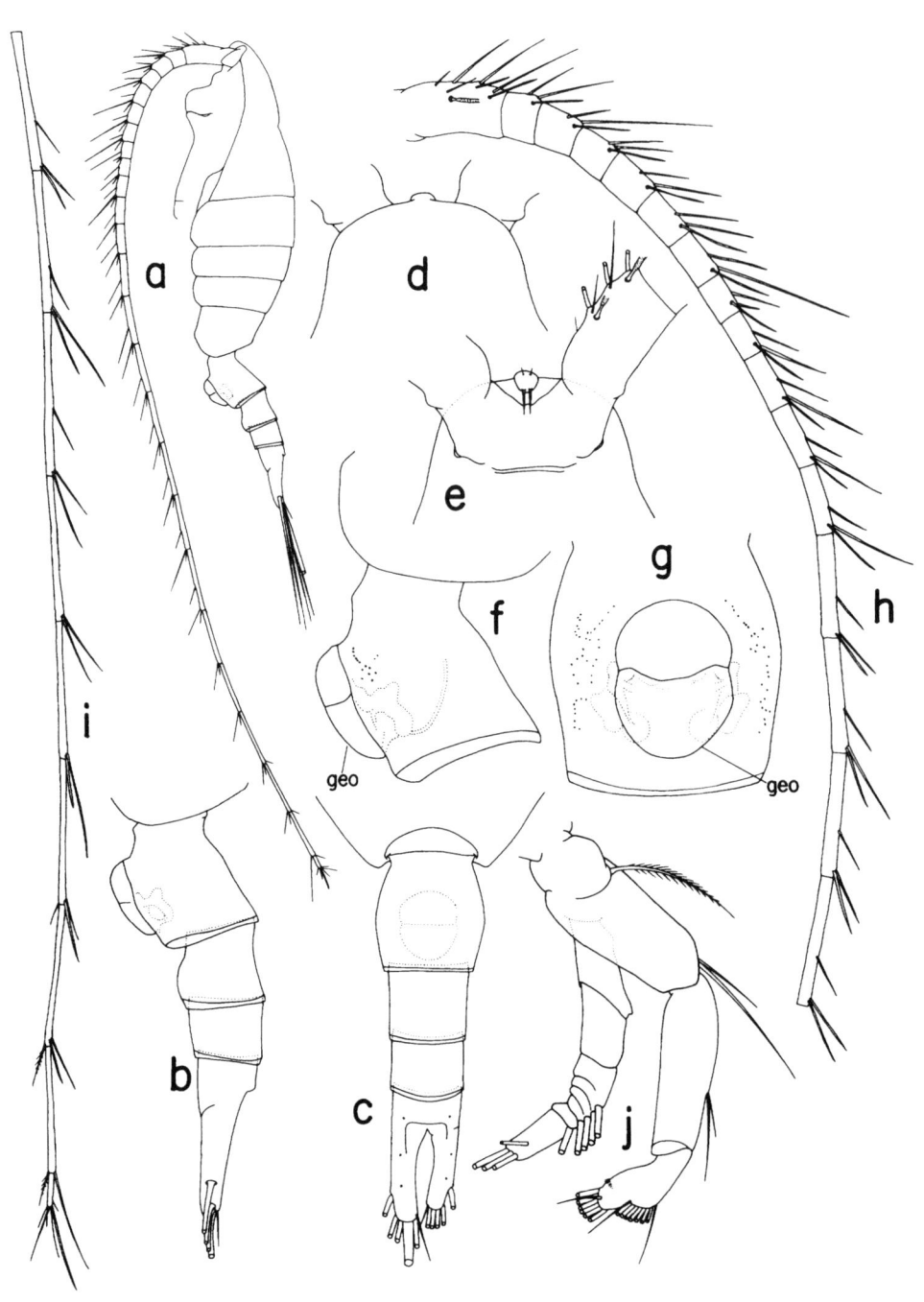

Figure 18. *Heterostylites major* female: **a**, habitus, left; **b**, urosome, left; **c**, do, dorsal; **d**, forehead, dorsal; **e**, do, ventral; **f**, genital somite, left; **g**, do, ventral; **h**, segments 1-16 of left antennule, ventral; **i**, segments 17-25 of left antennule, ventral; **j**, left antenna, posterior. geo = genital operculum.

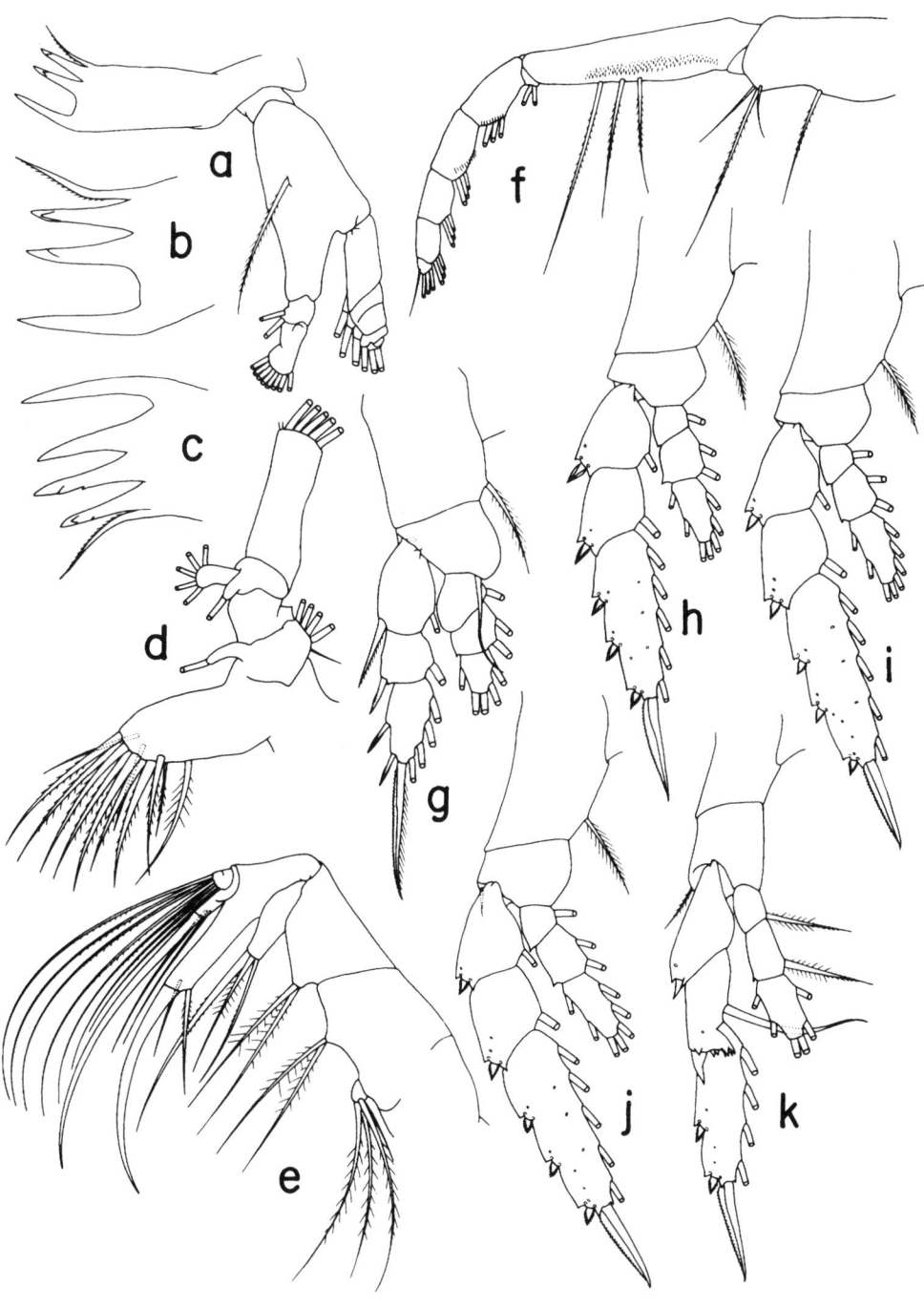

Figure 19. *Heterostylites major* female: **a**, left mandible, posterior; **b**, masticatory edge of left mandible, posterior; **c**, masticatory edge of right mandible, posterior; **d**, left maxillule, posterior; **e**, left maxilla, posterior; **f**, right maxilliped, anterior; **g**, first leg, anterior; **h**, second leg, anterior; **i**, third leg, anterior; **j**, fourth leg, anterior; **k**, fifth leg, anterior.

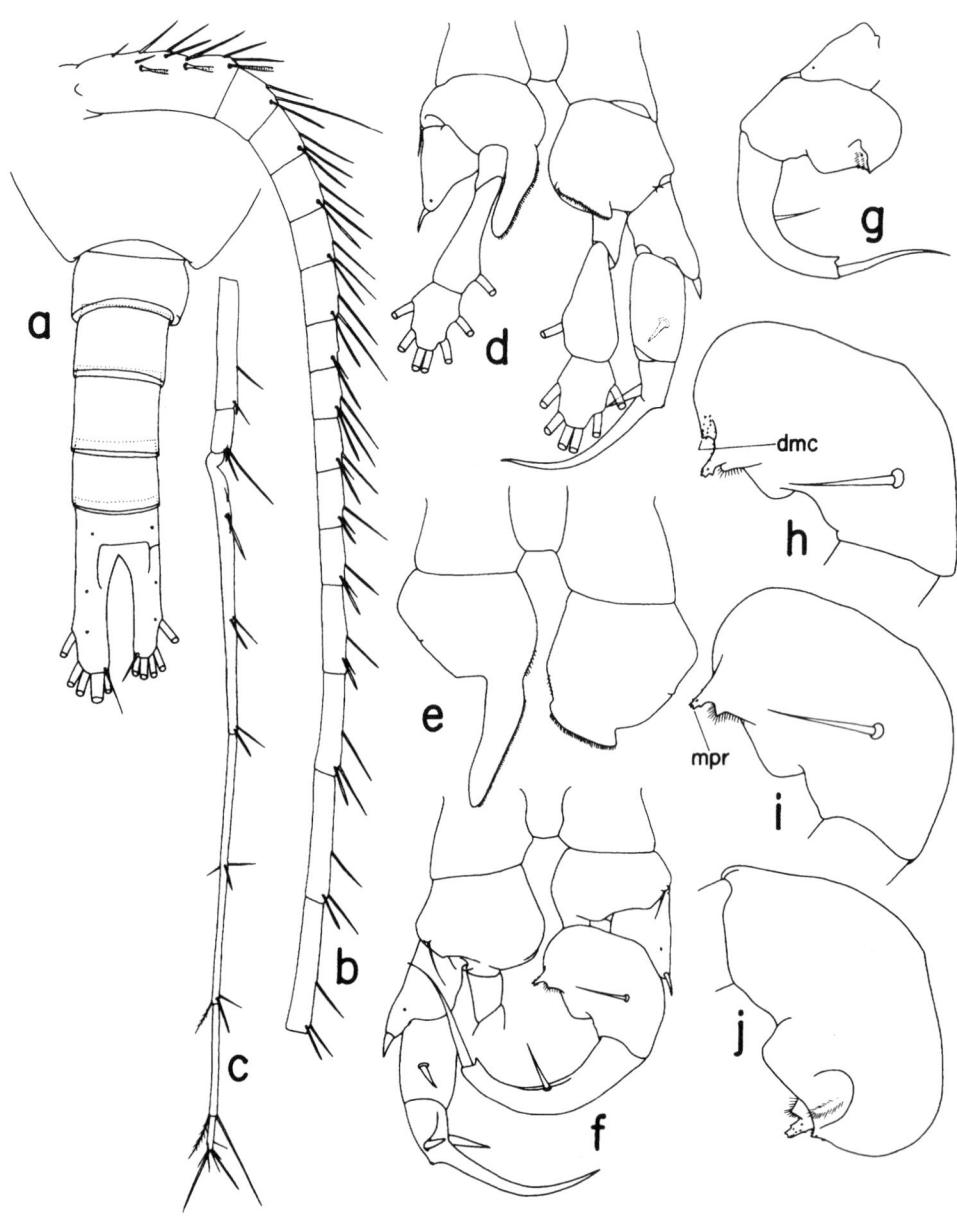

Figure 20. *Heterostylites major* male: **a**, urosome, dorsal; **b**, segments 1-16 of left antennule, ventral; **c**, segments 17-25 of left antennule, ventral; **d**, fifth pair of legs with distal segments of right exopod omitted, anterior; **e**, basipods of 5th pair of legs, anterior; **f**, fifth pair of legs with endopods omitted, posterior; **g**, exopod of right 5th leg, anterior; **h**, second exopodal segment of right 5th leg, posterior; **i**, do, tilted counterclockwise; **j**, do, anterior. dmc = distomedial corner; mpr = medial projection.

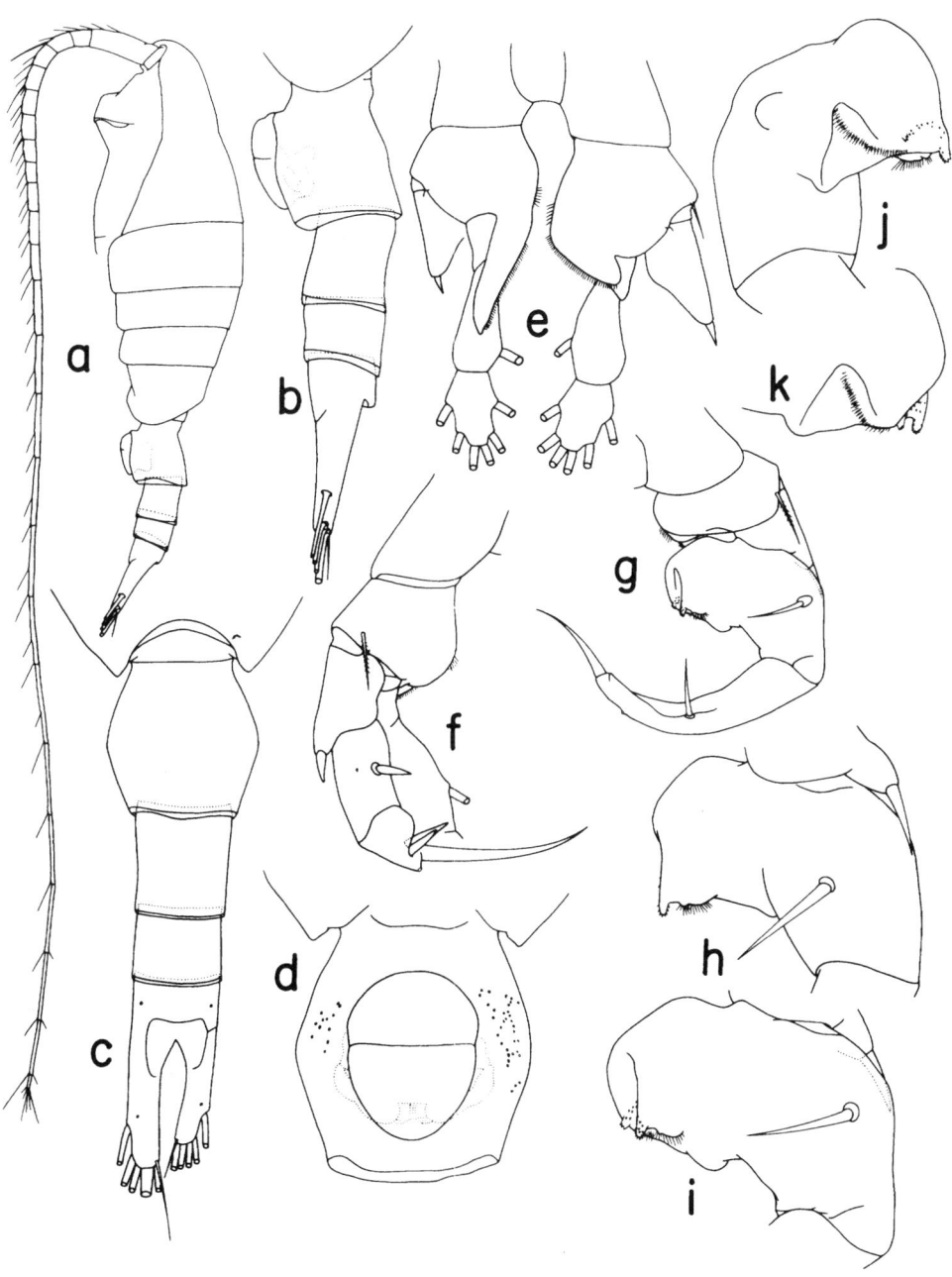

Figure 21. *Heterostylites submajor*, new species, female: **a**, habitus, left; **b**, urosome, left; **c**, do, dorsal; **d**, genital somite, ventral. Male: **e**, fifth pair of legs with distal exopodal segments omitted, anterior; **f**, left 5th leg, posterior; **g**, right 5th leg with endopod omitted, posterior; **h**, second exopodal segment of right 5th leg, posterior; **i**, do, tilted clockwise; **j**, do, anterior; **k**, do, tilted clockwise.

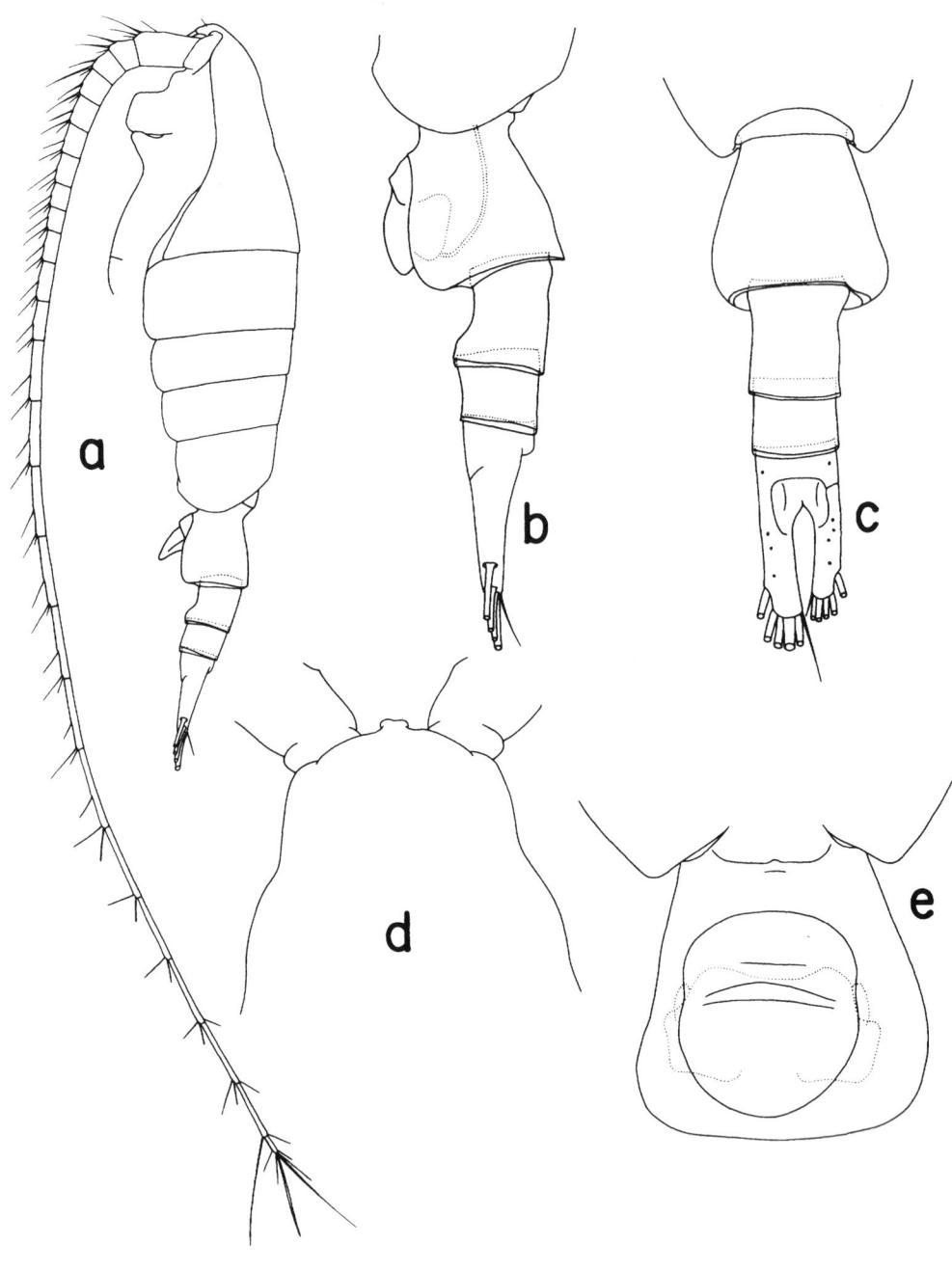

Figure 22. *Heterostylites nigrotinctus* female: **a**, habitus, left; **b**, urosome, left; **c**, do, dorsal; **d**, forehead, dorsal; **e**, genital somite, ventral.

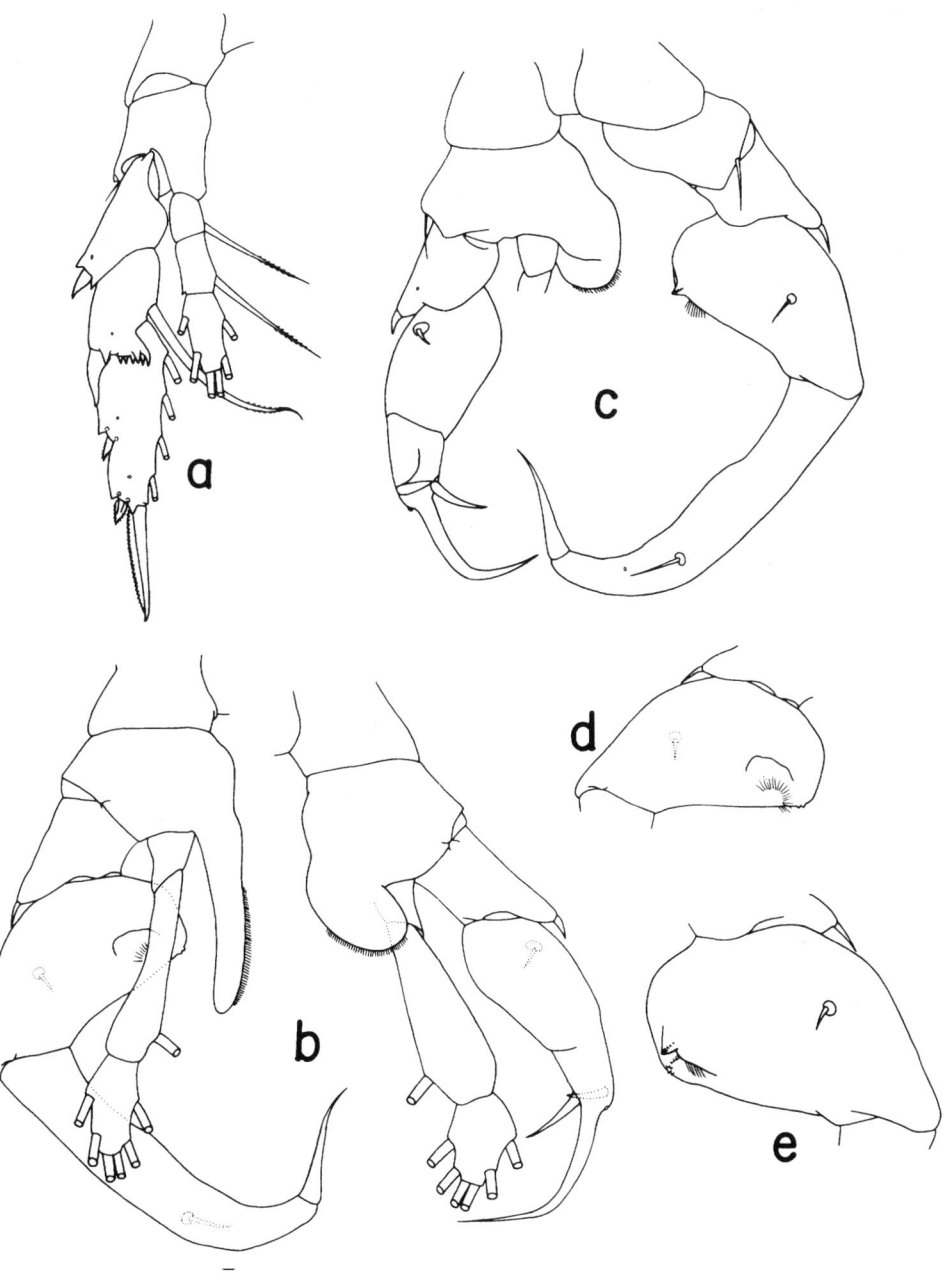

Figure 23. *Heterostylites nigrotinctus* female: **a**, fifth leg, anterior. Male: **b**, fifth pair of legs, anterior; **c**, fifth pair of legs with endopods omitted, posterior; **d**, second exopodal segment of right 5th leg, anterior; **e**, do, posterior.

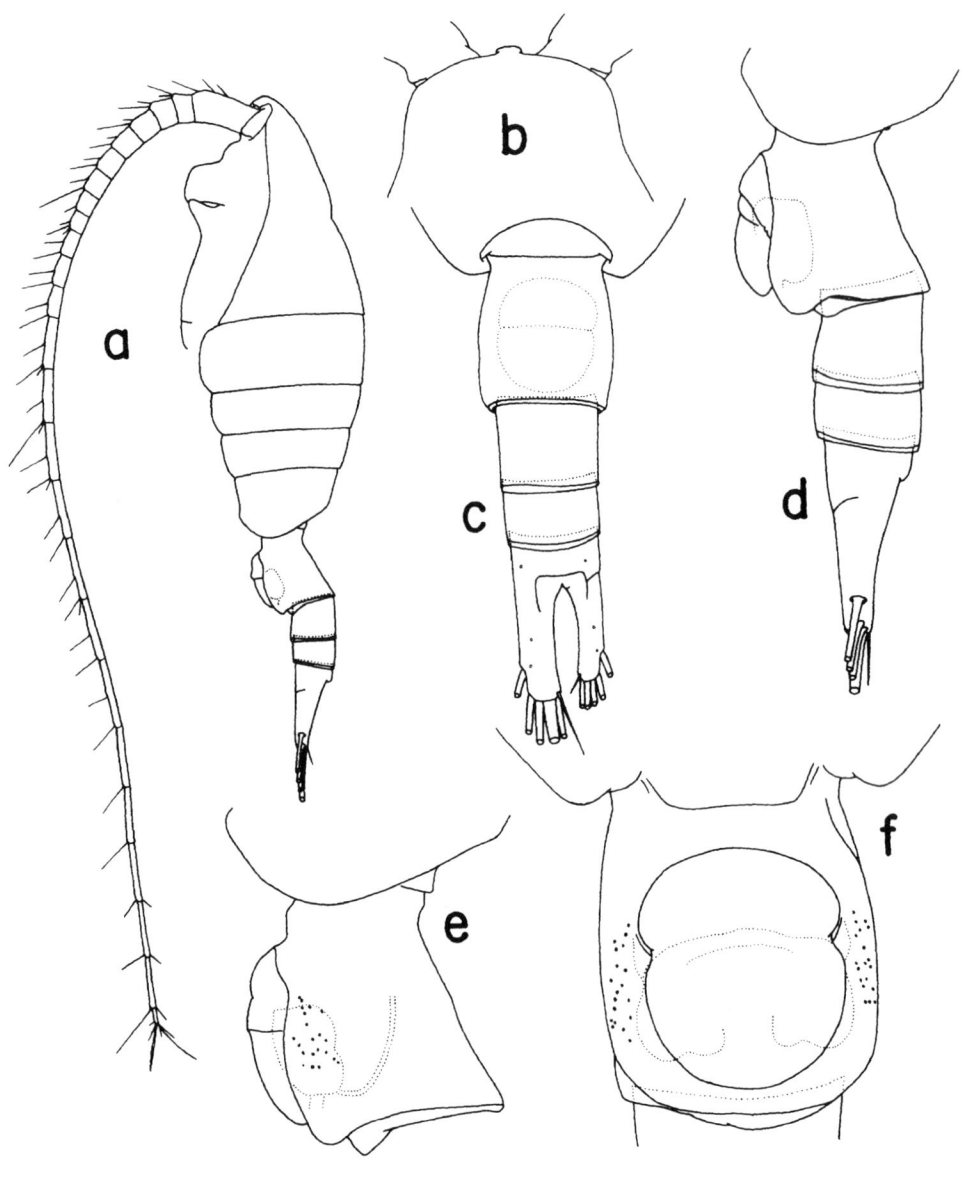

Figure 24. *Heterostylites longicornis* female: **a**, habitus, left; **b**, forehead, dorsal; **c**, urosome, dorsal; **d**, do, left; **e**, genital somite, left; **f**, do, ventral.

176

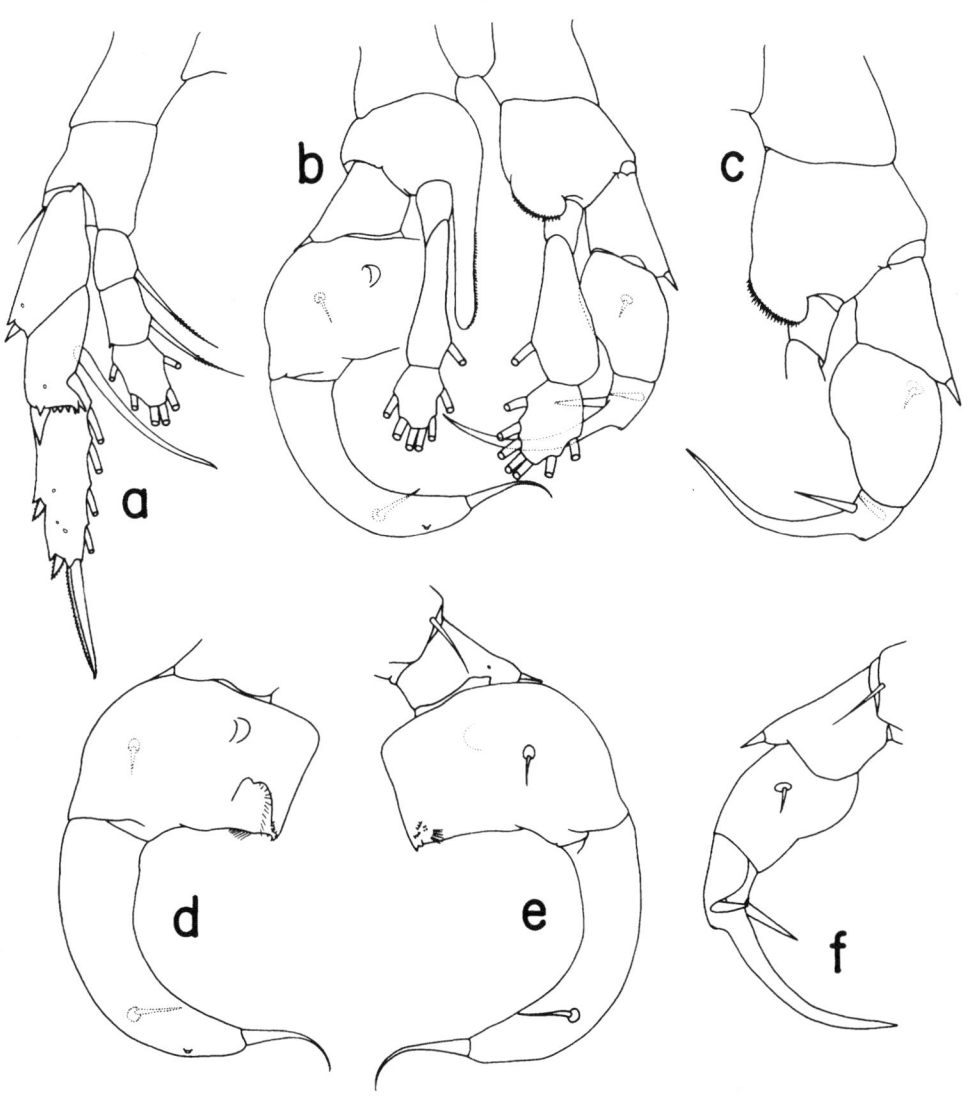

Figure 25. *Heterostylites longicornis* female: **a**, fifth leg, anterior. Male: **b**, fifth pair of legs, anterior; **c**, left 5th leg with distal endopodal segments omitted, anterior; **d**, exopod of right 5th leg, anterior; **e**, do, posterior; **f**, exopod of left 5th leg, posterior.

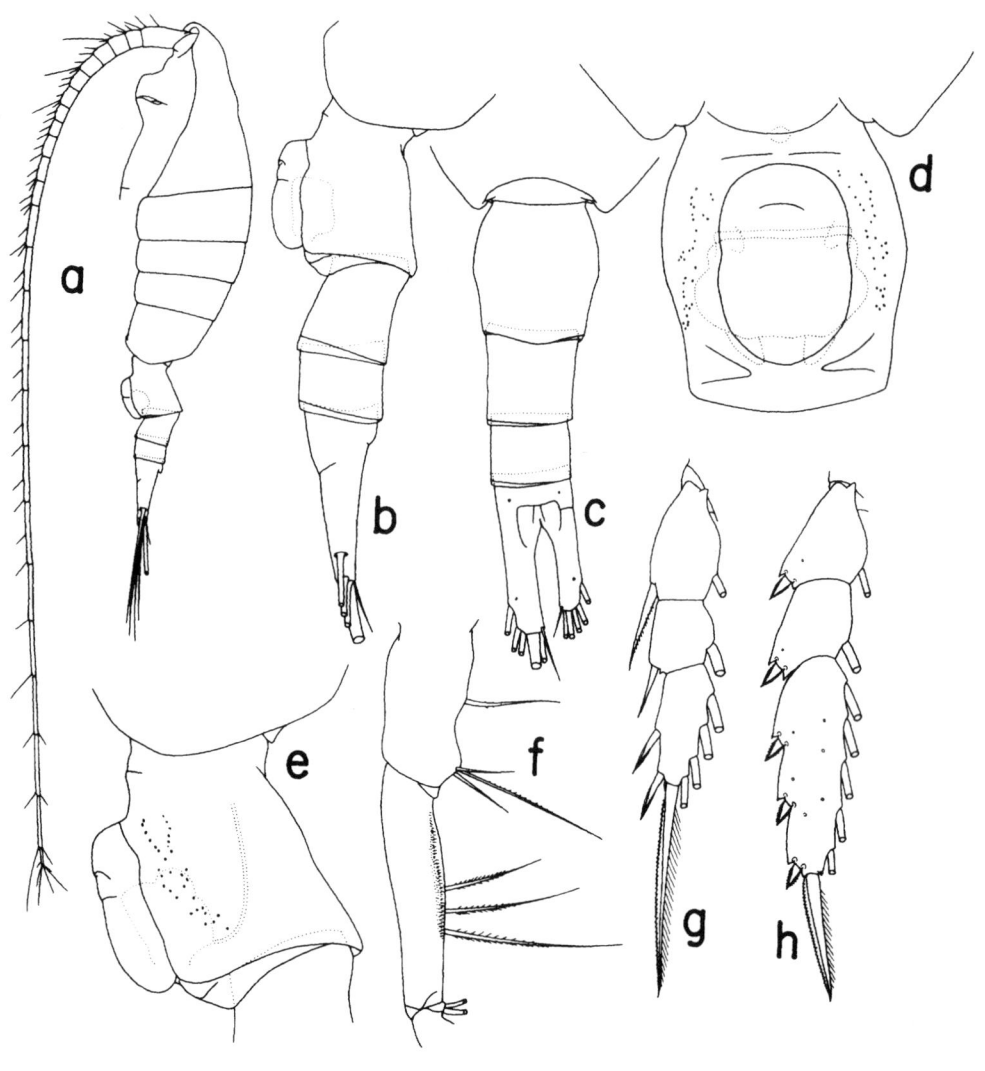

Figure 26. *Heterostylites longioperculis*, new species, female: **a**, habitus, left; **b**, urosome, left; **c**, do, dorsal; **d**, genital somite, ventral; **e**, do, left; **f**, basipod of right maxilliped, anterior; **g**, exopod of 1st leg, anterior; **h**, exopod of 2nd leg, anterior.

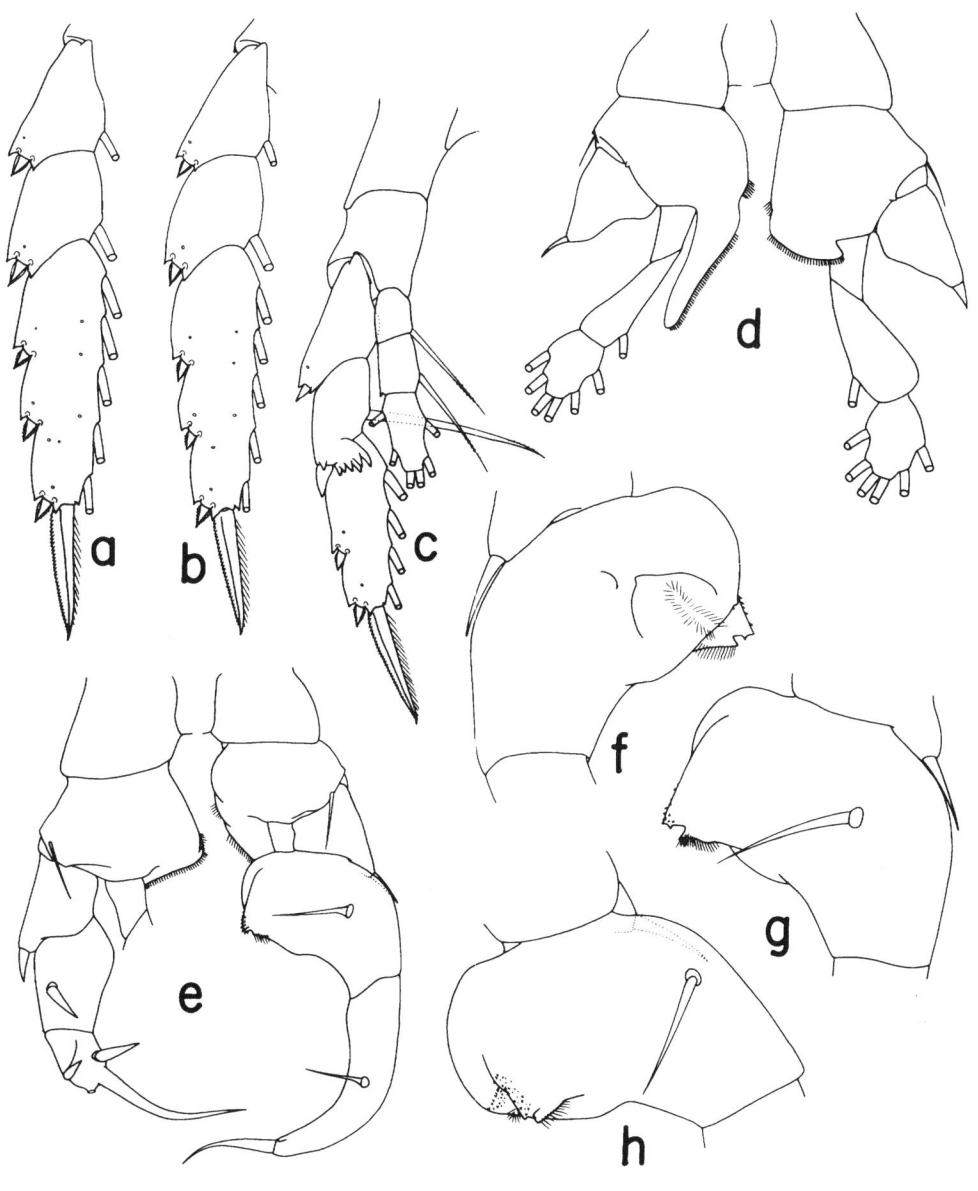

Figure 27. *Heterostylites longioperculis*, new species, female: **a**, exopod of 3rd leg, anterior; **b**, exopod of 4th leg, anterior; **c**, fifth leg, anterior. Male: **d**, fifth pair of legs with distal exopodal segments omitted, anterior; **e**, fifth pair of legs with endopods omitted, posterior; **f**, second exopodal segment of right 5th leg, anterior; **g**, do, posterior; **h**, do, tilted clockwise.

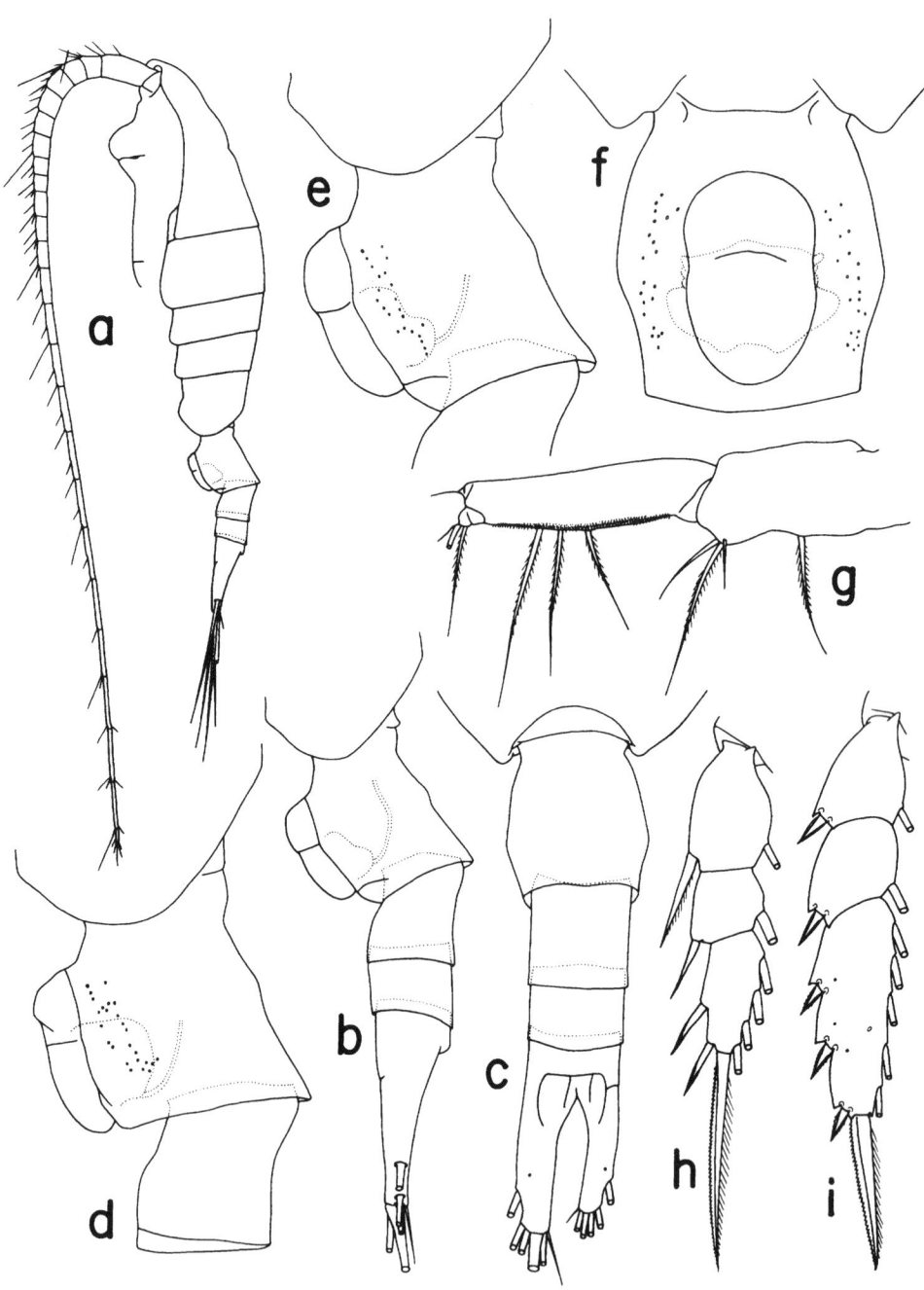

Figure 28. *Heterostylites echinatus*, new species, female: **a**, habitus, left; **b**, urosome, left; **c**, do, dorsal; **d**, first 2 urosomal somites, left; **e**, genital somite with arthrodial membrane on ventral side extended, left; **f**, genital somite, ventral; **g**, basipod of right maxilliped, anterior; **h**, exopod of 1st leg, anterior; **i**, exopod of 2nd leg, anterior.

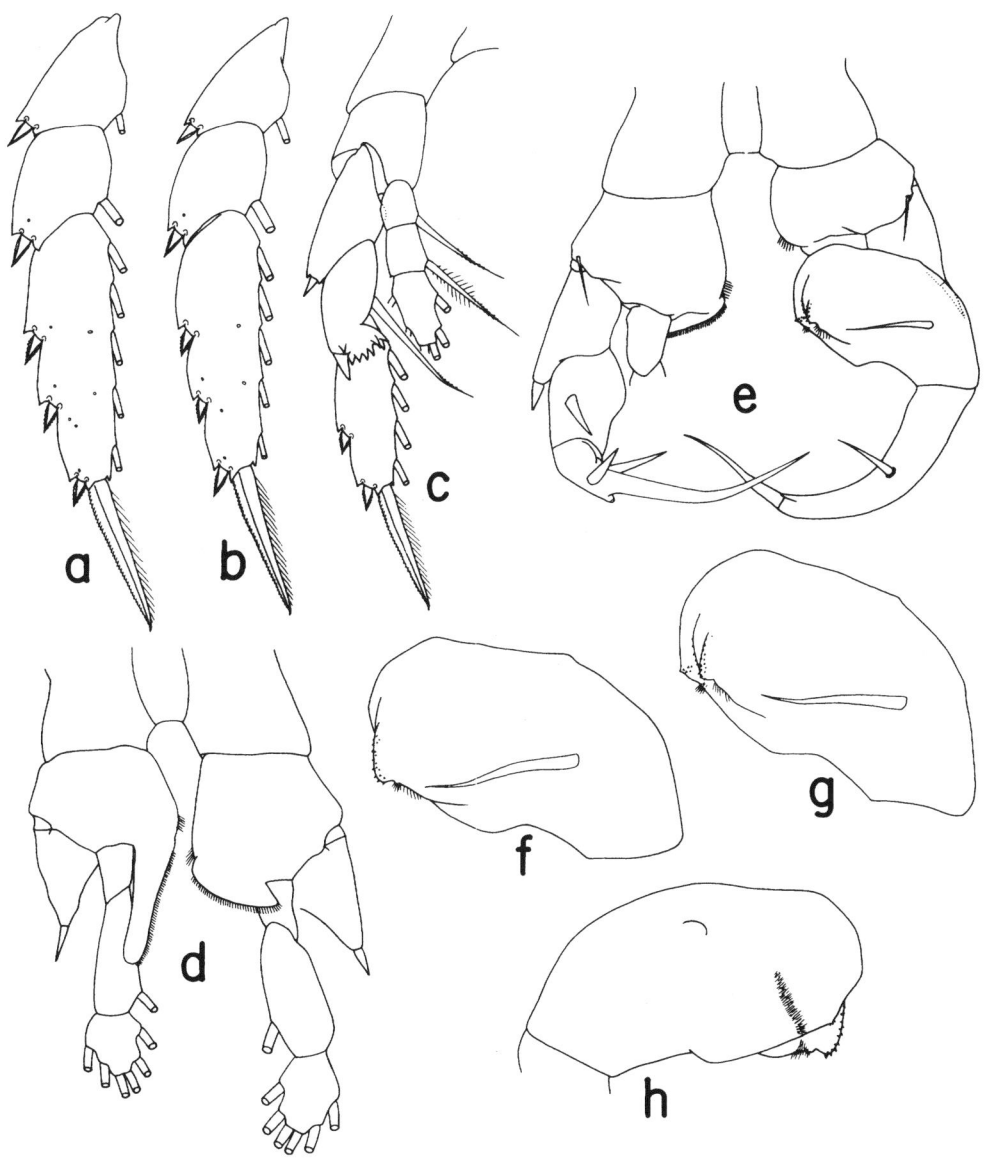

Figure 29. *Heterostylites echinatus*, new species, female: **a**, exopod of 3rd leg, anterior; **b**, exopod of 4th leg, anterior; **c**, fifth leg, anterior. Male: **d**, fifth pair of legs with distal exopodal segments omitted, anterior; **e**, fifth pair of legs with endopods omitted, posterior; **f**, second exopodal segment of right 5th leg, posterior; **g**, do, tilted clockwise; **h**, do, anterior.

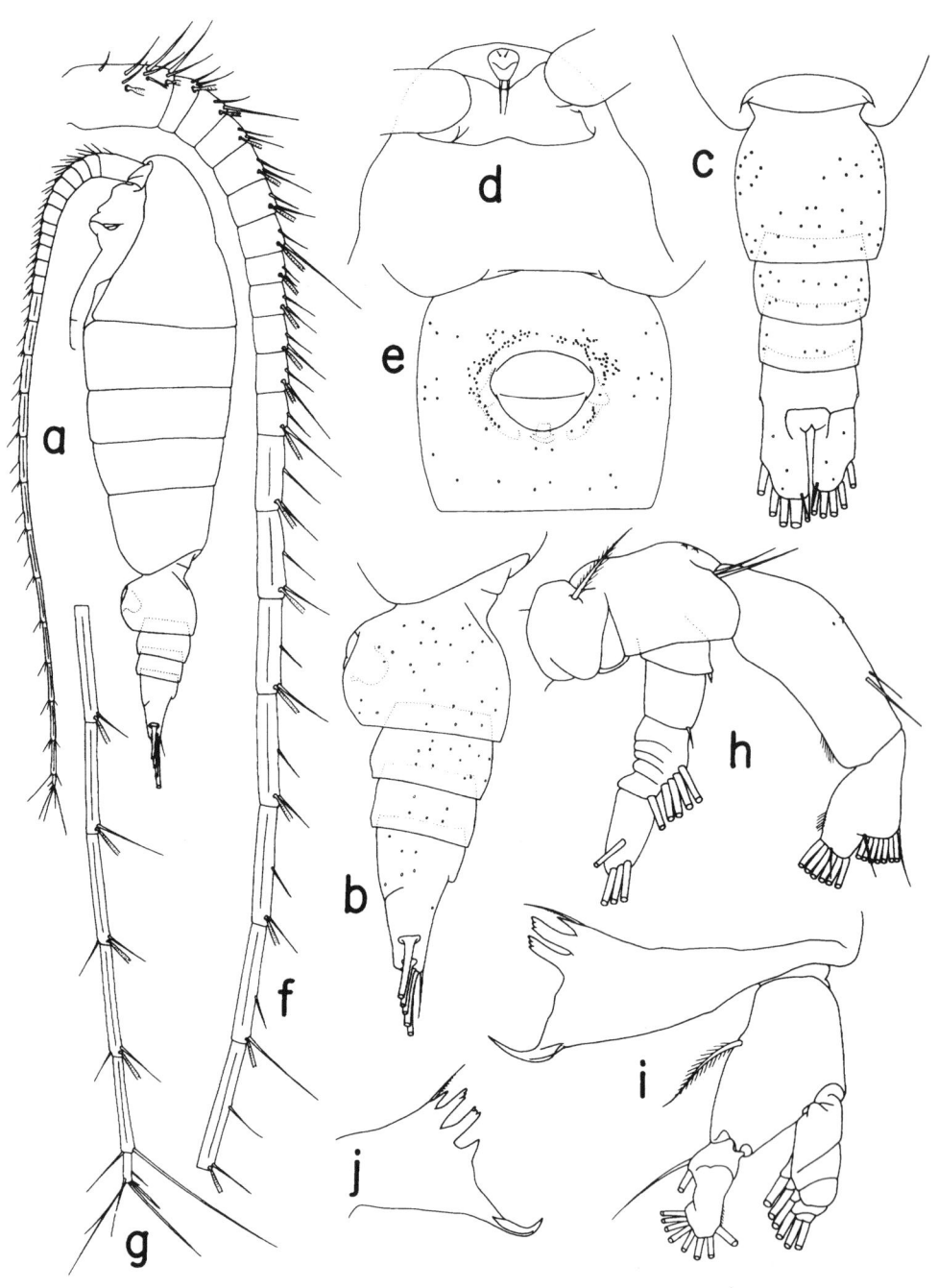

Figure 30. *Hemirhabdus grimaldii* female: **a**, habitus, left; **b**, urosome, left; **c**, do, dorsal; **d**, forehead, ventral; **e**, genital somite, ventral; **f**, segments 1-19 of left antennule, ventral; **g**, segments 20-25 of left antennule, ventral; **h**, left antenna, posterior; **i**, left mandible, posterior; **j**, masticatory edge of right mandible, posterior.

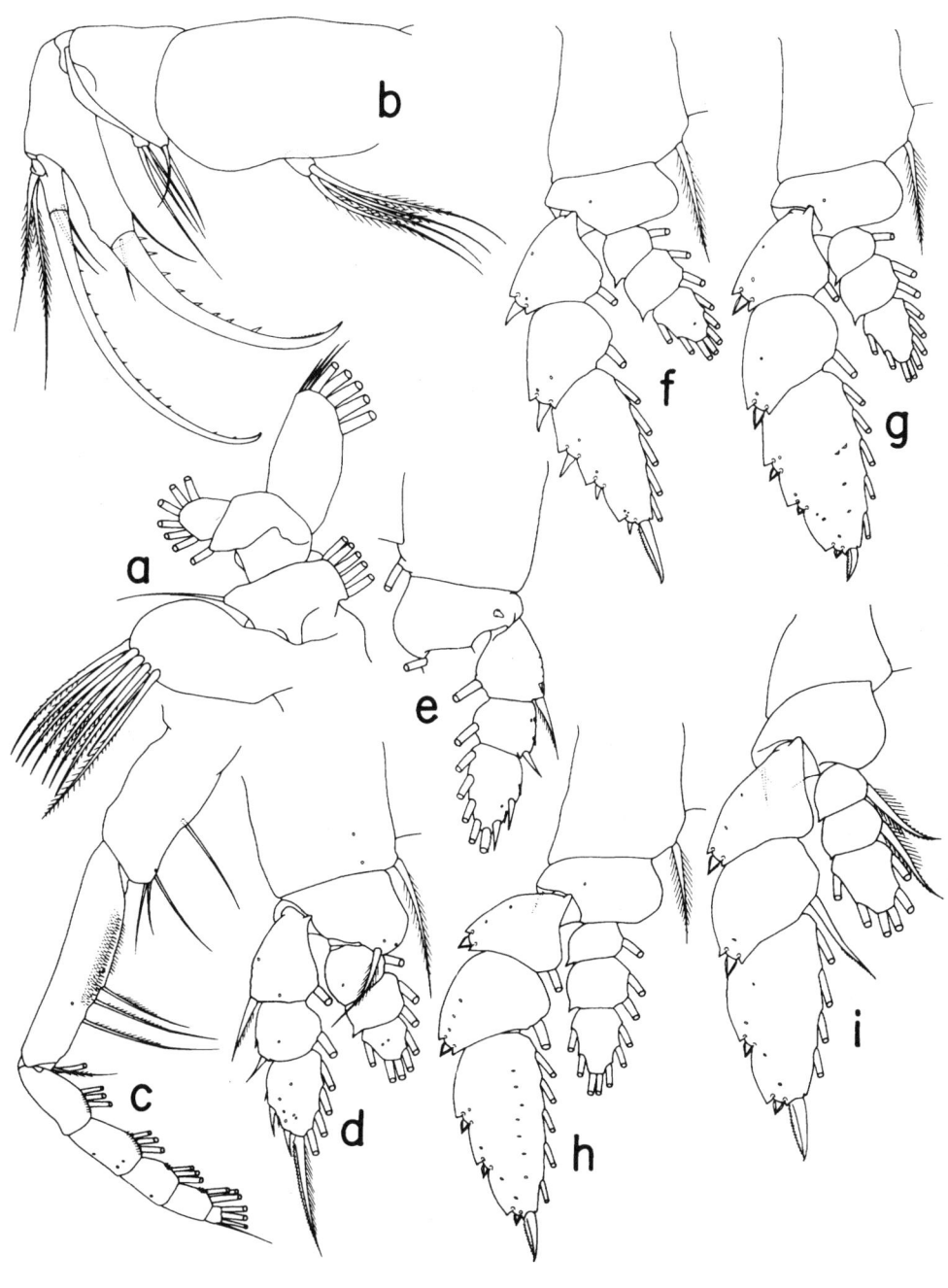

Figure 31. *Hemirhabdus grimaldii* female: **a**, left maxillule, posterior; **b**, left maxilla, posterior; **c**, right maxilliped, anterior; **d**, first leg, anterior; **e**, first leg with endopod omitted, posterior; **f**, second leg, anterior; **g**, third leg, anterior; **h**, fourth leg, anterior; **i**, fifth leg, anterior.

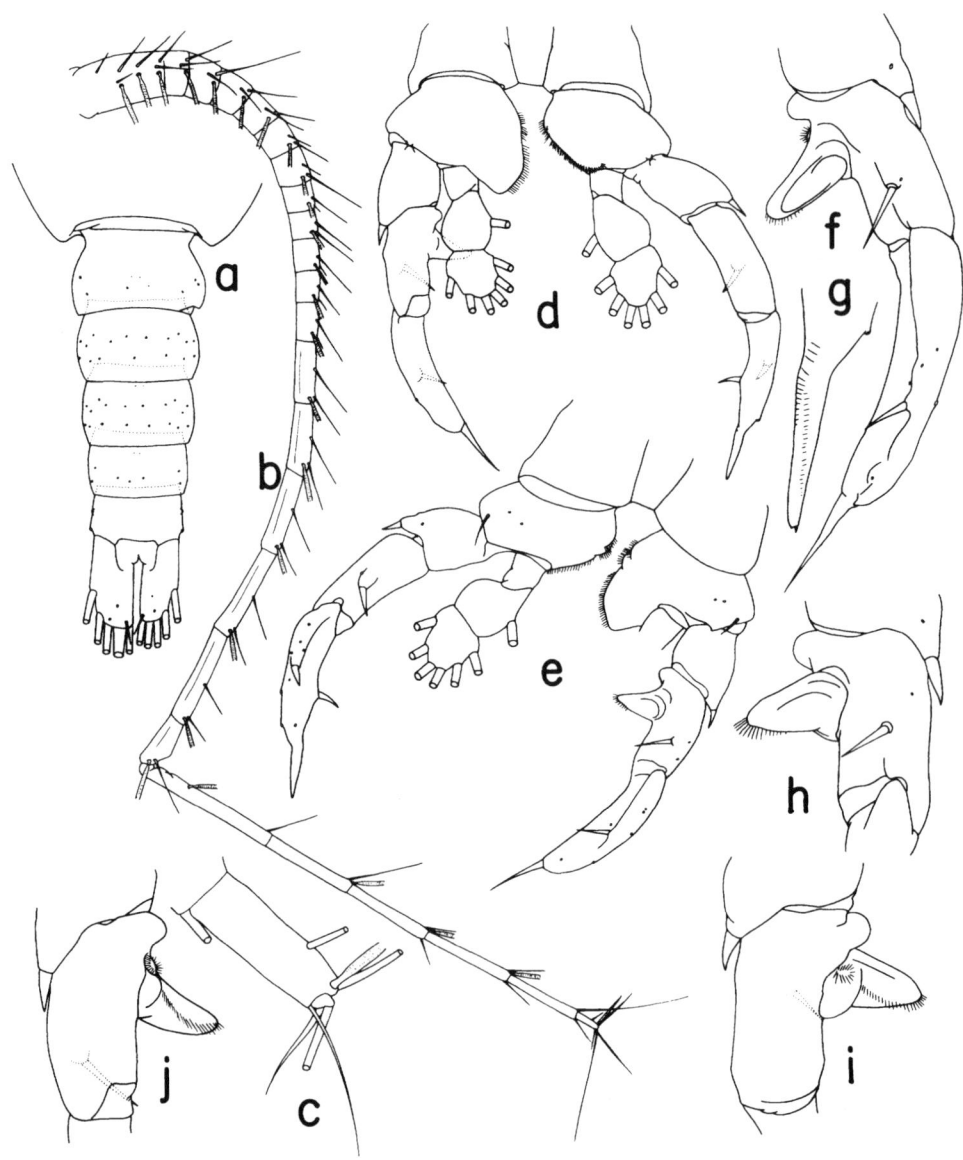

Figure 32. *Hemirhabdus grimaldii* male: **a**, urosome, dorsal; **b**, left antennule, ventral; **c**, distal segment of left antennule, ventral; **d**, fifth pair of legs, anterior; **e**, fifth pair of legs with right endopod omitted, posterior; **f**, exopod of right 5th leg, posterior; **g**, distal end of exopod of left 5th leg, anterior; **h**, second exopodal segment of right 5th leg, posterior; **i**, do, anterior; **j**, do, tilted clockwise.

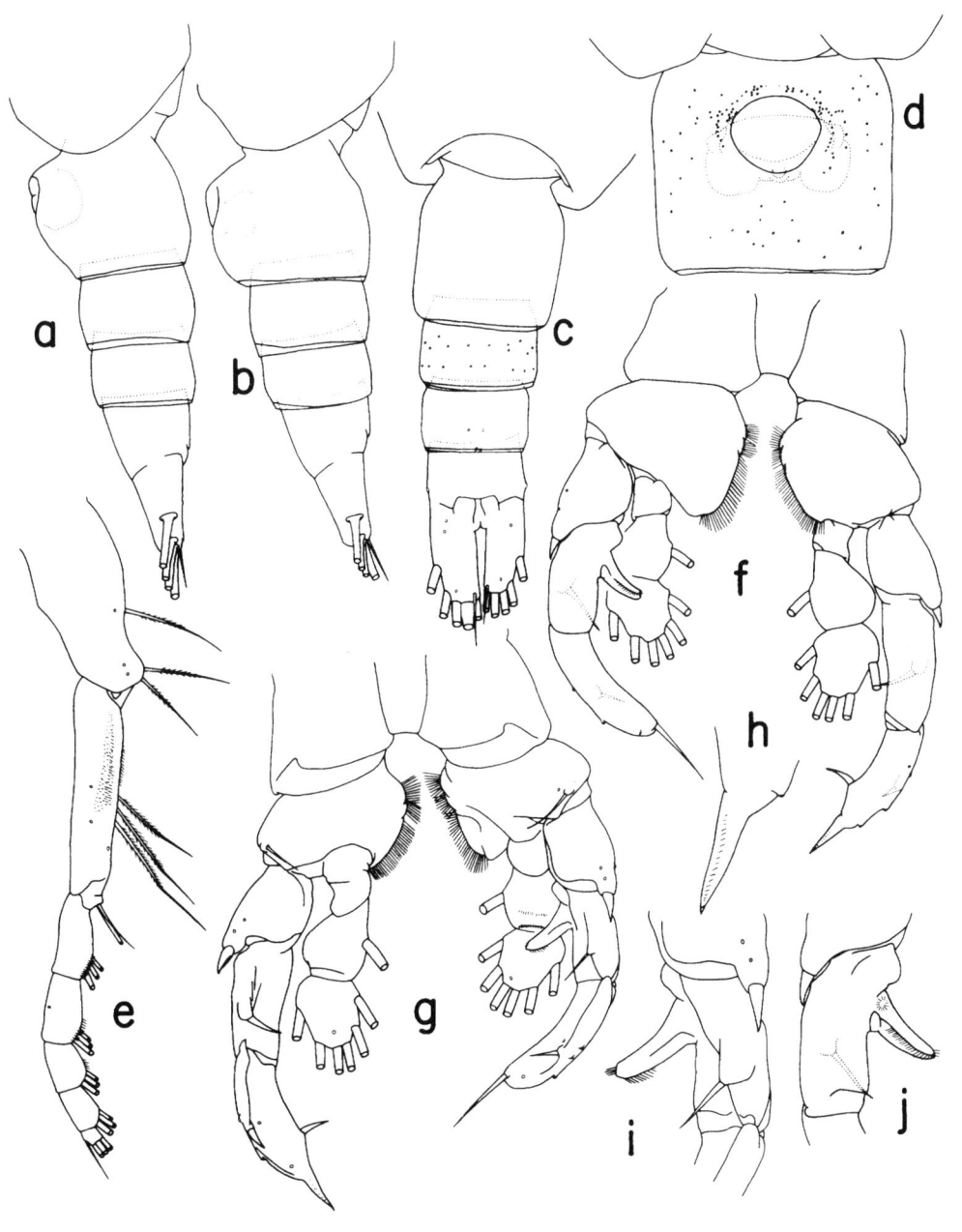

Figure 33. *Hemirhabdus amplus*, new species, female: **a**, urosome, left; **b**, urosome with genital somite somewhat inflated by deep telescoping of the following somite, left; **c**, urosome, dorsal; **d**, genital somite, ventral; **e**, right maxilliped, anterior. Male: **f**, fifth pair of legs, anterior; **g**, do, posterior; **h**, distal end of exopod of left 5th leg, anterior; **i**, second exopodal segment of right 5th leg, posterior; **j**, do, anterior.

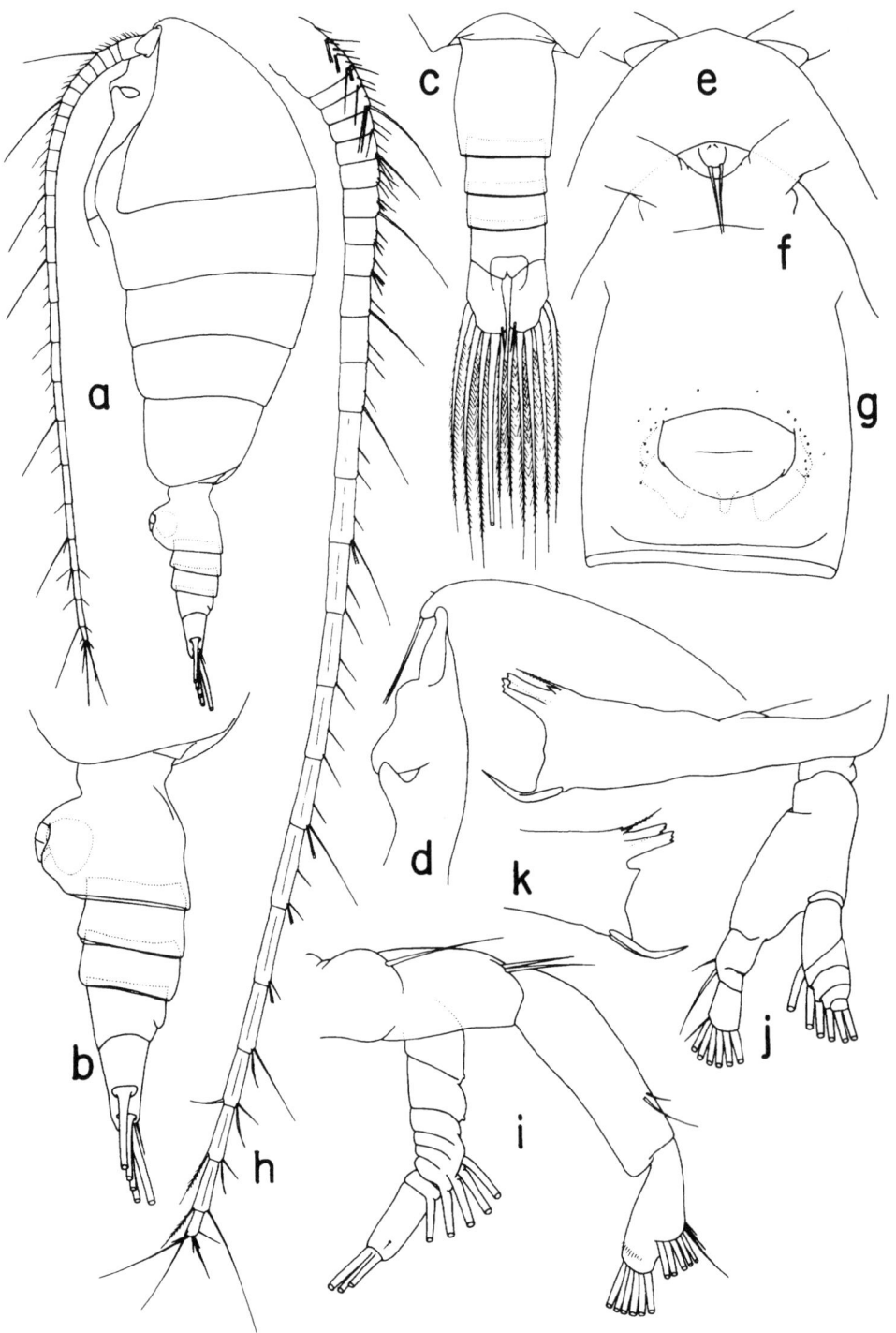

Figure 34. *Neorhabdus latus* female: **a**, habitus, left; **b**, urosome, left; **c**, do, dorsal; **d**, forehead, left; **e**, do, dorsal; **f**, do, ventral; **g**, genital somite, ventral; **h**, left antennule, ventral; **i**, left antenna, posterior; **j**, left mandible, posterior; **k**, masticatory edge of right mandible, posterior.

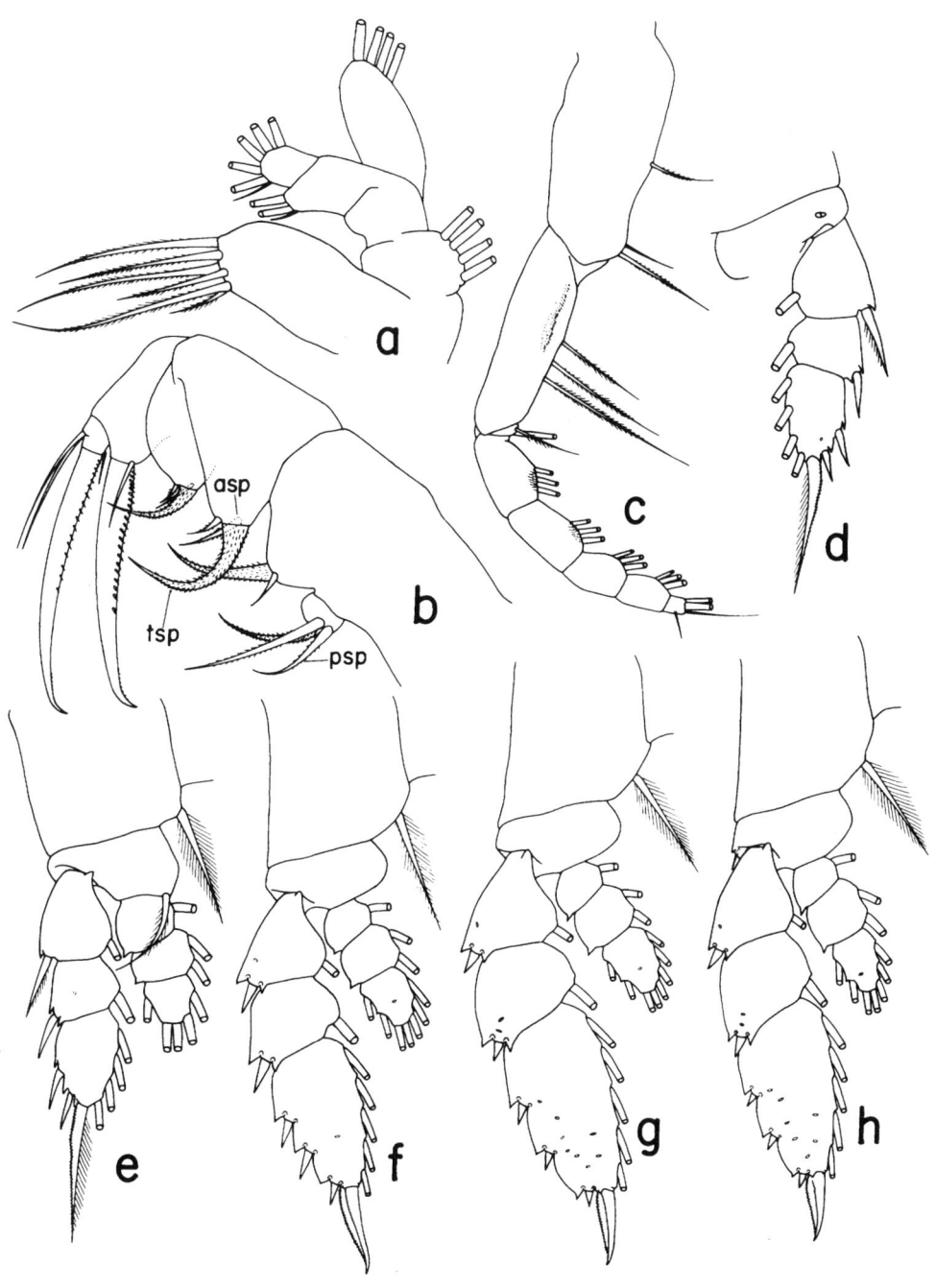

Figure 35. *Neorhabdus latus* female: **a**, left maxillule, posterior; **b**, left maxilla, posterior; **c**, right maxilliped, anterior; **d**, first leg with endopod omitted, posterior; **e**, first leg, anterior; **f**, second leg, anterior; **g**, third leg, anterior; **h**, fourth leg, anterior. asp = anterior spine; psp = posterior spine; tsp = terminal spine.

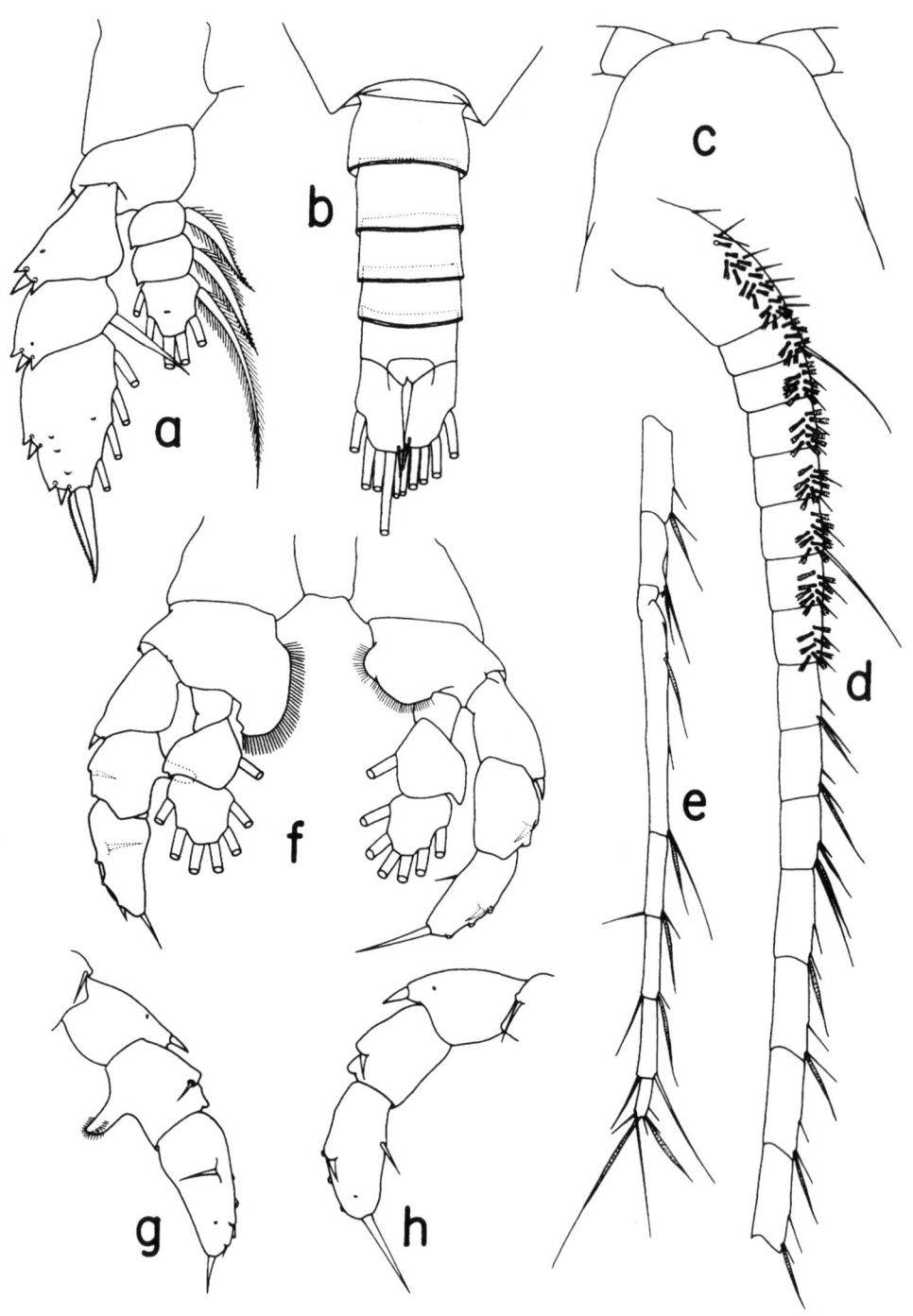

Figure 36. *Neorhabdus latus* female: **a**, fifth leg, anterior. Male: **b**, urosome, dorsal; **c**, forehead, dorsal; **d**, segments 1-16 of left antennule, ventral; **e**, segments 17-25 of left antennule, ventral; **f**, fifth pair of legs, anterior; **g**, exopod of right 5th leg, posterior; **h**, exopod of left 5th leg, posterior.

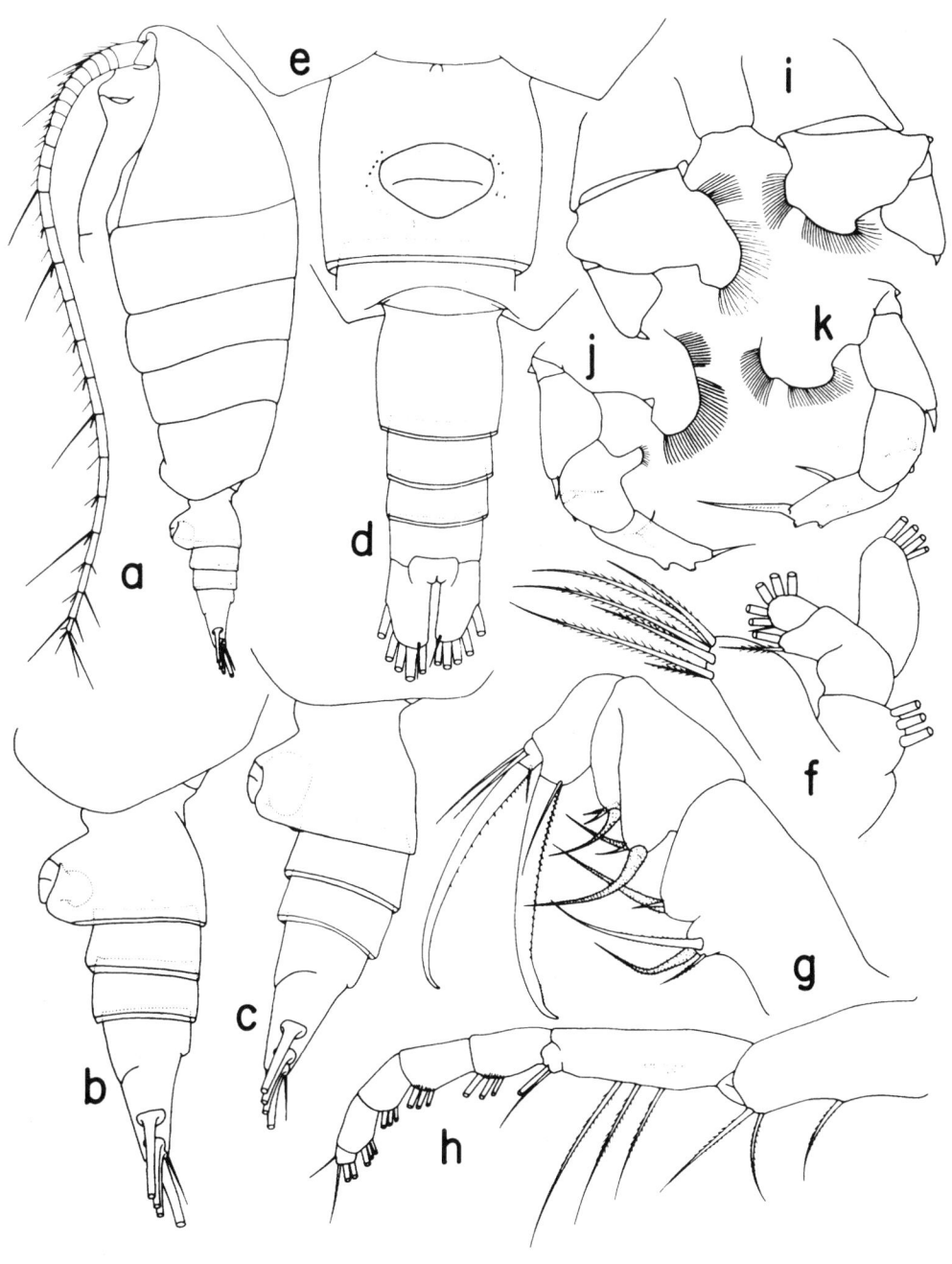

Figure 37. *Neorhabdus falciformis* female: **a**, habitus, left; **b**, urosome, left; **c**, urosome with 2nd somite deeply telescoped, left; **d**, urosome, dorsal; **e**, genital somite, ventral; **f**, left maxillule, posterior; **g**, left maxilla, posterior; **h**, right maxilliped, anterior; **i**, basipods of 5th pair of legs, anterior; **j**, basis and exopod of right 5th leg, anterior; **k**, basis and exopod of left 5th leg, anterior.

189

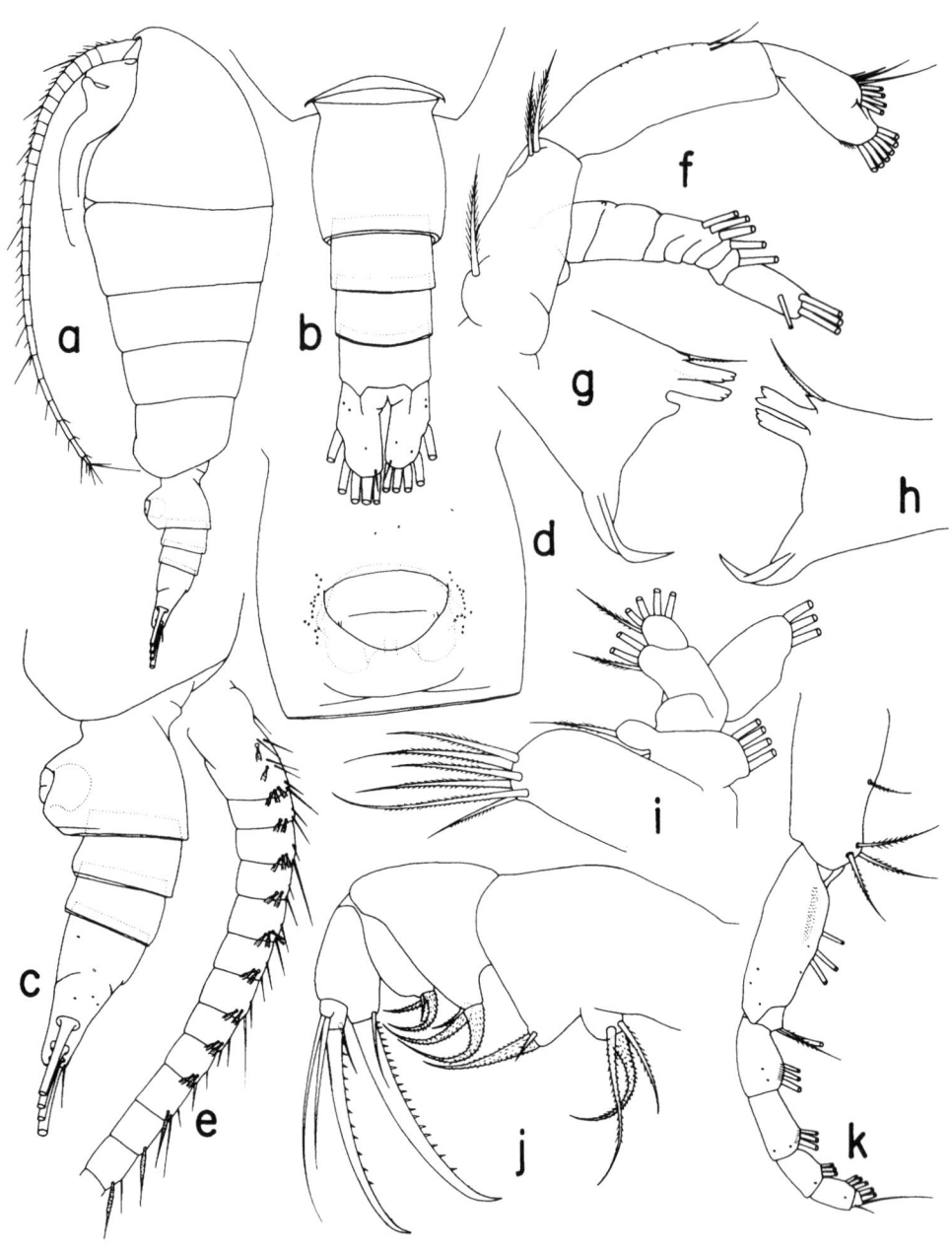

Figure 38. *Neorhabdus brevicornis*, new species, female: **a**, habitus, left; **b**, urosome, dorsal; **c**, do, left; **d**, genital somite, ventral; **e**, segments 1-12 of left antennule, ventral; **f**, left antenna, posterior; **g**, masticatory edge of right mandible, posterior; **h**, masticatory edge of left mandible, posterior; **i**, left maxillule, posterior; **j**, left maxilla, posterior; **k**, right maxilliped, anterior.

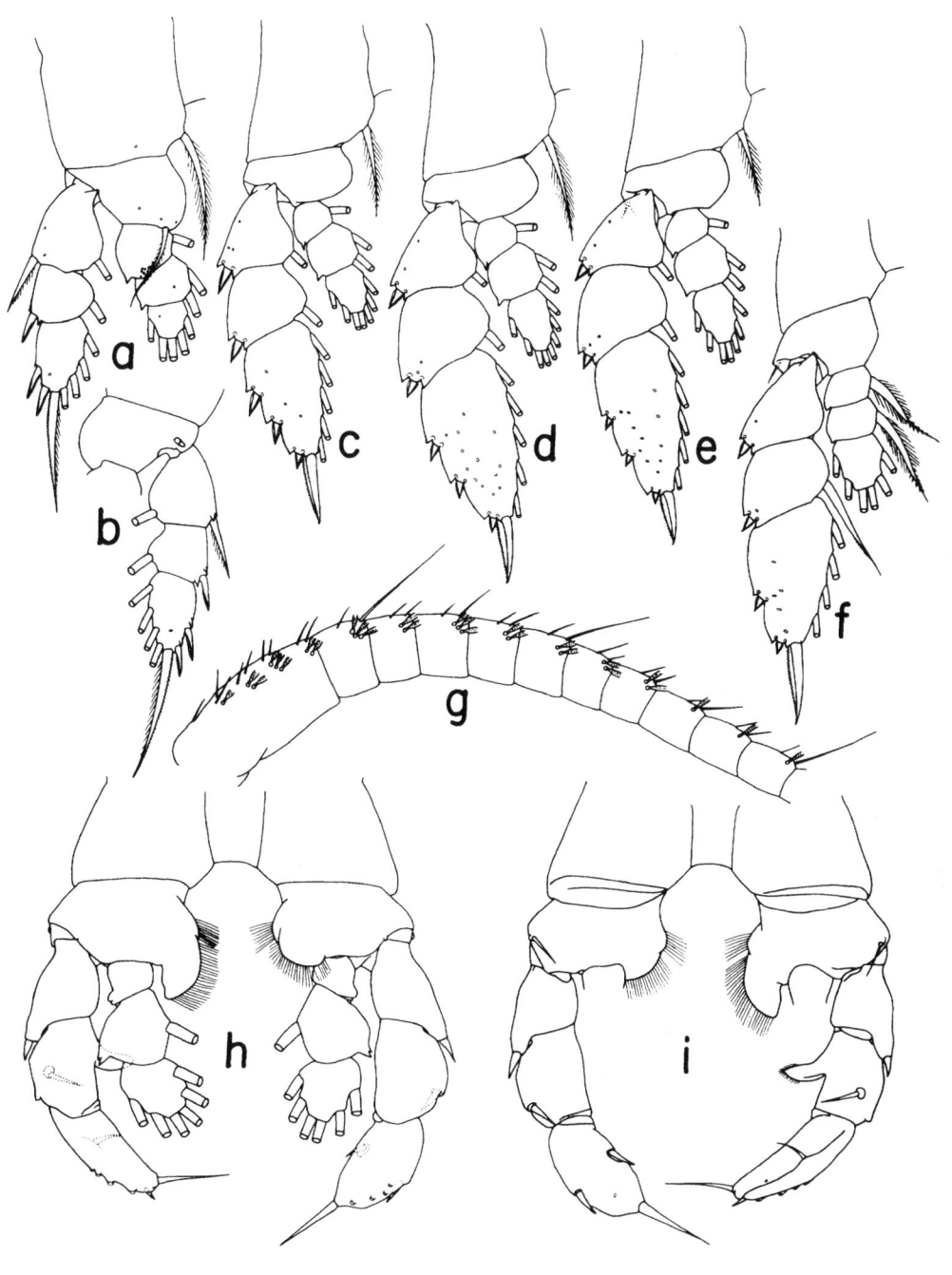

Figure 39. *Neorhabdus brevicornis*, new species, female: **a**, first leg, anterior; **b**, first leg with endopod omitted, posterior; **c**, second leg, anterior; **d**, third leg, anterior; **e**, fourth leg, anterior; **f**, fifth leg, anterior. Male: **g**, segments 1-12 of left antennule, ventral; **h**, fifth pair of legs, anterior; **i**, fifth pair of legs with endopods omitted, posterior.

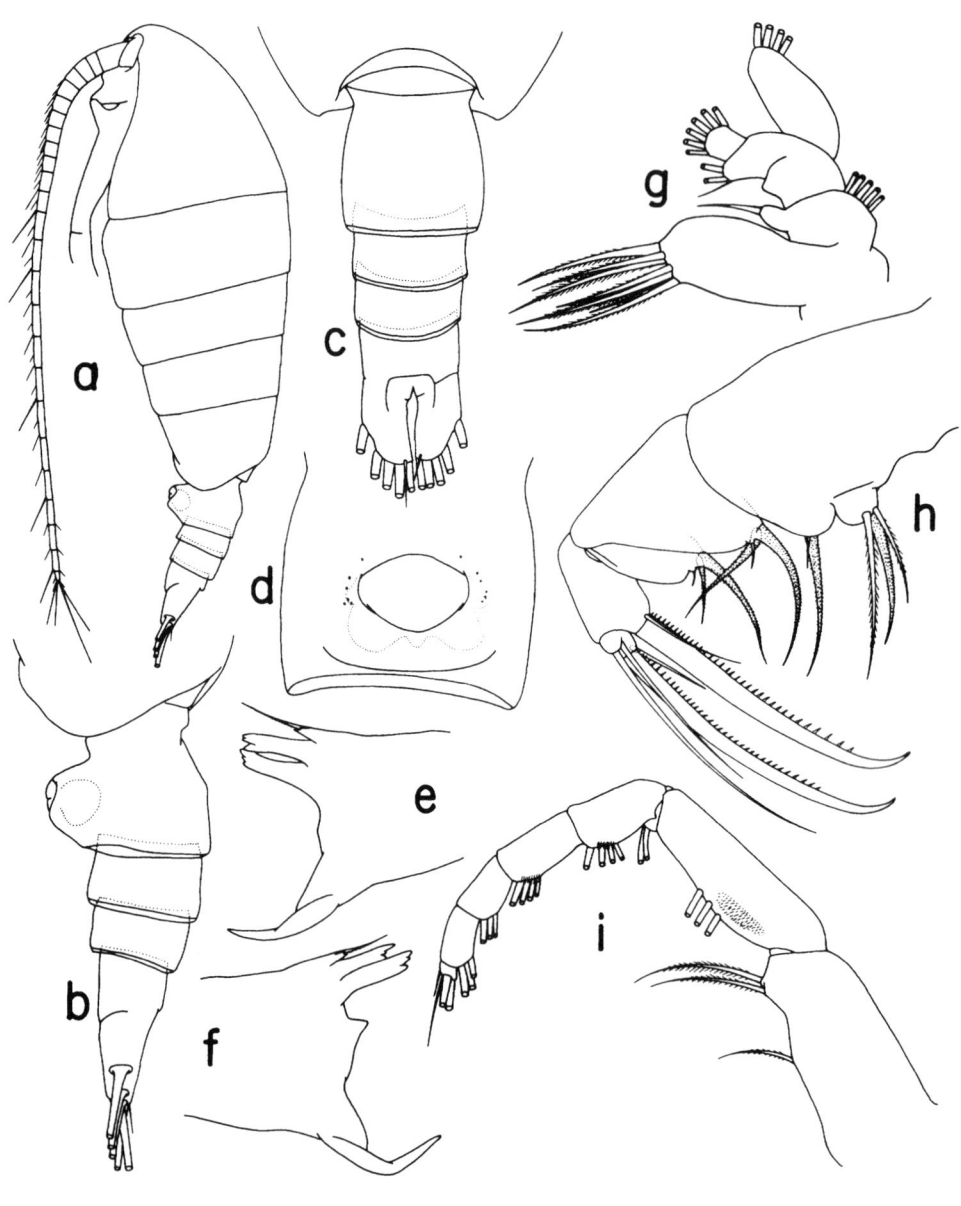

Figure 40. *Neorhabdus capitaneus*, new species, female: **a**, habitus, left; **b**, urosome, left; **c**, do, dorsal; **d**, genital somite, ventral; **e**, masticatory edge of left mandible, posterior; **f**, masticatory edge of right mandible, posterior; **g**, left maxillule, posterior; **h**, left maxilla, posterior; **i**, right maxilliped, anterior.

192

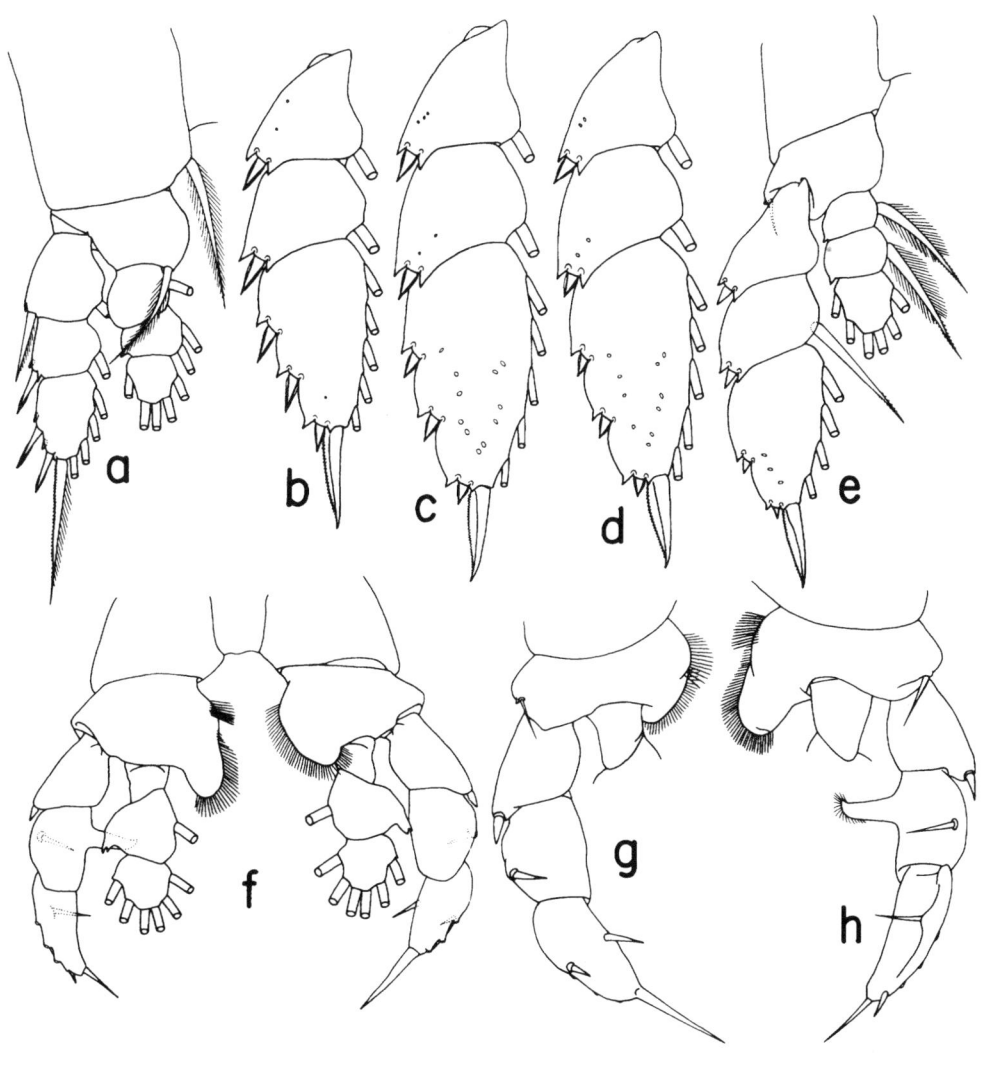

Figure 41. *Neorhabdus capitaneus*, new species, female: **a**, first leg, anterior; **b**, exopod of 2nd leg, anterior; **c**, exopod of 3rd leg, anterior; **d**, exopod of 4th leg, anterior; **e**, fifth leg, anterior. Male: **f**, fifth pair of legs, anterior; **g**, left 5th leg with distal endopodal segments omitted, posterior; **h**, right 5th leg with distal endopodal segments omitted, posterior.

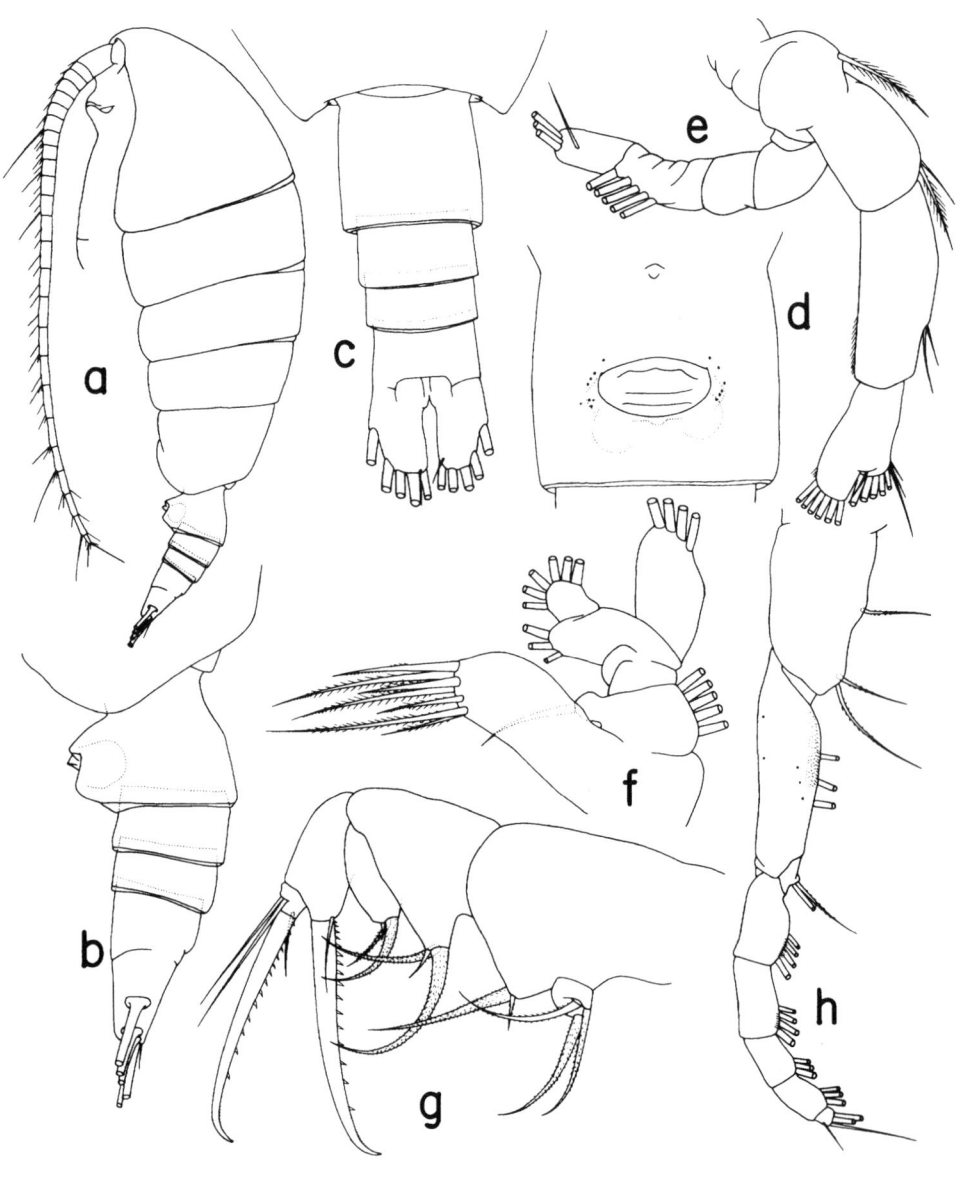

Figure 42. *Neorhabdus subcapitaneus*, new species, female: **a**, habitus, left; **b**, urosome, left; c, do, dorsal; **d**, genital somite, ventral; **e**, left antenna, posterior; **f**, left maxillule, posterior; **g**, left maxilla, posterior; **h**, right maxilliped, anterior.

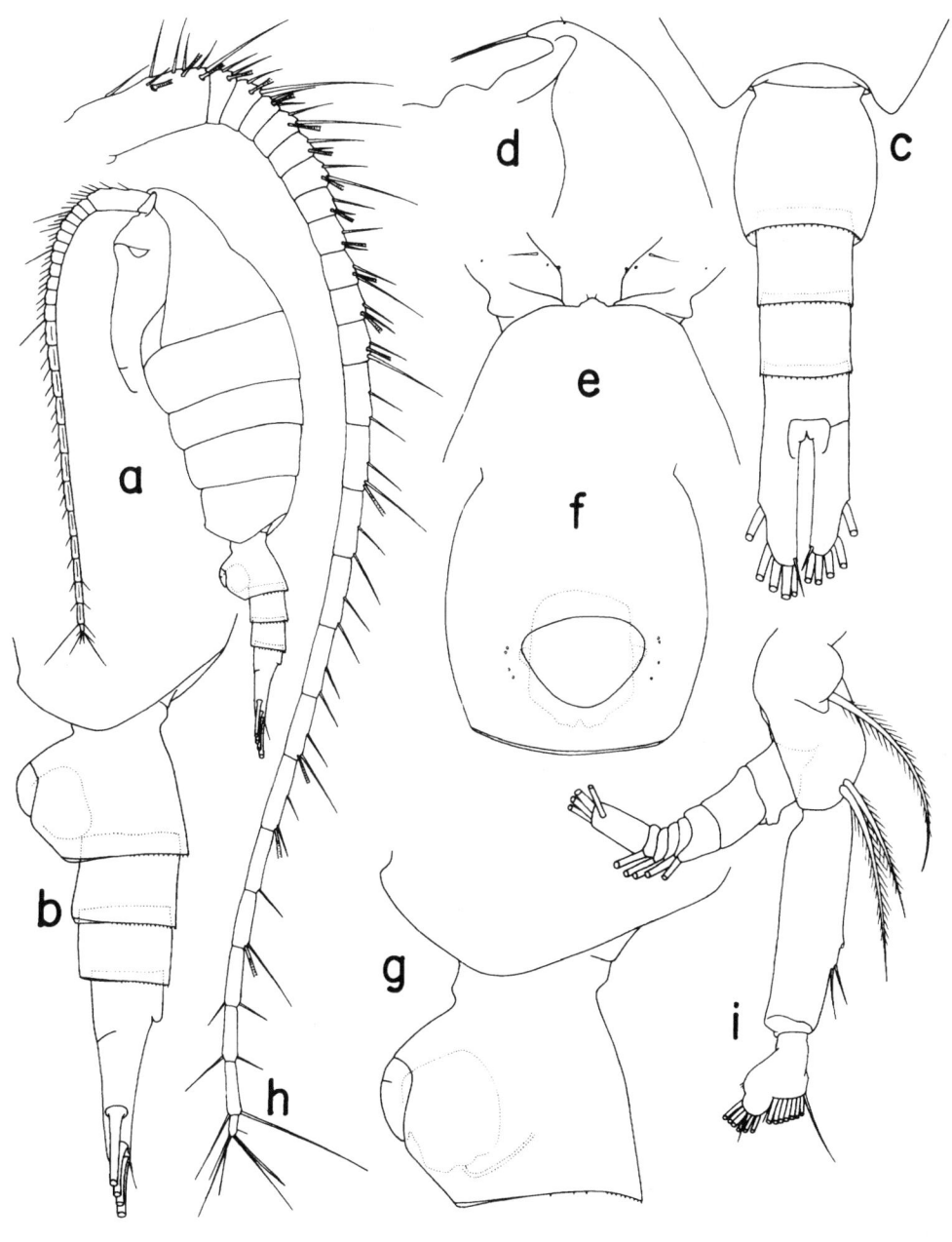

Figure 43. *Paraheterorhabdus (Paraheterorhabdus) robustus* female: **a**, habitus, left; **b**, urosome, left; **c**, do, dorsal; **d**, forehead, left; **e**, do, dorsal; **f**, genital somite, ventral; **g**, do, left; **h**, left antennule, ventral; **i**, left antenna, posterior.

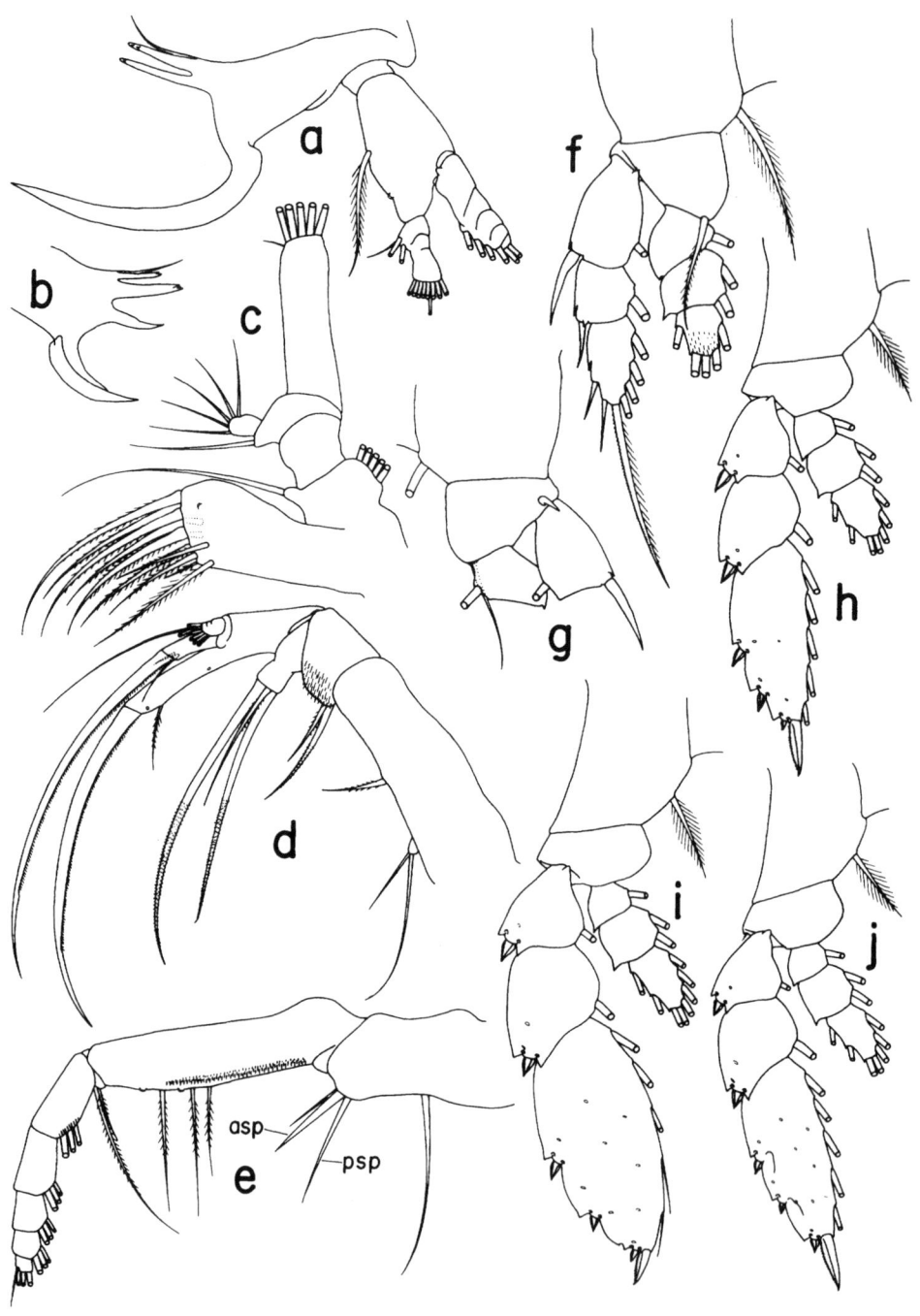

Figure 44. *Paraheterorhabdus (Paraheterorhabdus) robustus* female: **a**, left mandible, posterior; **b**, masticatory edge of right mandible, posterior; **c**, left maxillule, posterior; **d**, left maxilla, posterior; **e**, right maxilliped, anterior; **f**, first leg, anterior; **g**, first leg with distal segments of rami omitted, posterior; **h**, second leg, anterior; **i**, third leg, anterior; **j**, fourth leg, anterior. asp = anterior spine; psp = posterior spine.

196

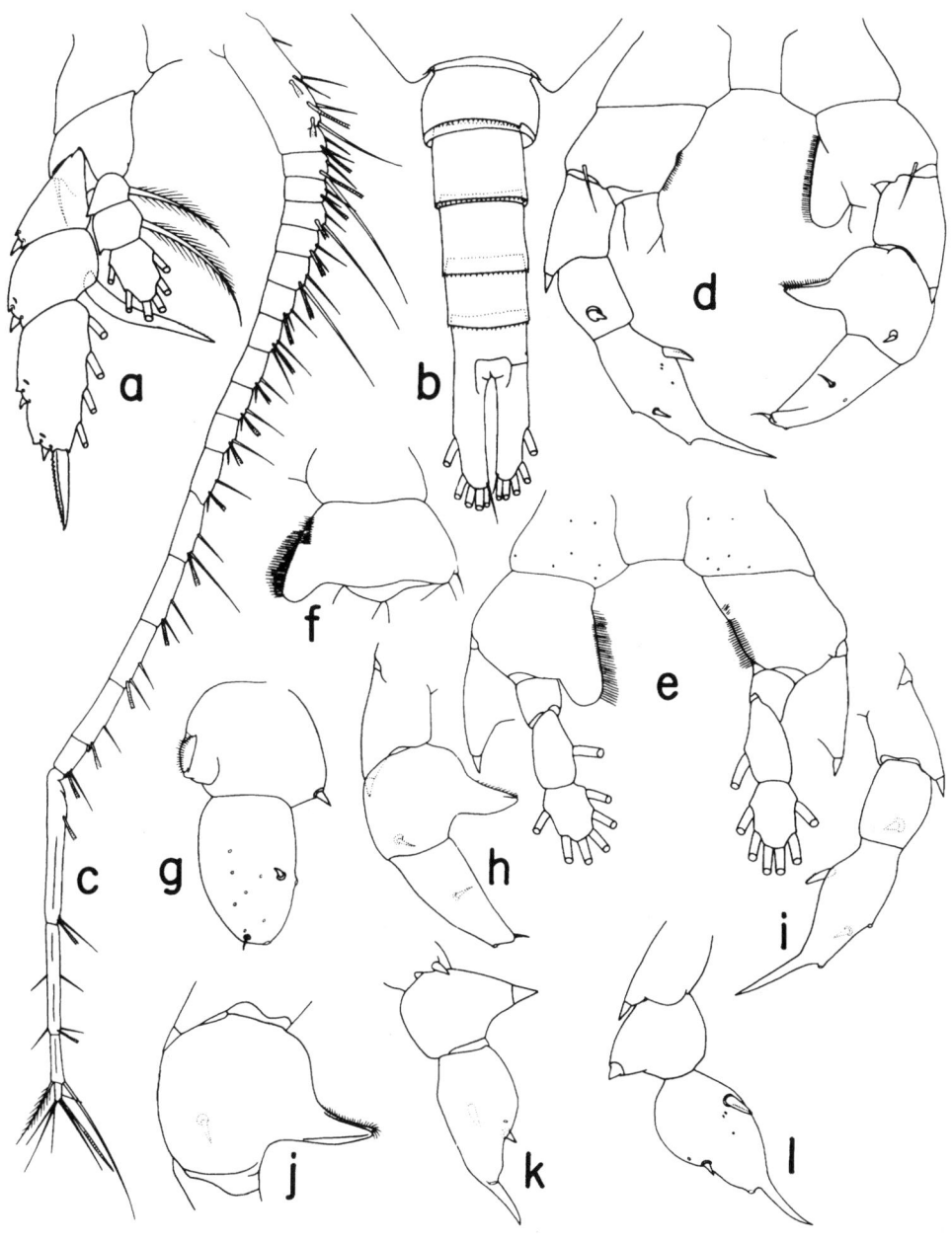

Figure 45. *Paraheterorhabdus (Paraheterorhabdus) robustus* female: **a**, fifth leg, anterior. Male: **b**, urosome, dorsal; **c**, left antennule, ventral; **d**, fifth pair of legs with endopods omitted, posterior; **e**, fifth pair of legs with distal exopodal segments omitted, anterior; **f**, basipod of right 5th leg, posterior; **g**, two distal exopodal segments of right 5th leg, posterior; **h**, exopod of right 5th leg, anterior; **i**, exopod of left 5th leg, anterior; **j**, second exopodal segment of right 5th leg, anterior; **k**, exopod of left 5th leg, anterior; **l**, do, posterior.

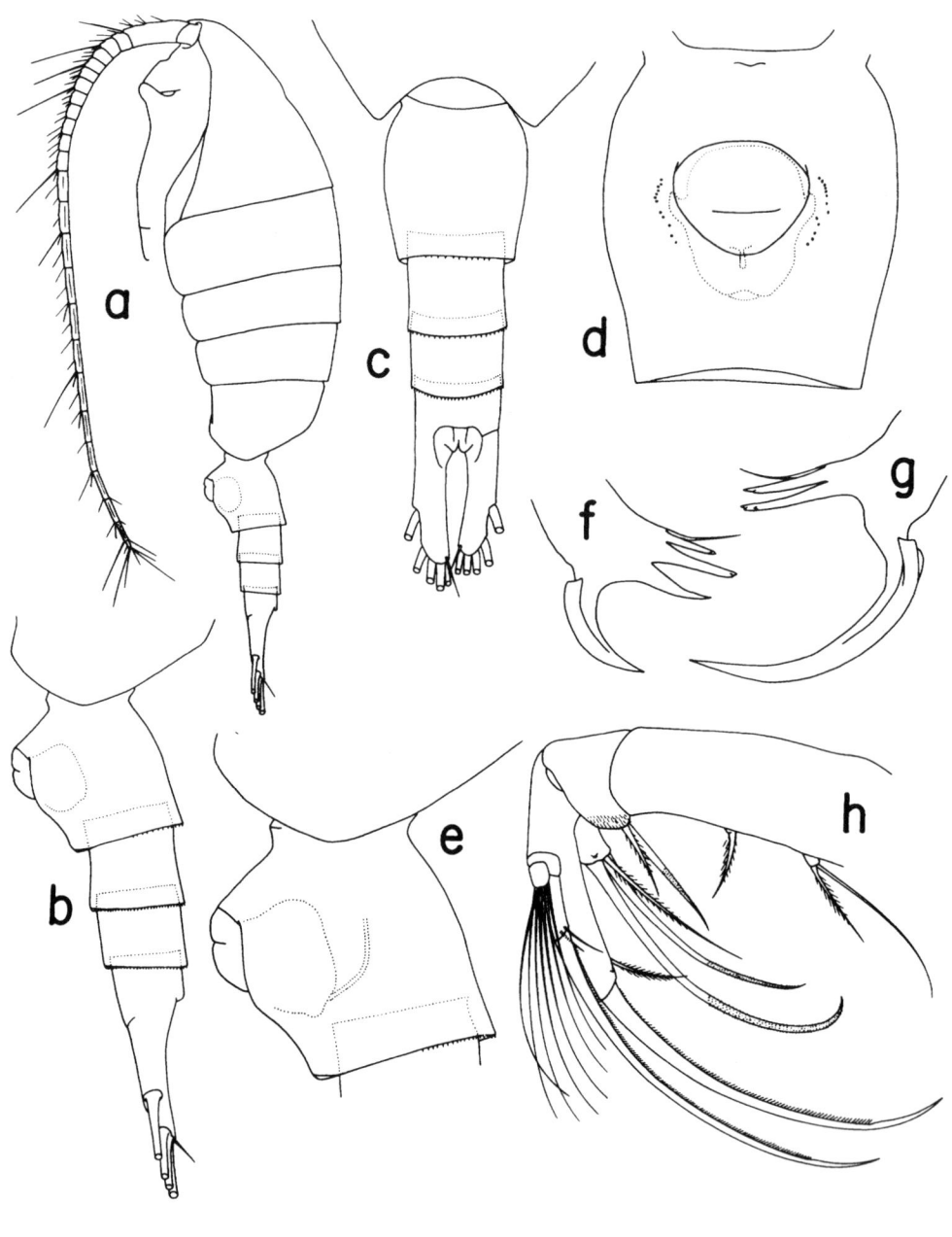

Figure 46. *Paraheterorhabdus* (*Paraheterorhabdus*) *longispinus* female: **a**, habitus, left; **b**, urosome, left; **c**, do, dorsal; **d**, genital somite, ventral; **e**, do, left; **f**, masticatory edge of right mandible, posterior; **g**, masticatory edge of left mandible, posterior; **h**, left maxilla, posterior.

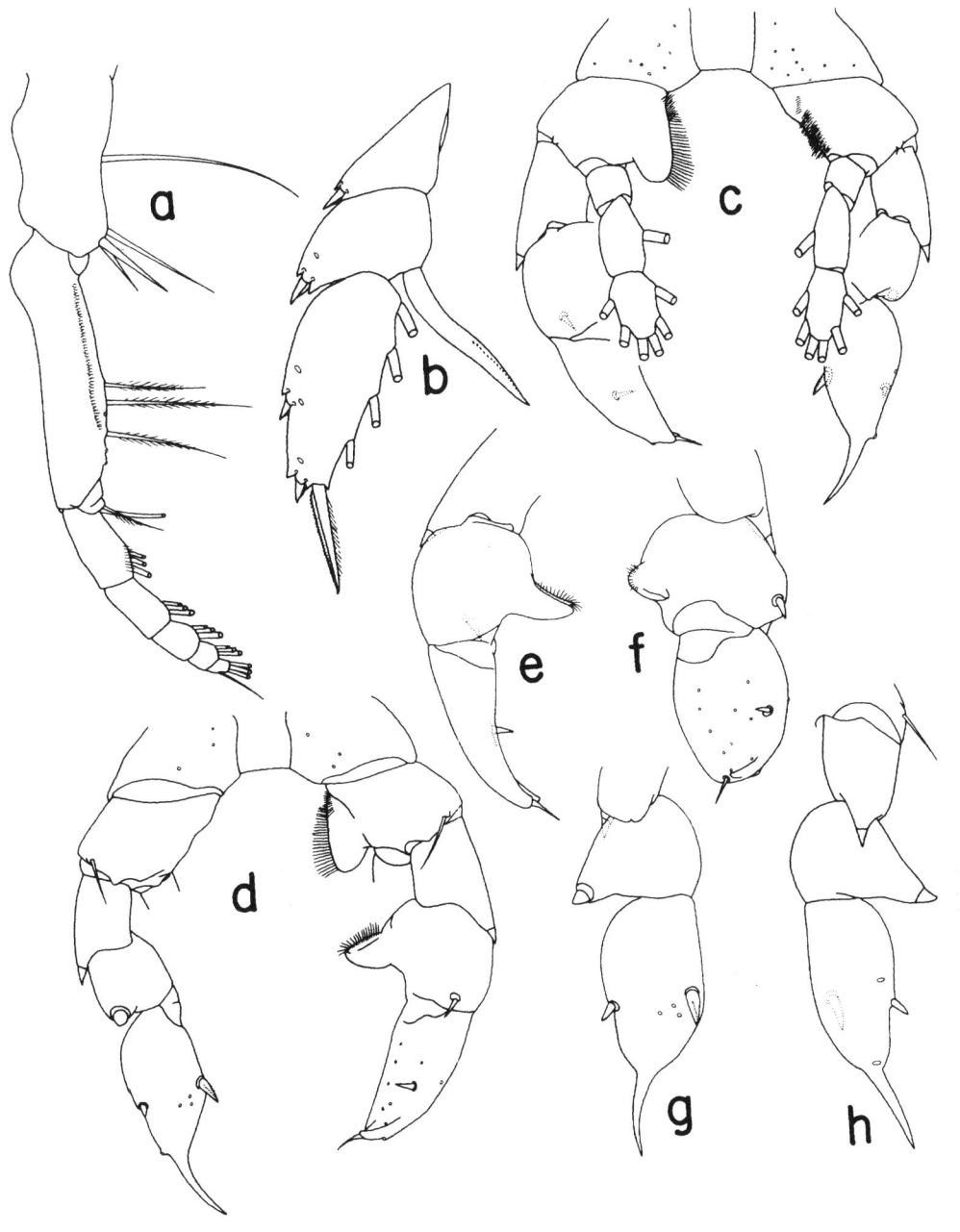

Figure 47. *Paraheterorhabdus* (*Paraheterorhabdus*) *longispinus* female: **a**, right maxilliped, anterior; **b**, exopod of 5th leg, anterior. Male: **c**, fifth pair of legs, anterior; **d**, fifth pair of legs with endopods omitted, posterior; **e**, exopod of right 5th leg, anterior; **f**, do, posterior, tilted clockwise; **g**, exopod of left 5th leg, posterior; **h**, do, anterior.

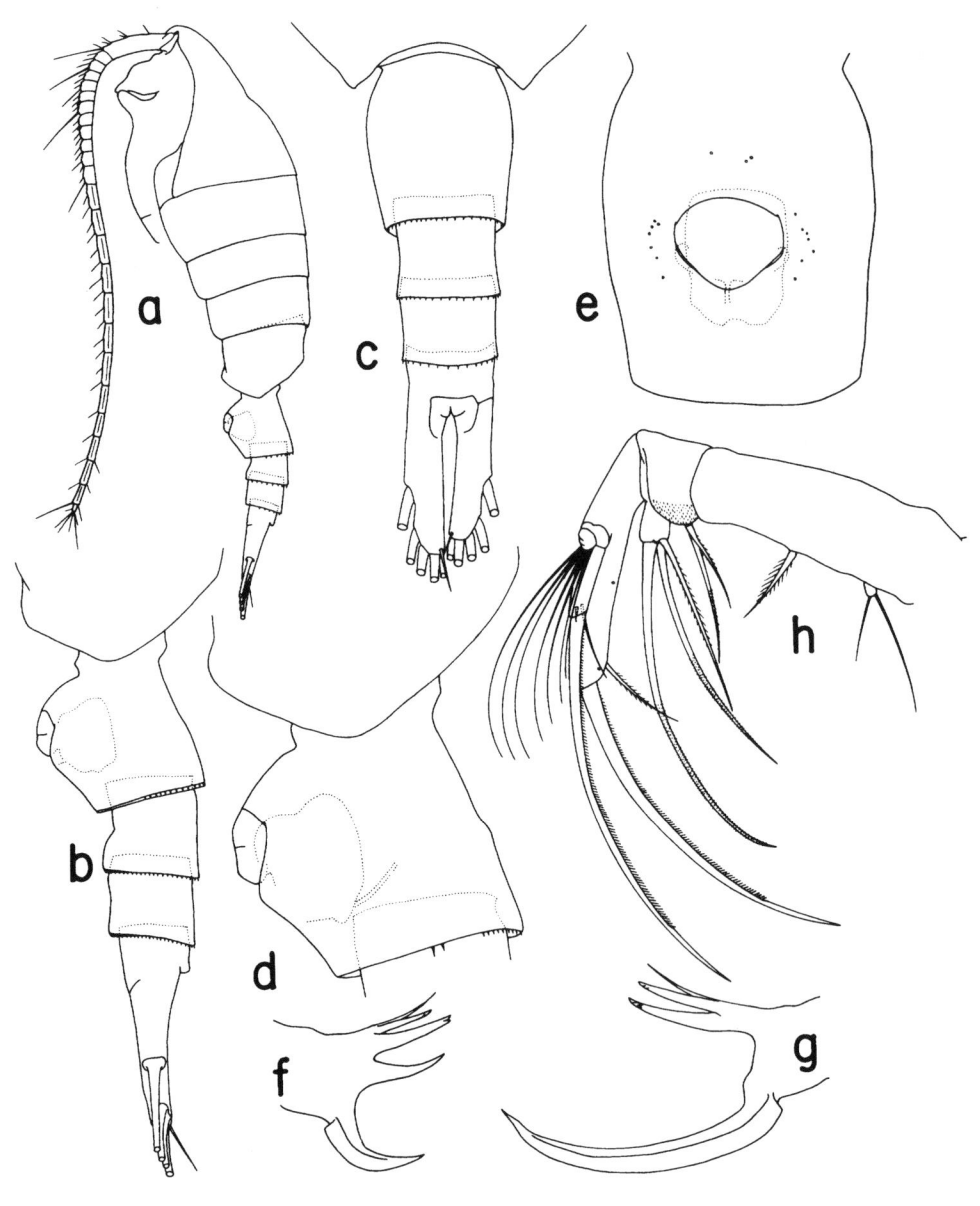

Figure 48. *Paraheterorhabdus* (*Paraheterorhabdus*) *farrani* female: **a**, habitus, left; **b**, urosome, left; **c**, do, dorsal; **d**, genital somite, left; **e**, do, ventral; **f**, masticatory edge of right mandible, posterior; **g**, masticatory edge of left mandible, posterior; **h**, left maxilla, posterior.

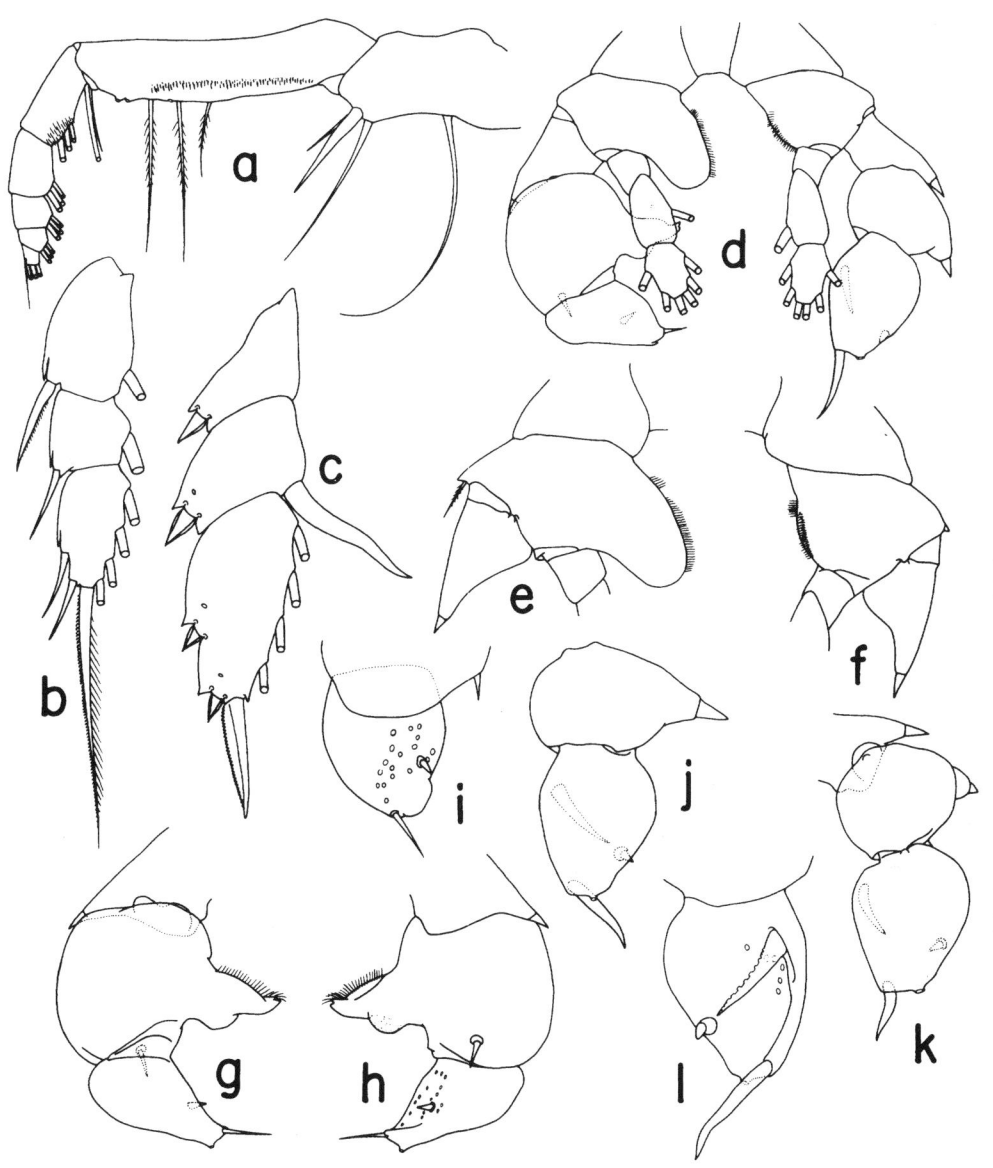

Figure 49. *Paraheterorhabdus (Paraheterorhabdus) farrani* female: **a**, right maxilliped, anterior; **b**, exopod of 1st leg, anterior; **c**, exopod of 5th leg, anterior. Male: **d**, fifth pair of legs, anterior; **e**, basipod of right 5th leg, anterior; **f**, basipod of left 5th leg, anterior; **g**, exopod of right 5th leg, anterior; **h**, do, posterior; **i**, distal exopodal segment of right 5th leg, posterior, tilted clockwise; **j**, distal exopodal segments of left 5th leg, anterior; **k**, exopod of left 5th leg, anterior; **l**, distal exopodal segment of left 5th leg, posterior.

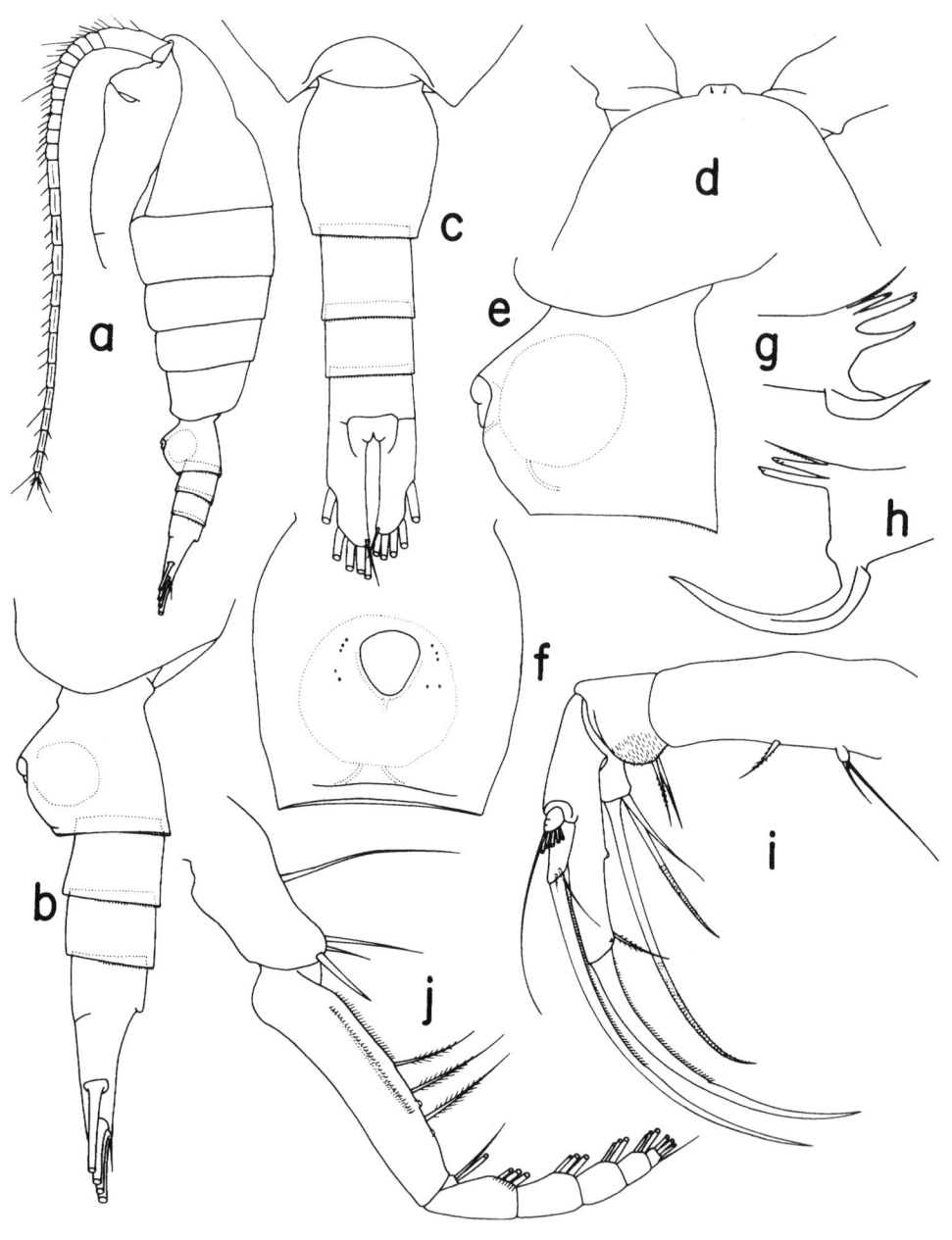

Figure 50. *Paraheterorhabdus (Paraheterorhabdus) illgi*, new species, female: **a**, habitus, left; **b**, urosome, left; **c**, do, dorsal; **d**, forehead, dorsal; **e**, genital somite, left; **f**, do, ventral; **g**, masticatory edge of right mandible, posterior; **h**, masticatory edge of left mandible, posterior; **i**, left maxilla, posterior; **j**, right maxilliped, anterior.

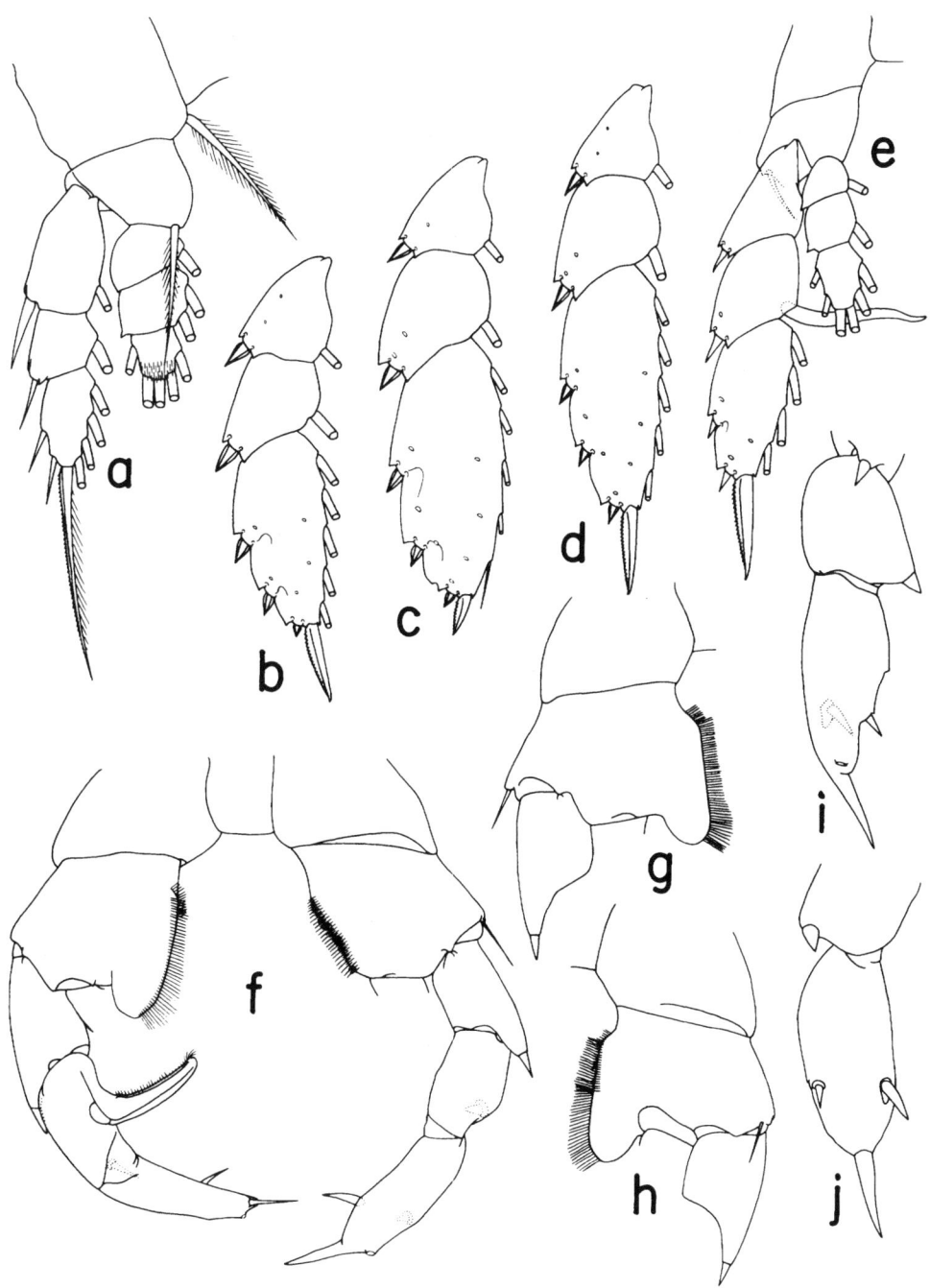

Figure 51. *Paraheterorhabdus (Paraheterorhabdus) illgi*, new species, female: **a**, first leg, anterior; **b**, exopod of 2nd leg, anterior; **c**, exopod of 3rd leg, anterior; **d**, exopod of 4th leg, anterior; **e**, fifth leg, anterior. Male: **f**, fifth pair of legs with endopods omitted, anterior; **g**, basipod of right 5th leg, anterior; **h**, do, posterior; **i**, exopod of left 5th leg, anterior; **j**, distal exopodal segments of left 5th leg, posterior.

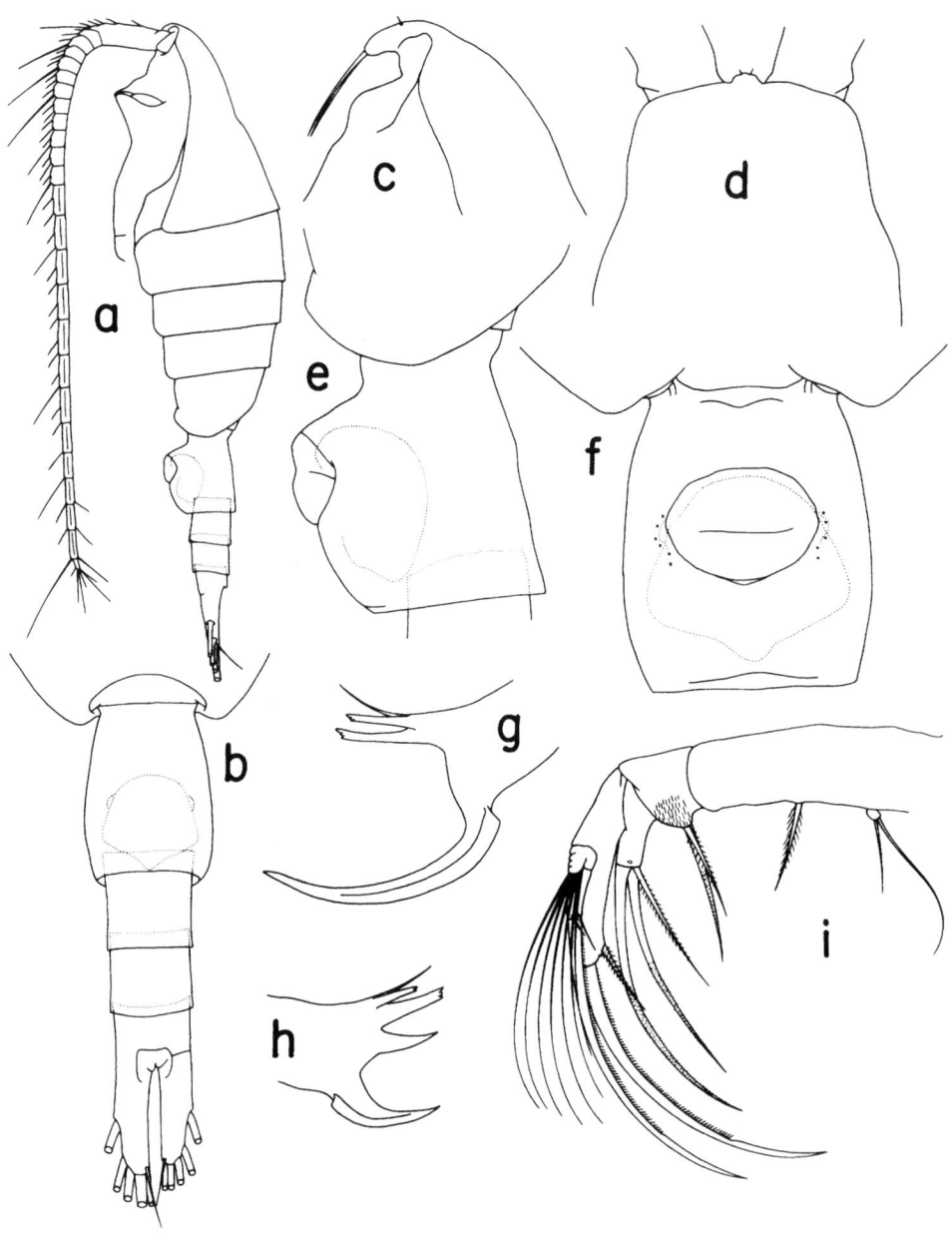

Figure 52. *Paraheterorhabdus* (*Paraheterorhabdus*) *medianus* female: **a**, habitus, left, **b**, urosome, dorsal; **c**, forehead, left; **d**, do, dorsal; **e**, genital somite, left; **f**, do, ventral; **g**, masticatory edge of left mandible, posterior; **h**, masticatory edge of right mandible; posterior; **i**, left maxilla, posterior.

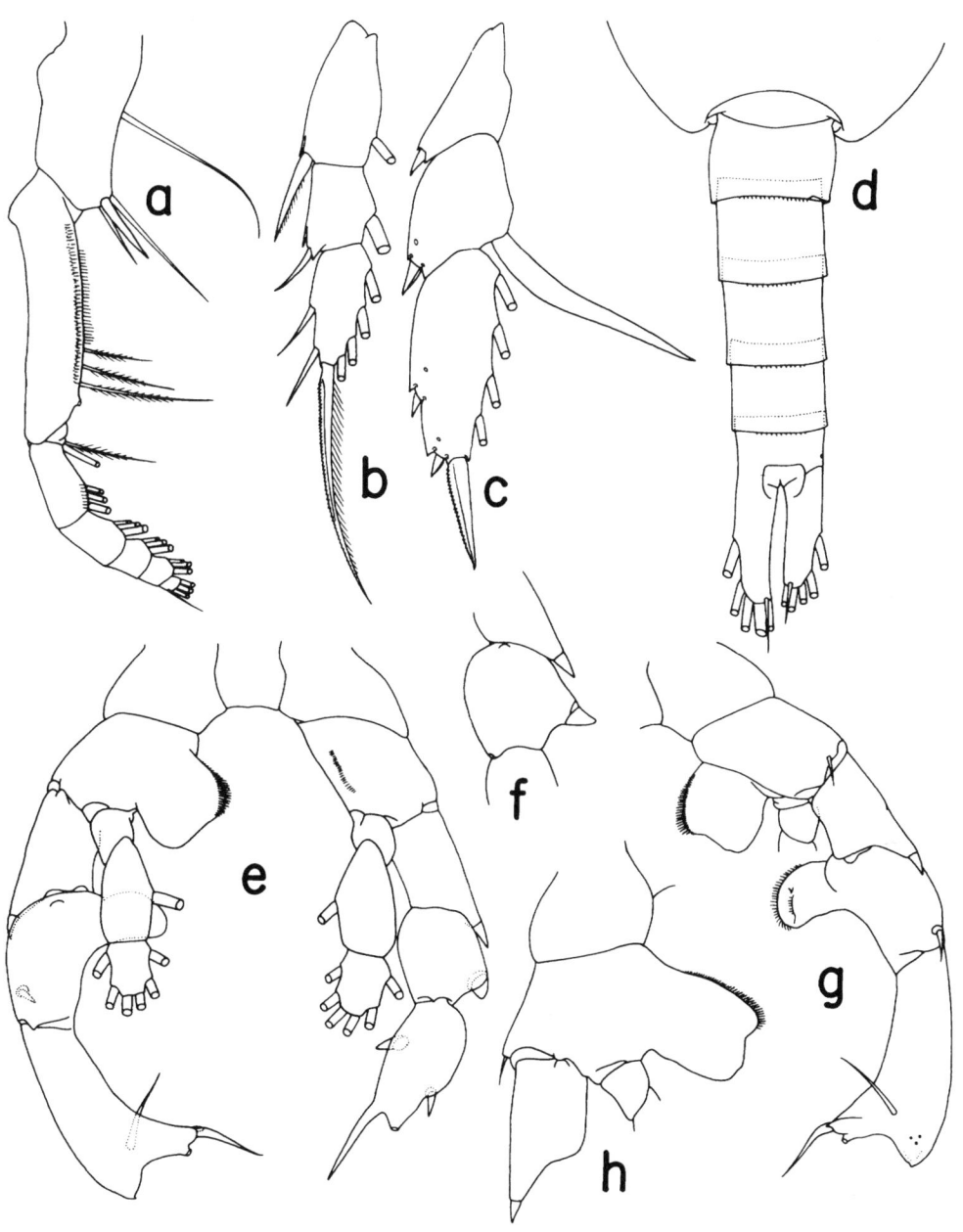

Figure 53. *Paraheterorhabdus* (*Paraheterorhabdus*) *medianus* female: **a**, right maxilliped, anterior; **b**, exopod of 1st leg, anterior; **c**, exopod of 5th leg, anterior. Male: **d**, urosome, dorsal; **e**, fifth pair of legs, anterior; **f**, second exopodal segment of left 5th leg, anterior; **g**, right 5th leg with endopod omitted, posterior; **h**, basipod of right 5th leg, anterior.

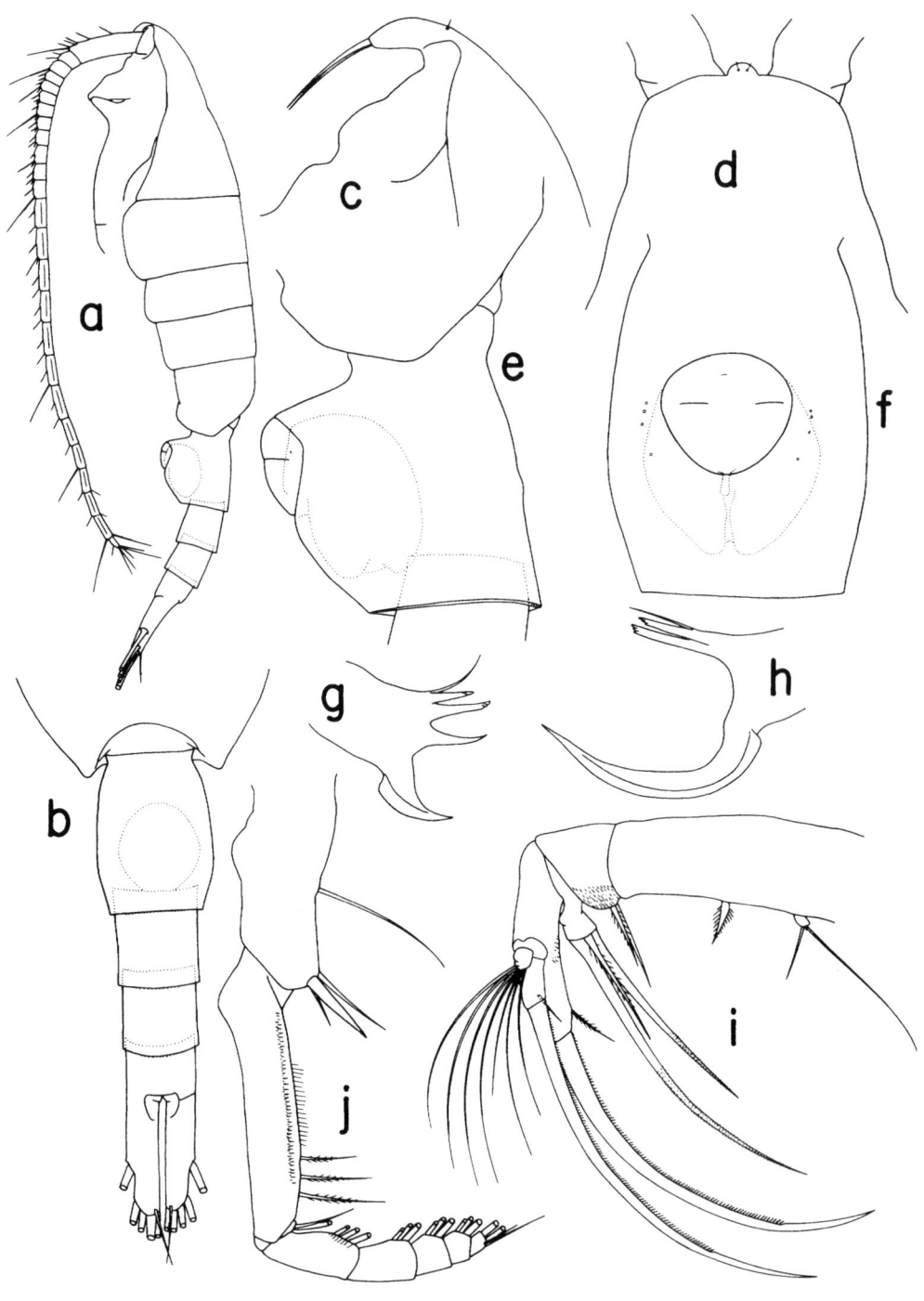

Figure 54. *Paraheterorhabdus (Paraheterorhabdus) vipera* female: **a**, habitus, left; **b**, urosome, dorsal; **c**, forehead, left; **d**, do, dorsal; **e**, genital somite, left; **f**, do, ventral; **g**, masticatory edge of right mandible, posterior; **h**, masticatory edge of left mandible, posterior; **i**, left maxilla, posterior; **j**, right maxilliped, anterior.

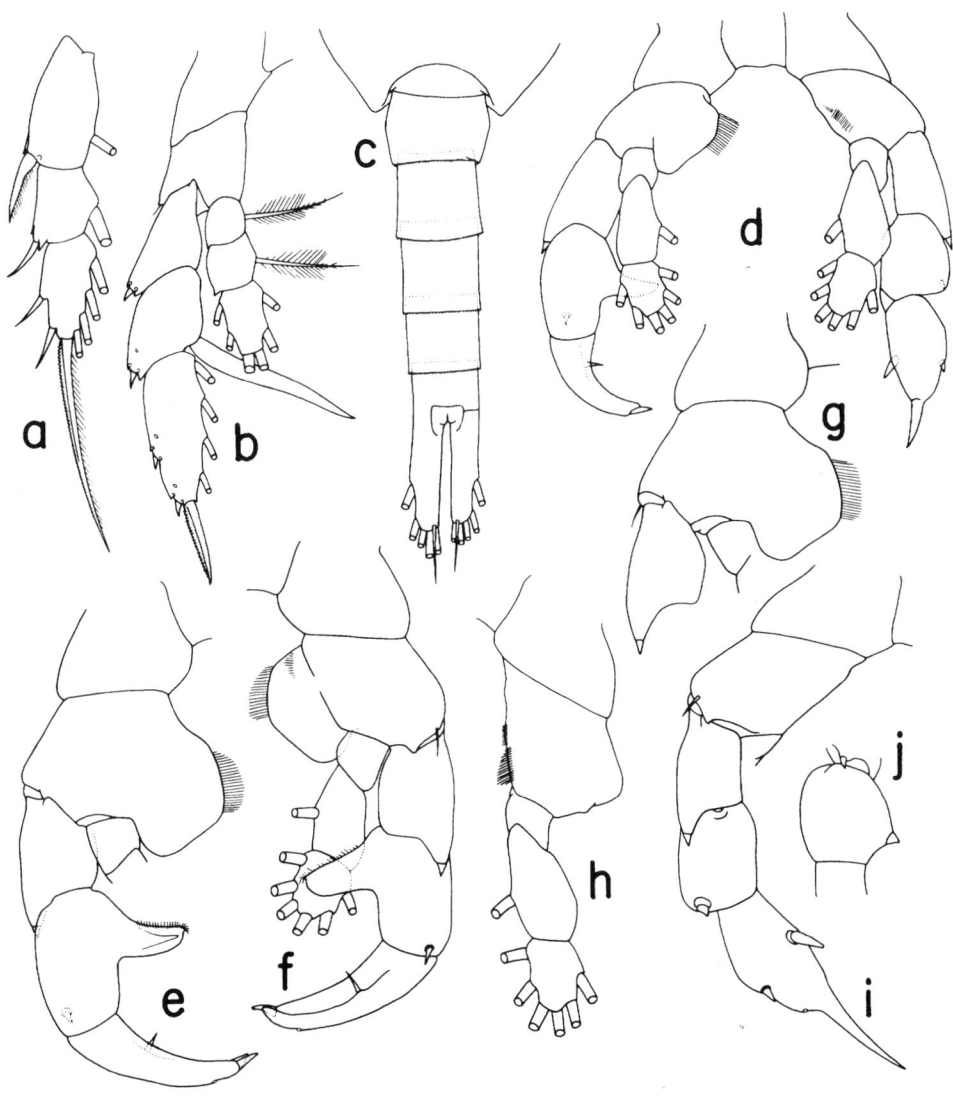

Figure 55. *Paraheterorhabdus (Paraheterorhabdus) vipera* female: **a**, exopod of 1st leg, anterior; **b**, fifth leg, anterior. Male: **c**, urosome, dorsal; **d**, fifth pair of legs, anterior; **e**, right 5th leg with endopod omitted, anterior; **f**, right 5th leg, posterior; **g**, basipod of right 5th leg, anterior; **h**, left 5th leg with exopod omitted, anterior; **i**, left 5th leg with endopod omitted, posterior; **j**, second exopodal segment of left 5th leg, anterior.

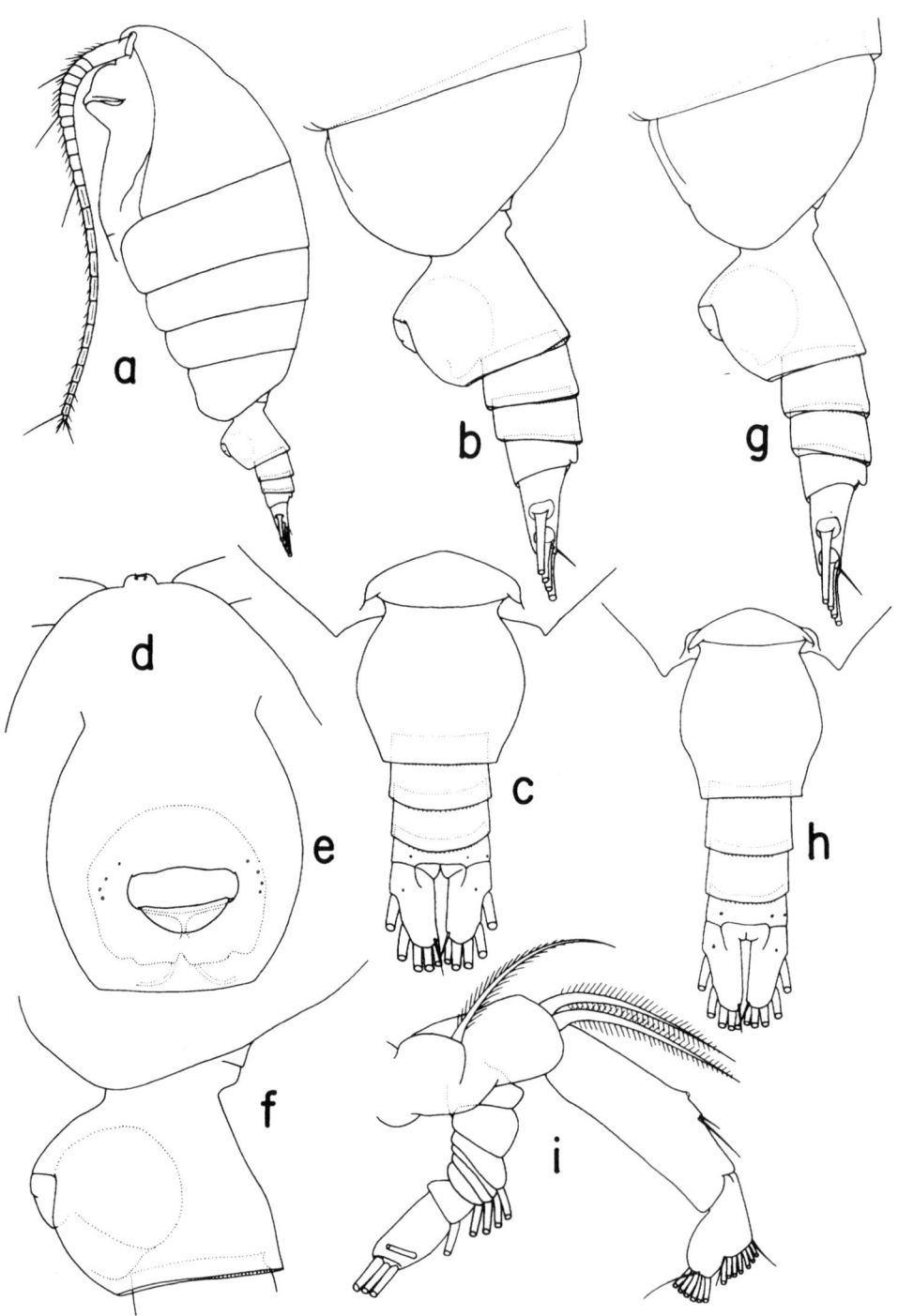

Figure 56. *Paraheterorhabdus* (*Antirhabdus*) *compactus* female: **a**, habitus, left; **b**, urosome of a small specimen (PL = 1.96), left; **c**, do, dorsal; **d**, forehead, dorsal; **e**, genital somite, ventral; **f**, do, left; **g**, urosome of a large specimen (PL = 2.32), left; **h**, do, dorsal; **i**, left antenna, posterior.

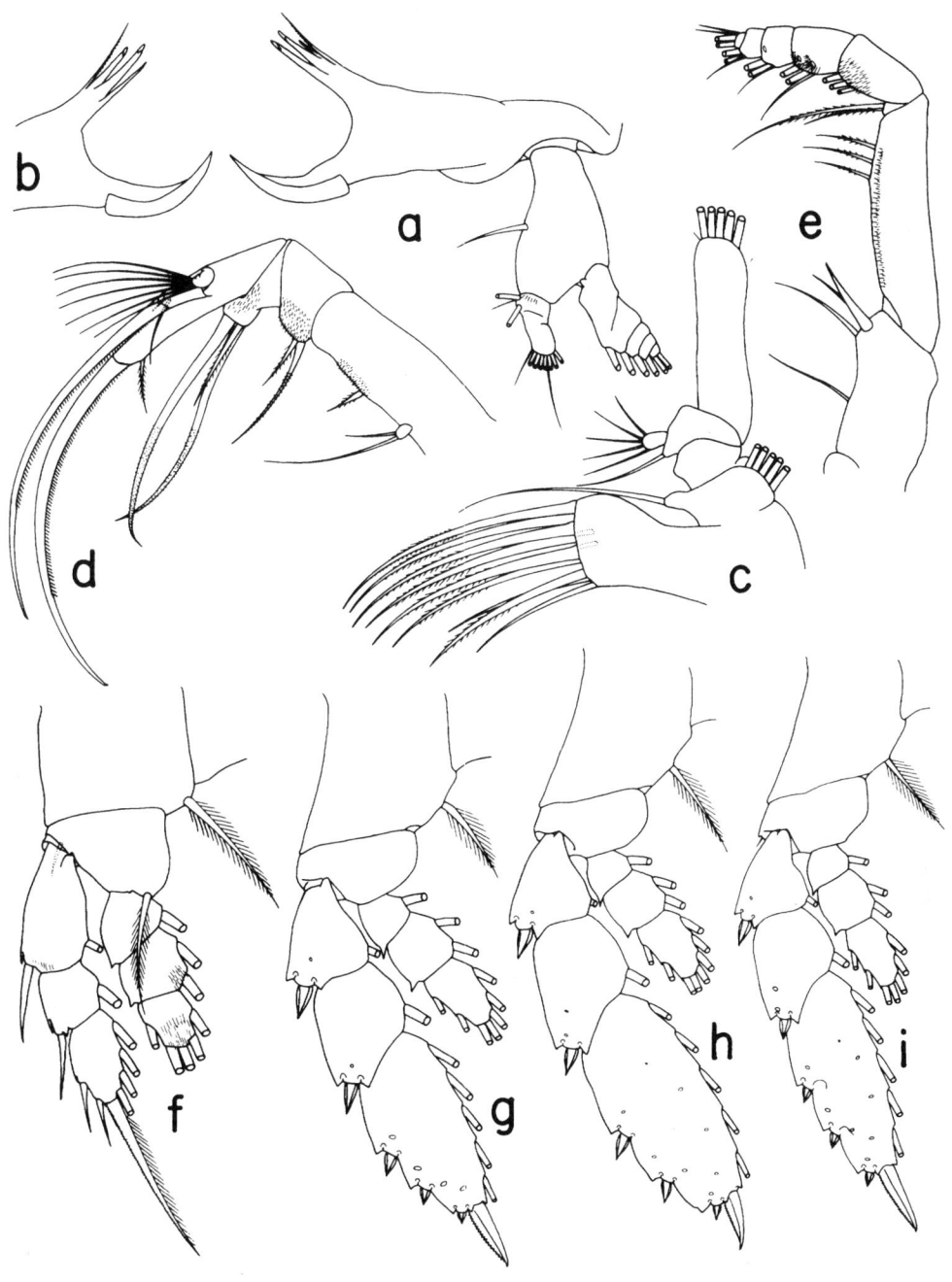

Figure 57. *Paraheterorhabdus* (*Antirhabdus*) *compactus* female: **a**, left mandible, posterior; **b**, masticatory edge of right mandible, posterior; **c**, left maxillule, posterior; **d**, left maxilla, posterior; **e**, right maxilliped, anterior; **f**, first leg, anterior; **g**, second leg, anterior; **h**, third leg, anterior; **i**, fourth leg, anterior.

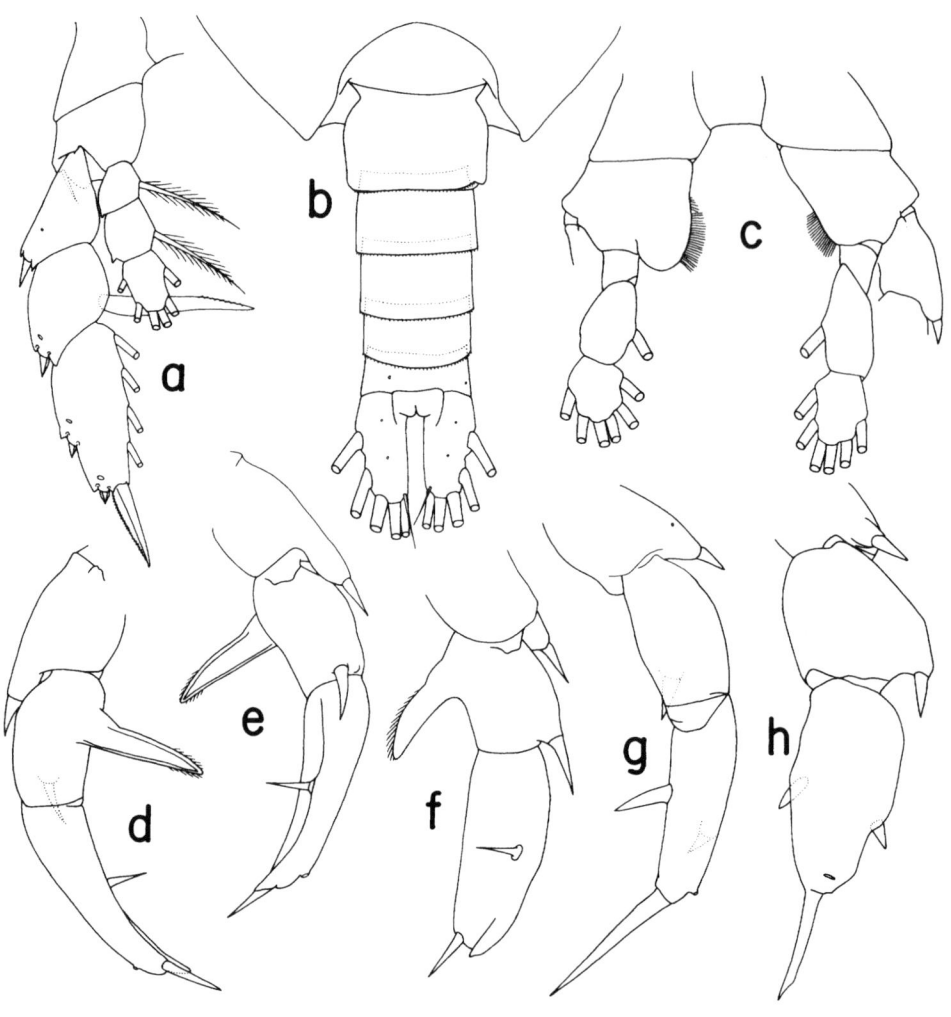

Figure 58. *Paraheterorhabdus (Antirhabdus) compactus* female: **a**, fifth leg, anterior. Male: **b**, urosome, dorsal; **c**, fifth pair of legs with exopods omitted, anterior; **d**, exopod of right 5th leg, anterior; **e**, do, posterior; **f**, do, tilted clockwise; **g**, exopod of left 5th leg, anterior; **h**, do, tilted counterclockwise.

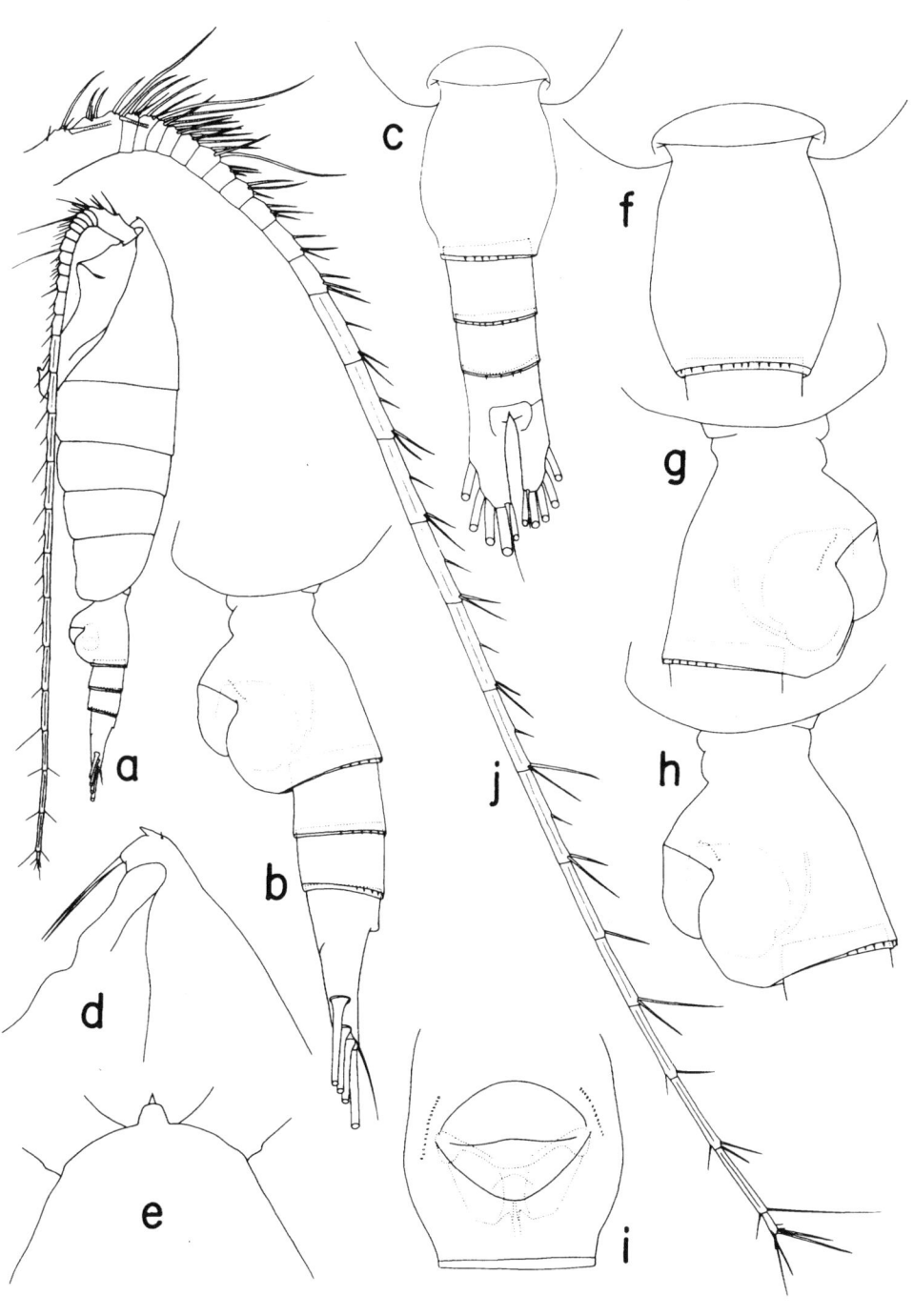

Figure 59. *Heterorhabdus spinifrons* female (Atlantic specimen): **a**, habitus, left; **b**, urosome, left; **c**, do, dorsal; **d**, forehead, left; **e**, do, dorsal; **f**, genital somite, dorsal; **g**, do, right; **h**, do, left; **i**, do, ventral; **j**, left antennule, ventral.

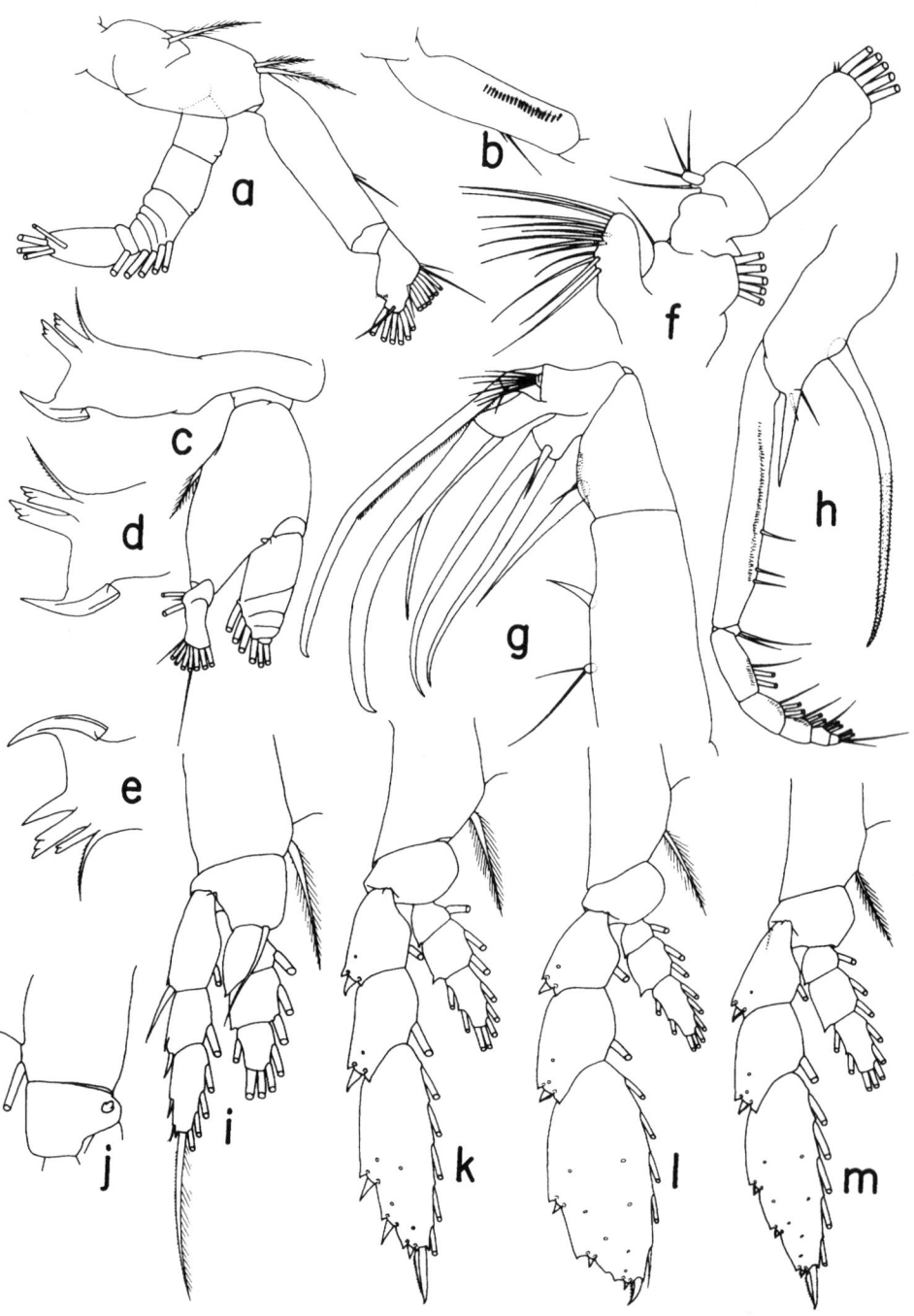

Figure 60. *Heterorhabdus spinifrons* female: **a**, left antenna, posterior; **b**, first endopodal segment of left antenna, anterior; **c**, left mandible, posterior; **d**, masticatory edge of left mandible, posterior; **e**, masticatory edge of right mandible, posterior; **f**, left maxillule, posterior; **g**, left maxilla, posterior; **h**, right maxilliped, anterior; **i**, first leg, anterior; **j**, basipod of 1st leg, posterior; **k**, second leg, anterior; **l**, third leg, anterior; **m**, fourth leg, anterior.

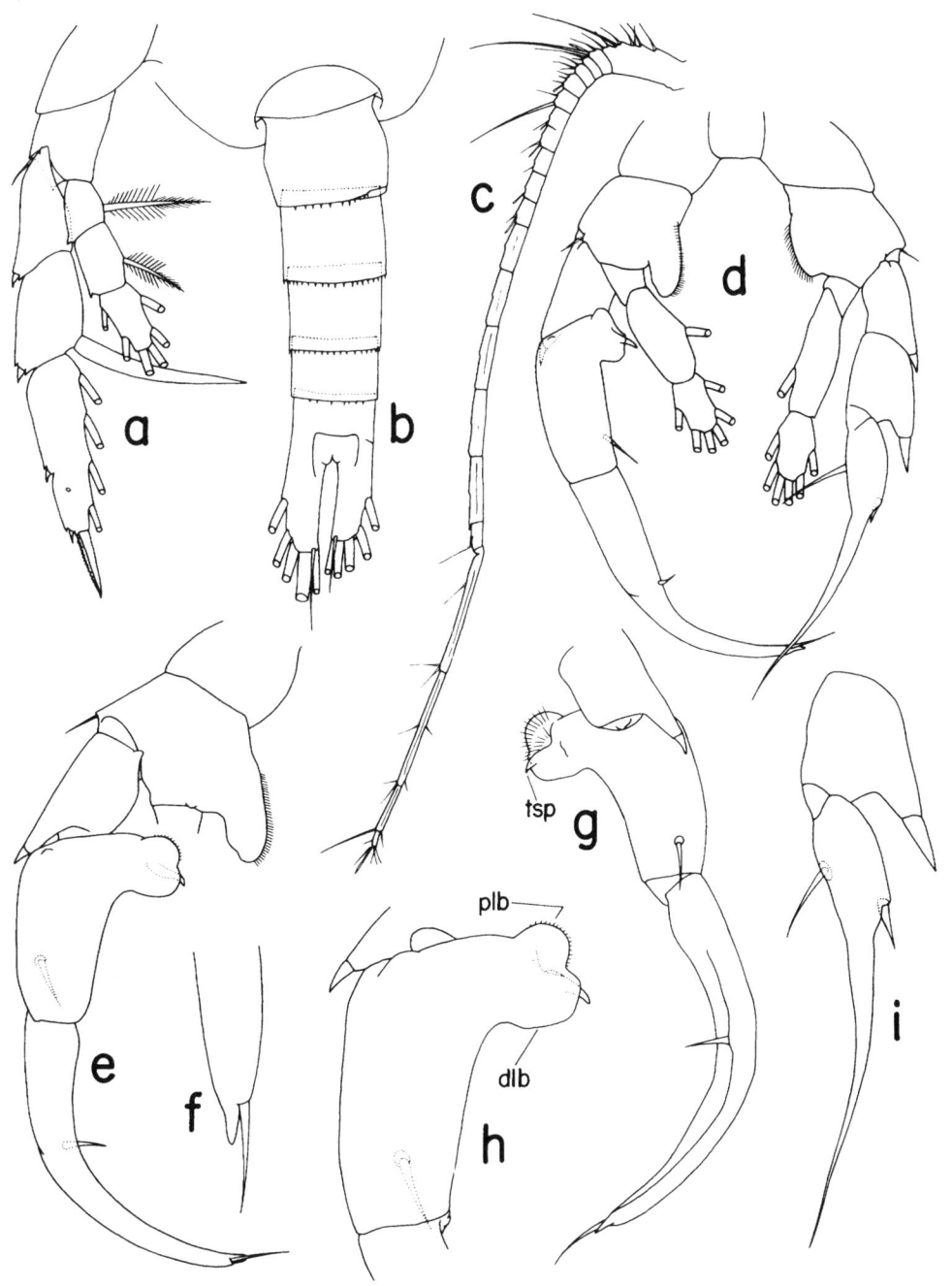

Figure 61. *Heterorhabdus spinifrons* female: **a**, fifth leg, anterior. Male: **b**, urosome, dorsal; **c**, left antennule, dorsal; **d**, fifth pair of legs, anterior; **e**, right 5th leg with endopod omitted, anterior; **f**, distal end of exopod of right 5th leg, lateral; **g**, exopod of right 5th leg, posterior; **h**, second exopodal segment of right 5th leg, anterior; **i**, distal exopodal segments of left 5th leg, anterior. dlb = distal lobe; plb = proximal lobe; tsp = terminal spiniform process.

213

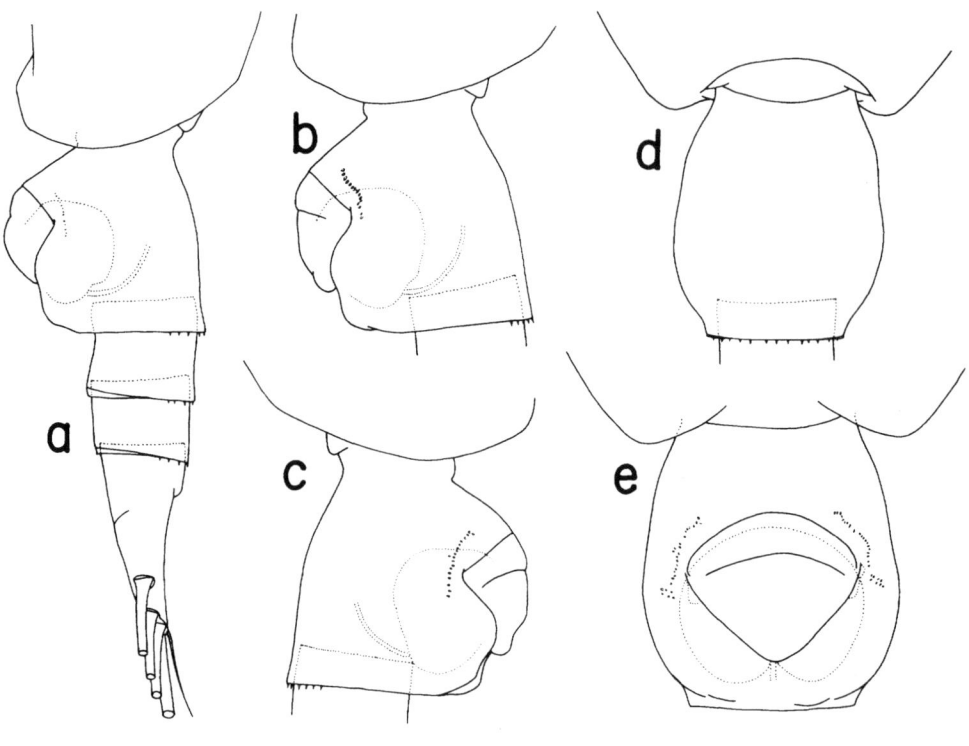

Figure 62. *Heterorhabdus spinifrons* female (specimen from off west coast of North America): **a**, urosome, left; **b**, genital somite, left; **c**, do, right; **d**, do, dorsal; **e**, do, ventral.

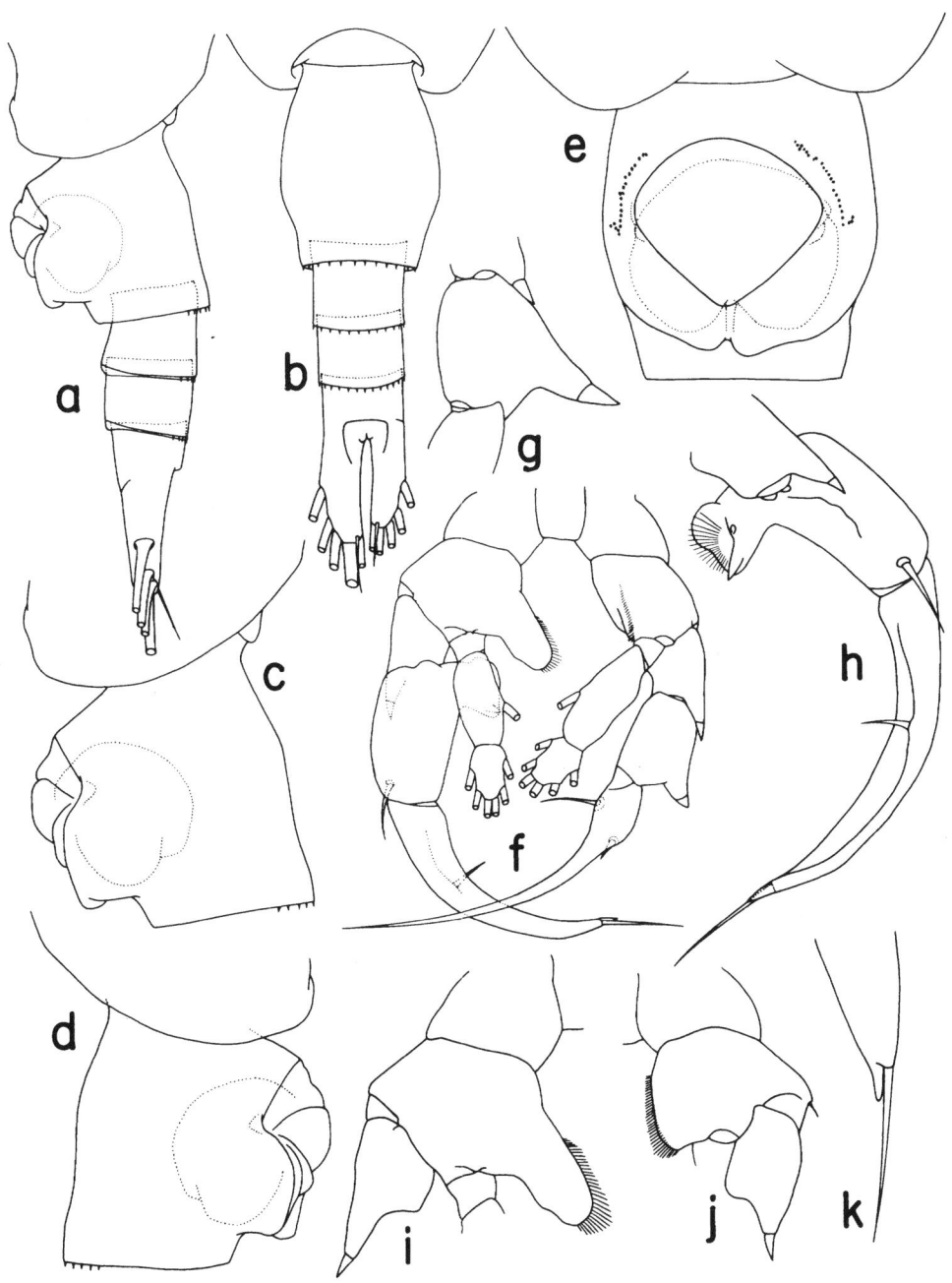

Figure 63. *Heterorhabdus heterolobus*, new species, female: **a**, urosome, left; **b**, do, dorsal; **c**, genital somite, left; **d**, do, right; **e**, do, ventral. Male: **f**, fifth pair of legs, anterior; **g**, second exopodal segment of left 5th leg, anterior; **h**, exopod of right 5th leg, posterior; **i**, basipod of right 5th leg, anterior; **j**, basipod of left 5th leg, anterior; **k**, distal end of exopod of right 5th leg, lateral.

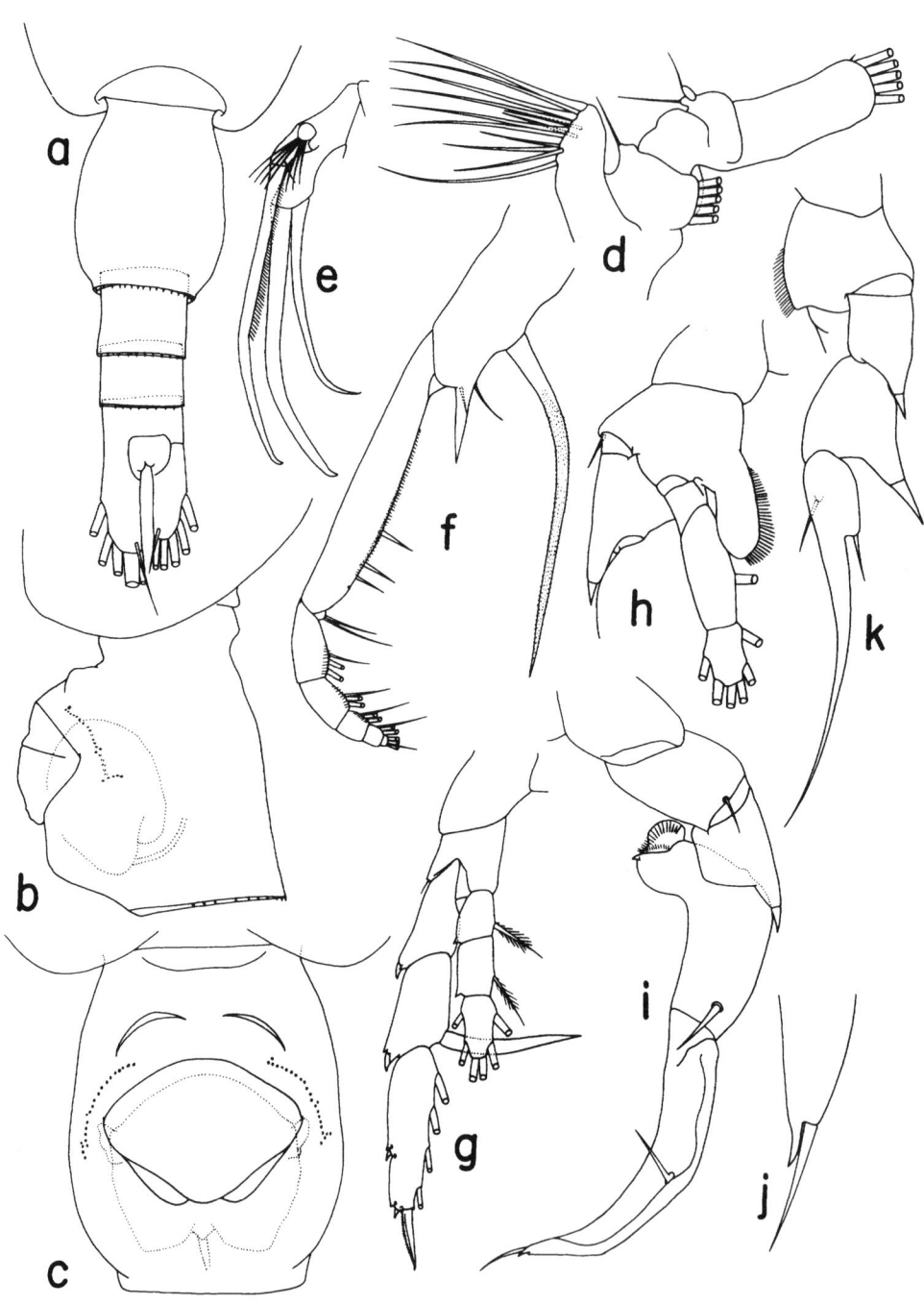

Figure 64. *Heterorhabdus subspinifrons* female: **a**, urosome, dorsal; **b**, genital somite, left; **c**, do, ventral; **d**, left maxillule, posterior; **e**, distal end of left maxilla, posterior; **f**, right maxilliped, anterior; **g**, fifth leg, anterior. Male: **h**, right 5th leg with distal exopodal segments omitted, anterior; **i**, right 5th leg with endopod omitted, posterior; **j**, distal end of exopod of right 5th leg, lateral; **k**, left 5th leg with endopod omitted, anterior.

216

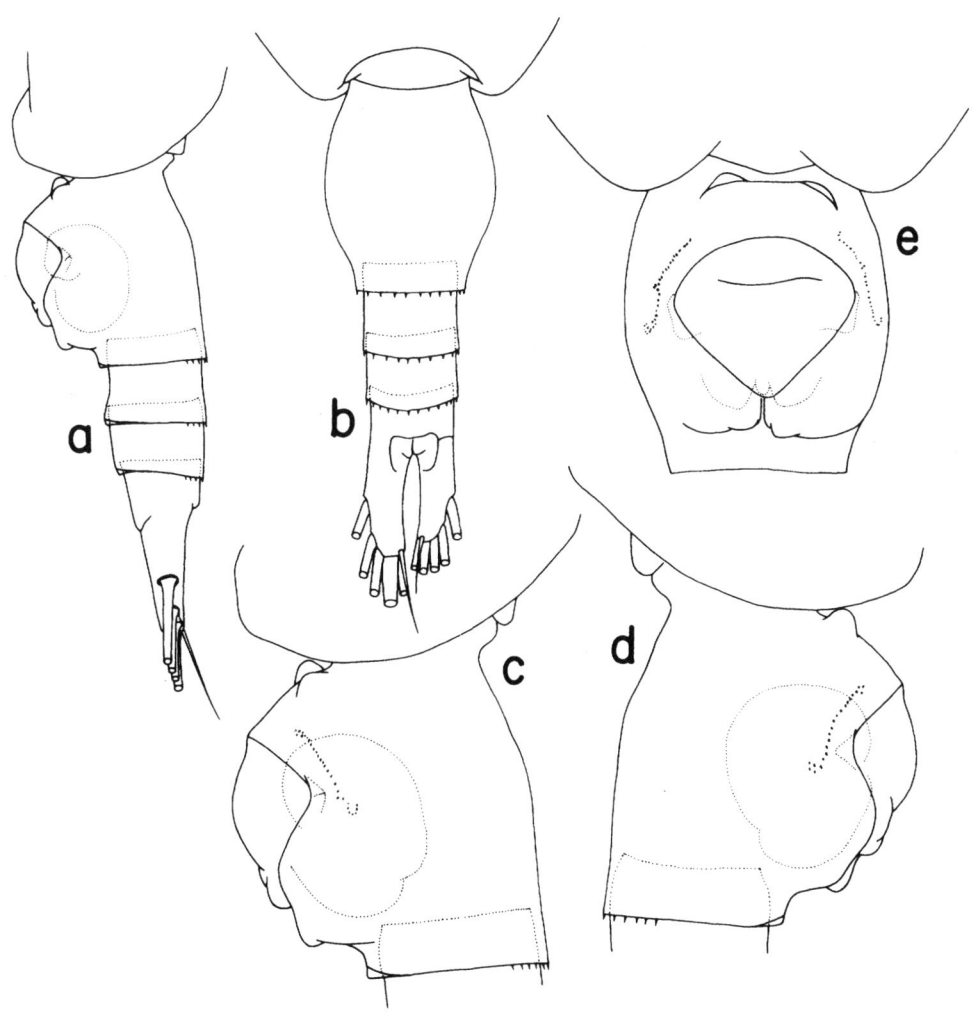

Figure 65. *Heterorhabdus quadrilobus*, new species, female: **a**, urosome, left; **b**, do, dorsal; **c**, genital somite, left; **d**, do, right; **e**, do, ventral.

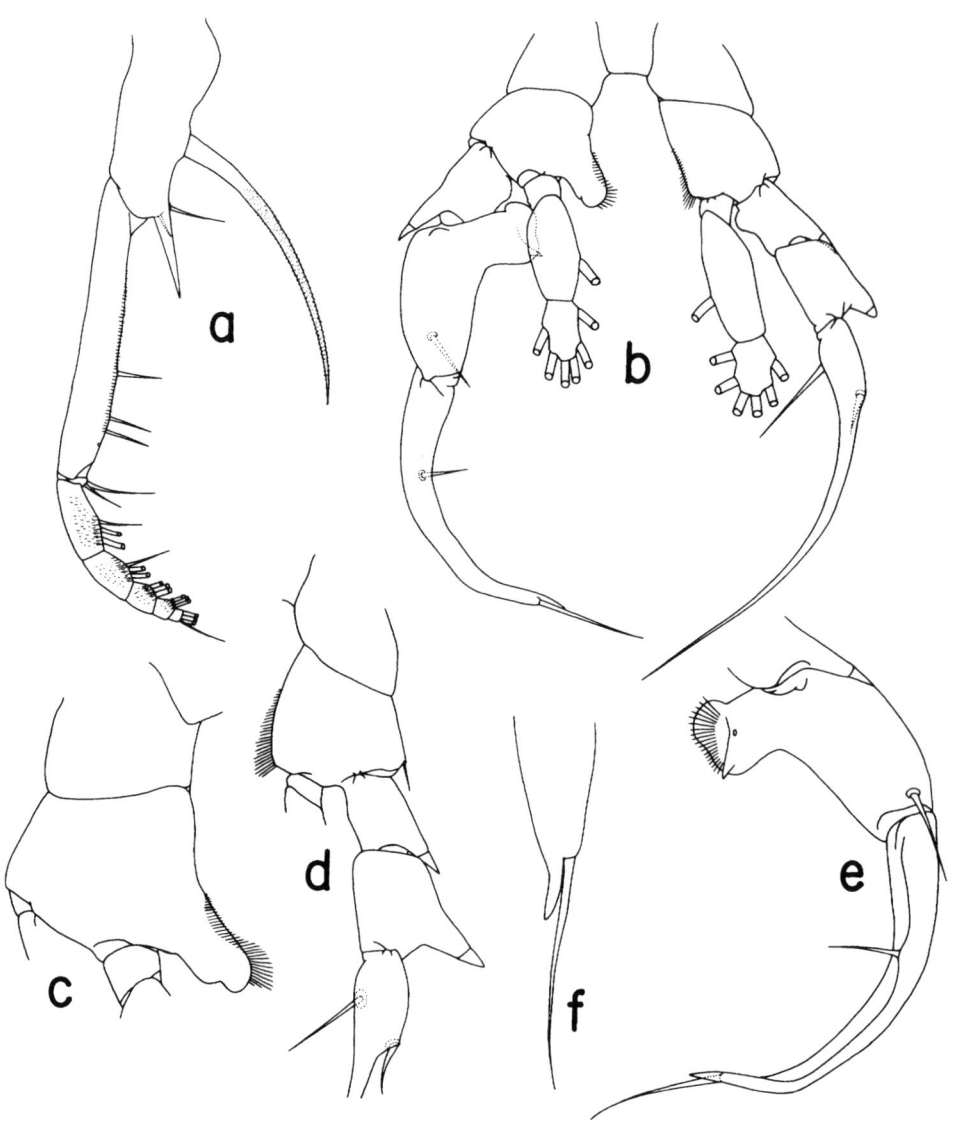

Figure 66. *Heterorhabdus quadrilobus*, new species, female: **a**, right maxilliped, anterior. Male: **b**, fifth pair of legs, anterior; **c**, basipod of right 5th leg, anterior; **d**, left 5th leg with endopod omitted, anterior; **e**, exopod of right 5th leg, posterior; **f**, distal end of exopod of right 5th leg, lateral.

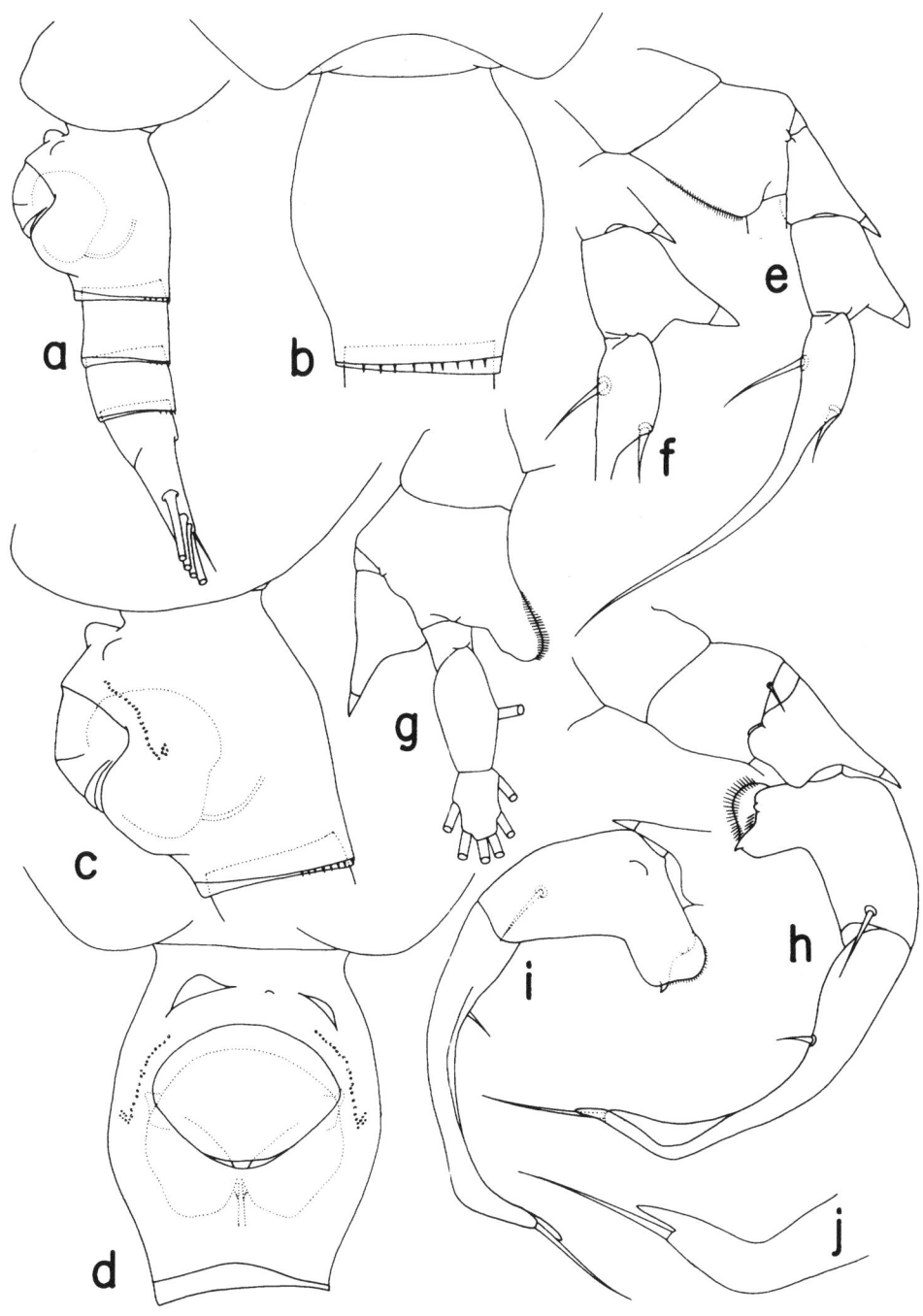

Figure 67. *Heterorhabdus ankylocolus*, new species, female: **a**, urosome, left; **b**, genital somite, dorsal; **c**, do, left; **d**, do, ventral. Male: **e**, left 5th leg with endopod omitted, anterior; **f**, exopod of left 5th leg, anterior; **g**, right 5th leg with distal exopodal segments omitted, anterior; **h**, right 5th leg with endopod omitted, posterior; **i**, exopod of right 5th leg, anterior; **j**, distal end of exopod of right 5th leg, lateral.

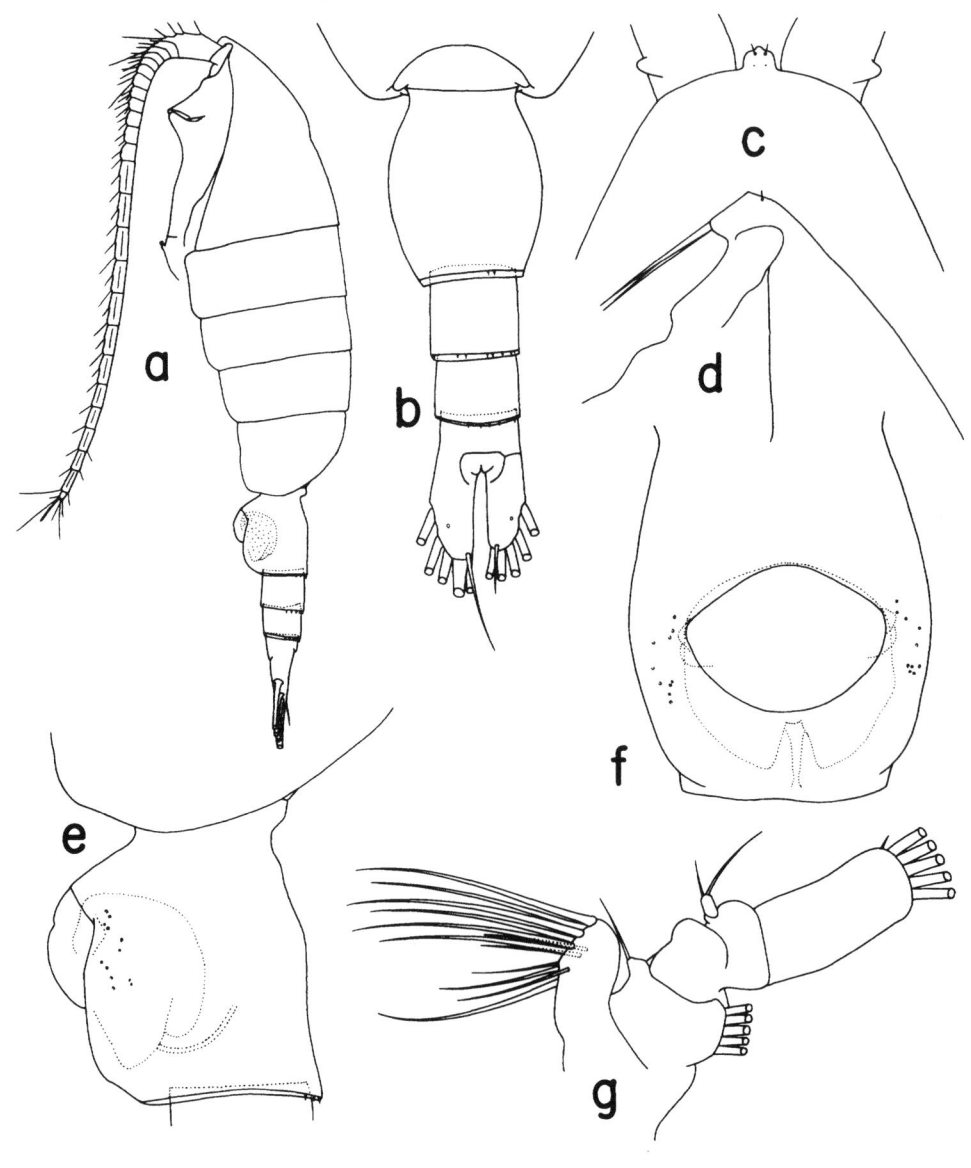

Figure 68. *Heterorhabdus caribbeanensis* female: **a**, habitus, left; **b**, urosome, dorsal; **c**, forehead, dorsal; **d**, do, left; **e**, genital somite, left; **f**, do, ventral; **g**, left maxillule, posterior.

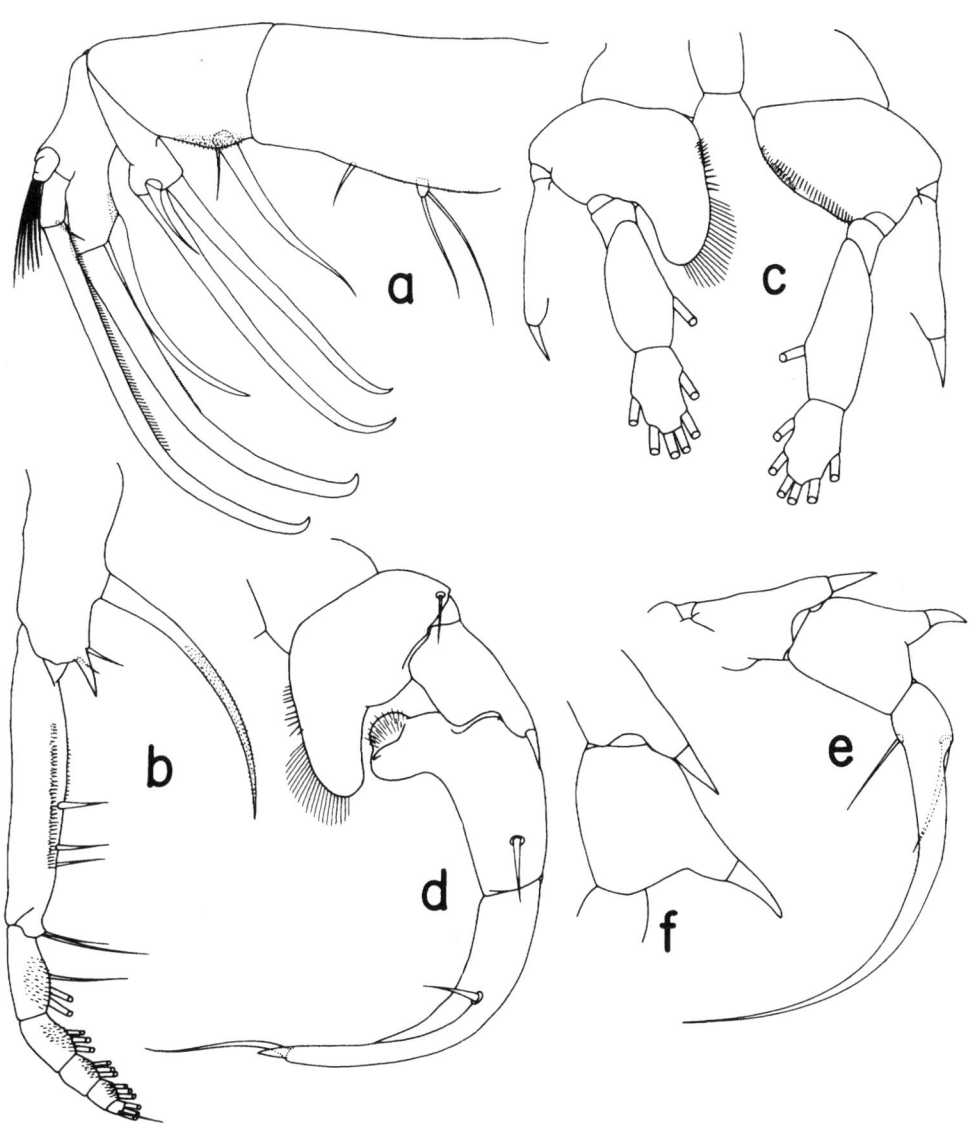

Figure 69. *Heterorhabdus caribbeanensis* female: **a**, left maxilla, posterior; **b**, right maxilliped, anterior. Male: **c**, fifth pair of legs with distal exopodal segments omitted, anterior; **d**, right 5th leg with endopod omitted, posterior; **e**, exopod of left 5th leg, anterior; **f**, first 2 exopodal segments of left 5th leg, anterior.

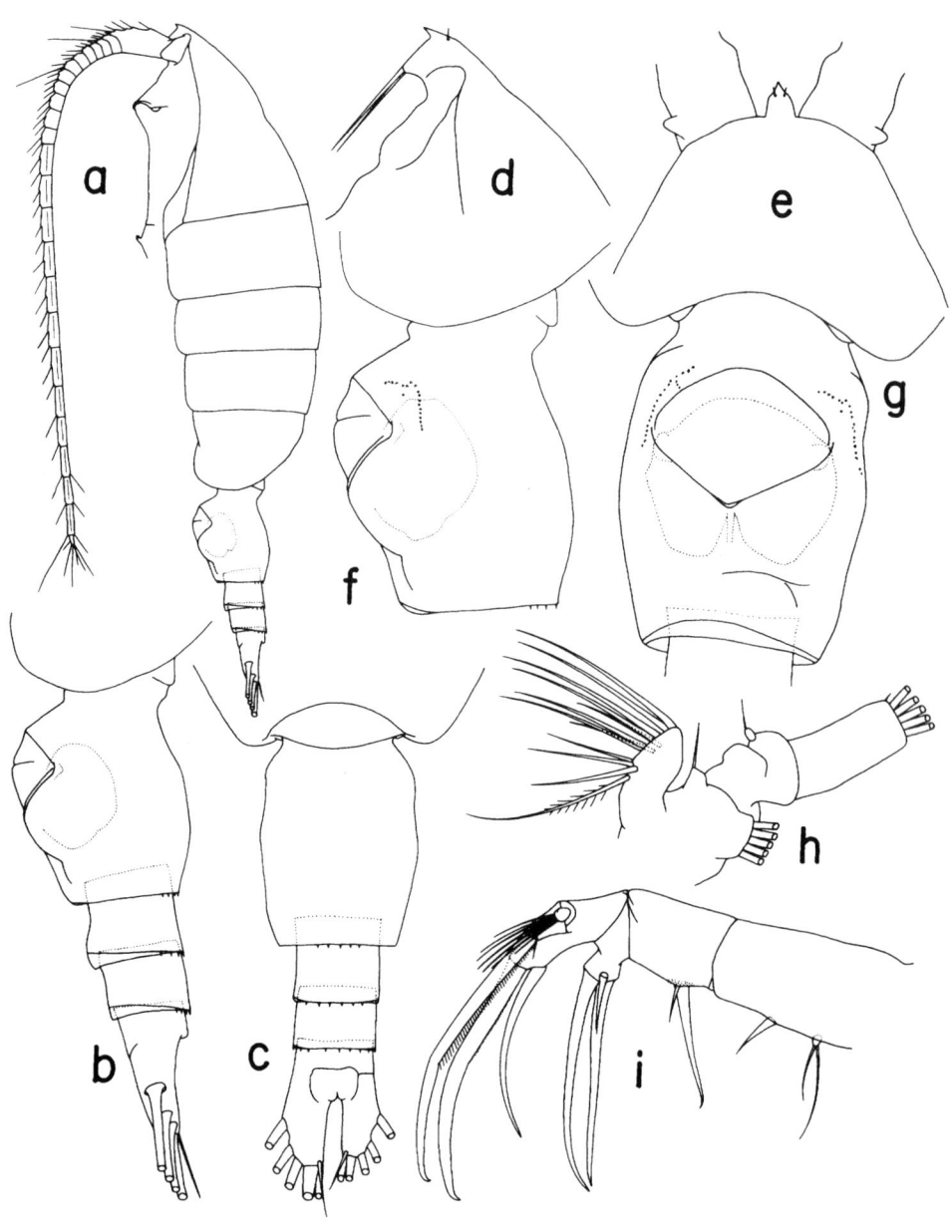

Figure 70. *Heterorhabdus insukae*, new species, female: **a**, habitus, left; **b**, urosome, left; **c**, do, dorsal; **d**, forehead, left; **e**, do, dorsal; **f**, genital somite, left; **g**, do, ventral; **h**, left maxillule, posterior; **i**, left maxilla, posterior.

222

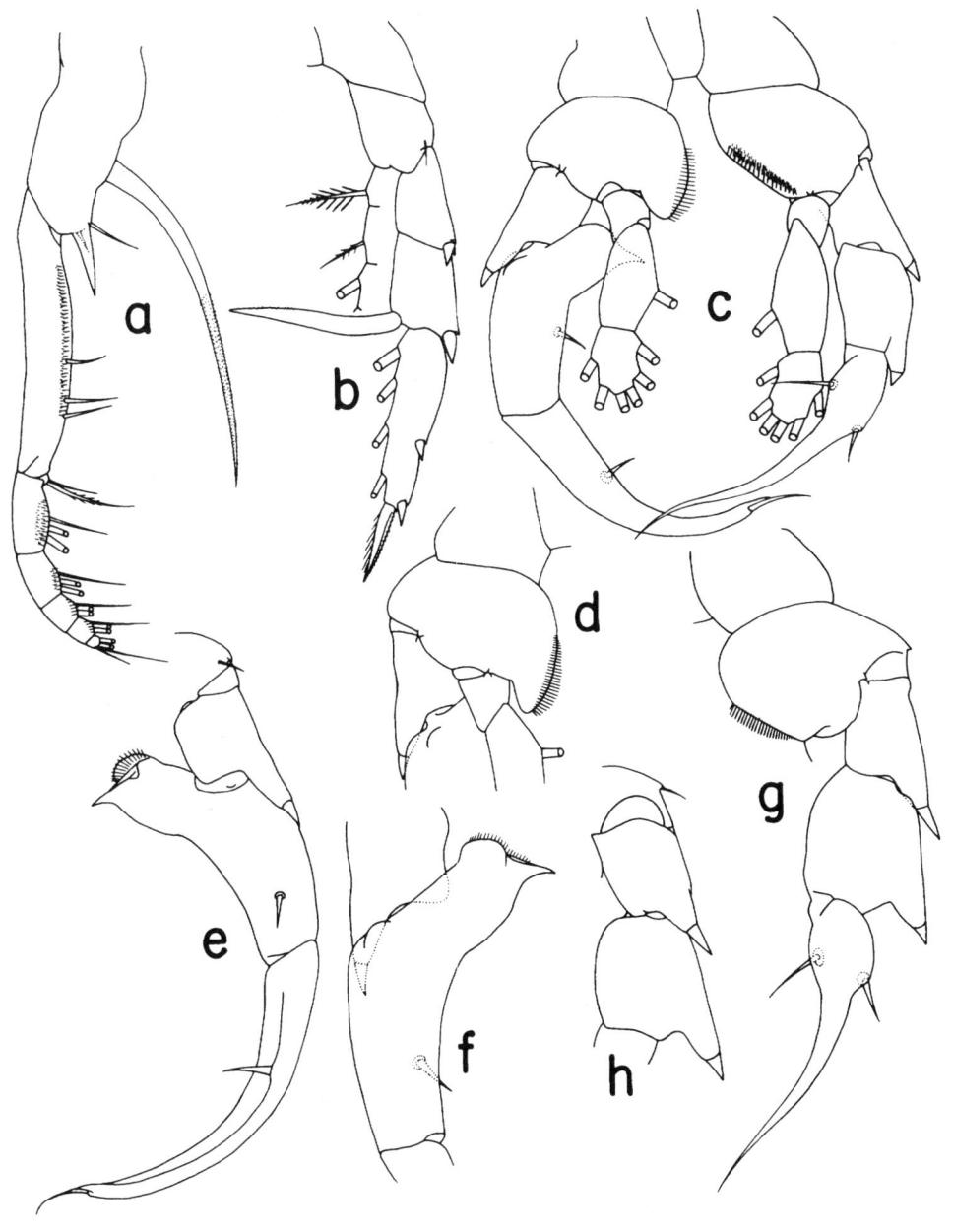

Figure 71. *Heterorhabdus insukae*, new species, female: **a**, right maxilliped, anterior; **b**, fifth leg, posterior. Male: **c**, fifth pair of legs, anterior; **d**, basipod of right 5th leg, anterior; **e**, exopod of right 5th leg, posterior; **f**, second exopodal segment of right 5th leg, anterior; **g**, left 5th leg with endopod omitted, anterior; **h**, first 2 exopodal segments of left 5th leg, anterior.

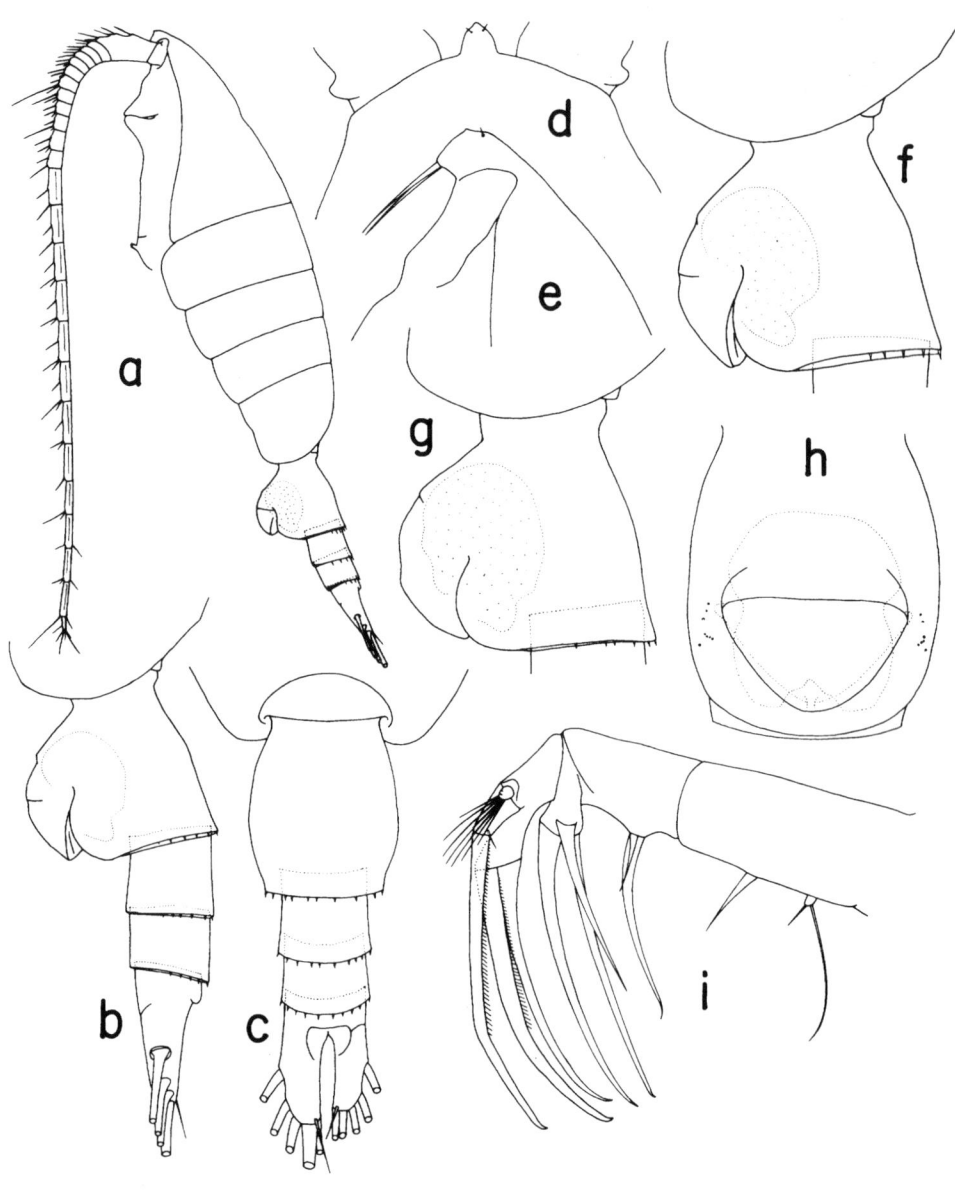

Figure 72. *Heterorhabdus guineanensis*, new species, female: **a**, habitus, left; **b**, urosome, left; **c**, do, dorsal; **d**, forehead, dorsal; **e**, do, left; **f**, genital somite with operculum partially open, left; **g**, genital somite with operculum closed, left; **h**, genital somite, ventral; **i**, left maxilla, posterior.

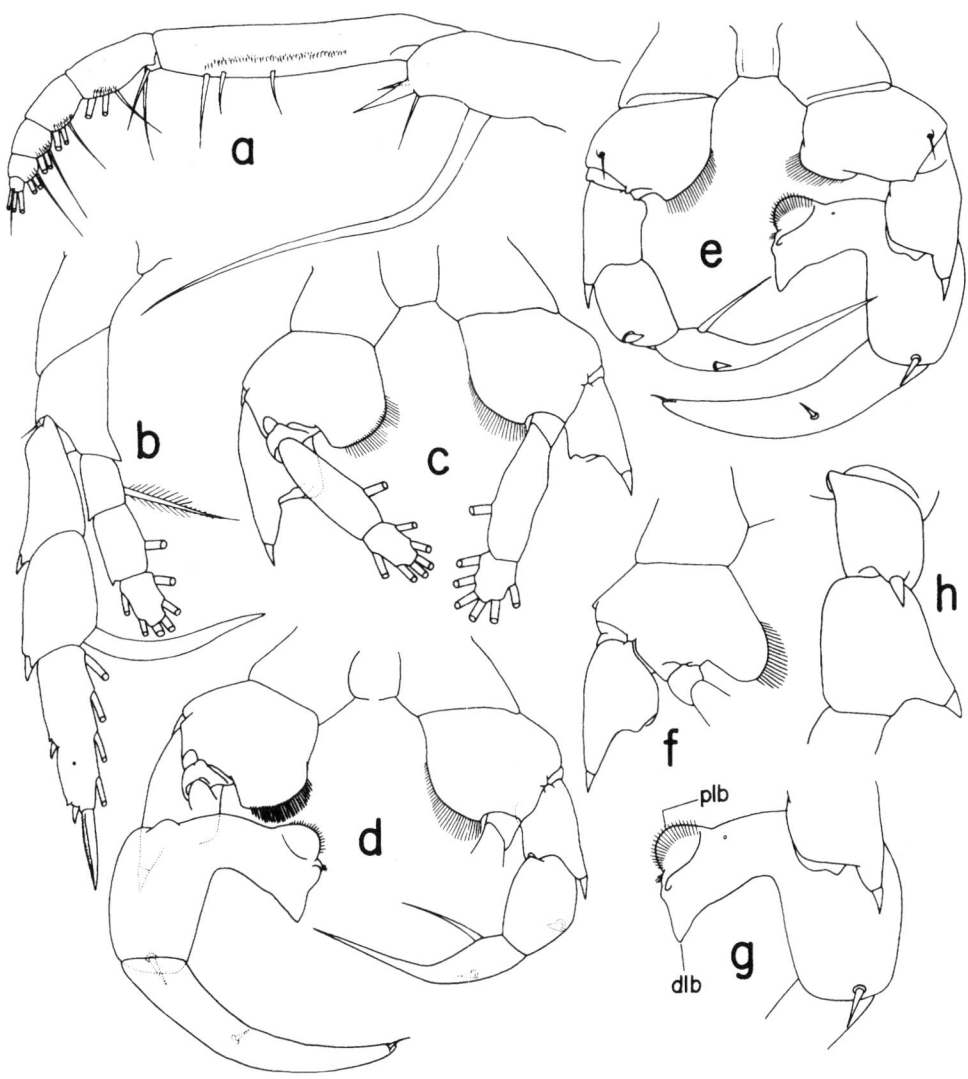

Figure 73. *Heterorhabdus guineanensis*, new species, female: **a**, right maxilliped, anterior; **b**, fifth leg, anterior. Male: **c**, fifth pair of legs with distal exopodal segments omitted, anterior; **d**, fifth pair of legs with endopods omitted, anterior; **e**, fifth pair of legs with endopods omitted, posterior; **f**, basipod of right 5th leg, anterior; **g**, second exopodal segment of right 5th leg, posterior; **h**, first 2 exopodal segments of left 5th leg, anterior. dlb = distal lobe; plb = proximal lobe.

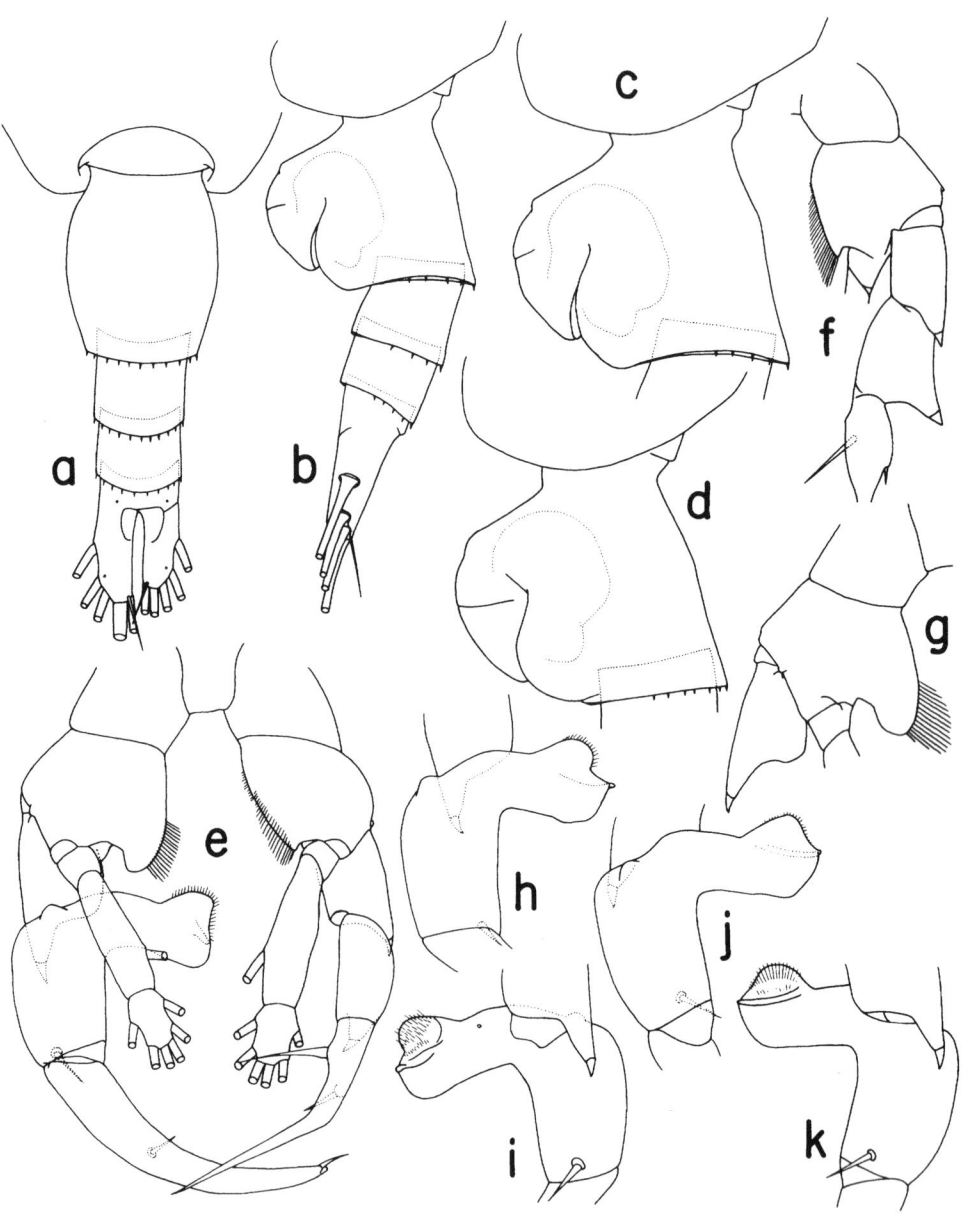

Figure 74. *Heterorhabdus lobatus* female: **a**, urosome, dorsal; **b**, do, left; **c**, genital somite with posteroventral arthrodial membrane folded in, left; **d**, genital somite with posteroventral arthrodial membrane somewhat extended, left. Male: **e**, fifth pair of legs, anterior; **f**, left 5th leg with endopod omitted; anterior; **g**, basipod of right 5th leg, anterior; **h**, second exopodal segment of right 5th leg, anterior; **i**, do, posterior; **j**, second exopodal segment of right 5th leg from different specimen, anterior; **k**, do, posterior.

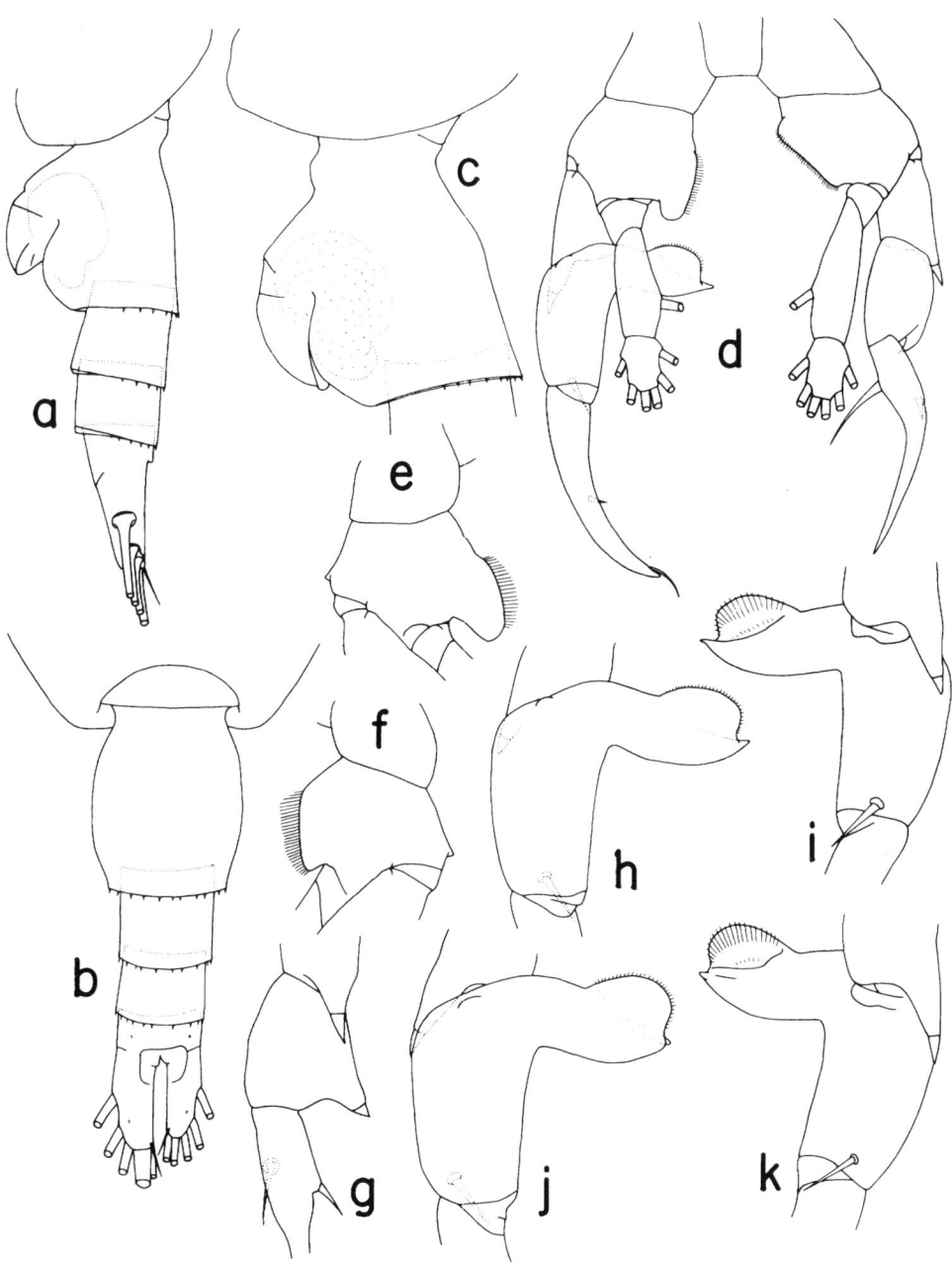

Figure 75. *Heterorhabdus papilliger* female: **a**, urosome, left; **b**, do, dorsal; **c**, genital somite, left. Male: **d**, fifth pair of legs, anterior; **e**, basipod of right 5th leg, anterior; **f**, basipod of left 5th leg, anterior; **g**, exopod of left 5th leg, anterior; **h**, second exopodal segment of right 5th leg, anterior; **i**, do, posterior; **j**, second exopodal segment of right 5th leg from different specimen, anterior; **k**, do, posterior.

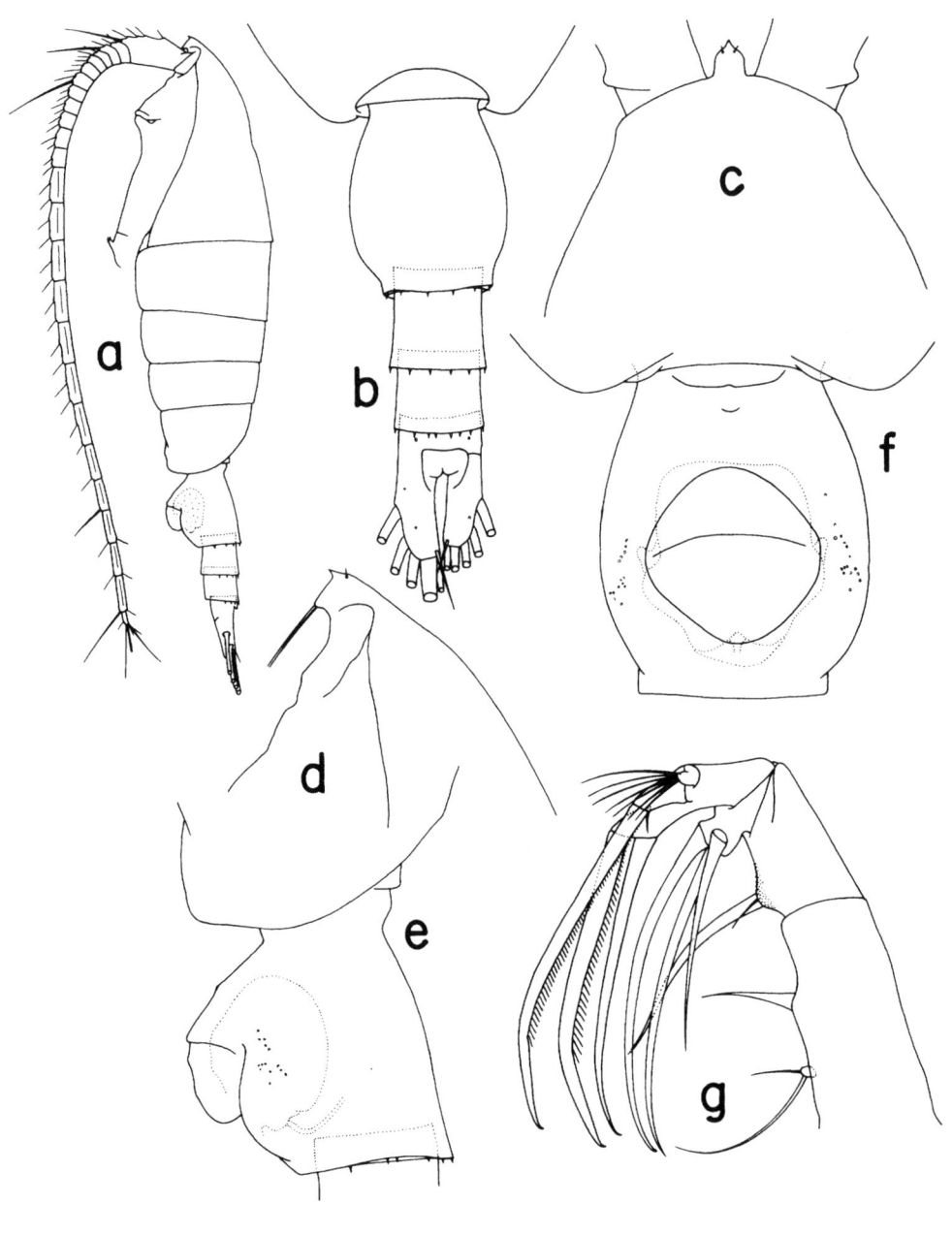

Figure 76. *Heterorhabdus spinifer* female: **a**, habitus, left; **b**, urosome, dorsal; **c**, forehead, dorsal; **d**, do, left; **e**, genital somite, left; **f**, do, ventral; **g**, left maxilla, posterior.

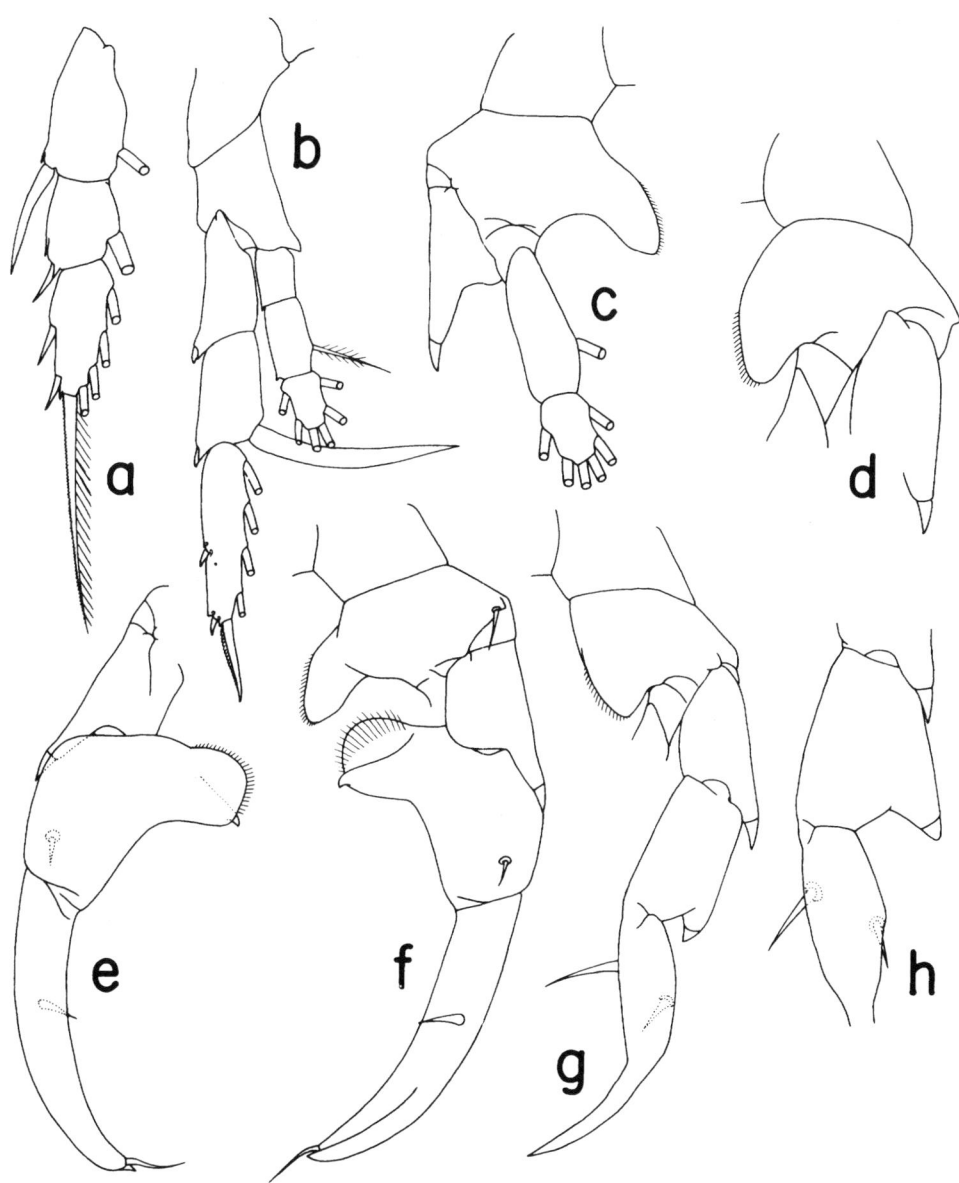

Figure 77. *Heterorhabdus spinifer* female: **a**, exopod of 1st leg, anterior; **b**, fifth leg, anterior. Male: **c**, right 5th leg with exopod omitted, anterior; **d**, basipod of left 5th leg, anterior; **e**, exopod of right 5th leg, anterior; **f**, right 5th leg with endopod omitted, posterior; **g**, left 5th leg with endopod omitted, anterior; **h**, exopod of left 5th leg, anterior.

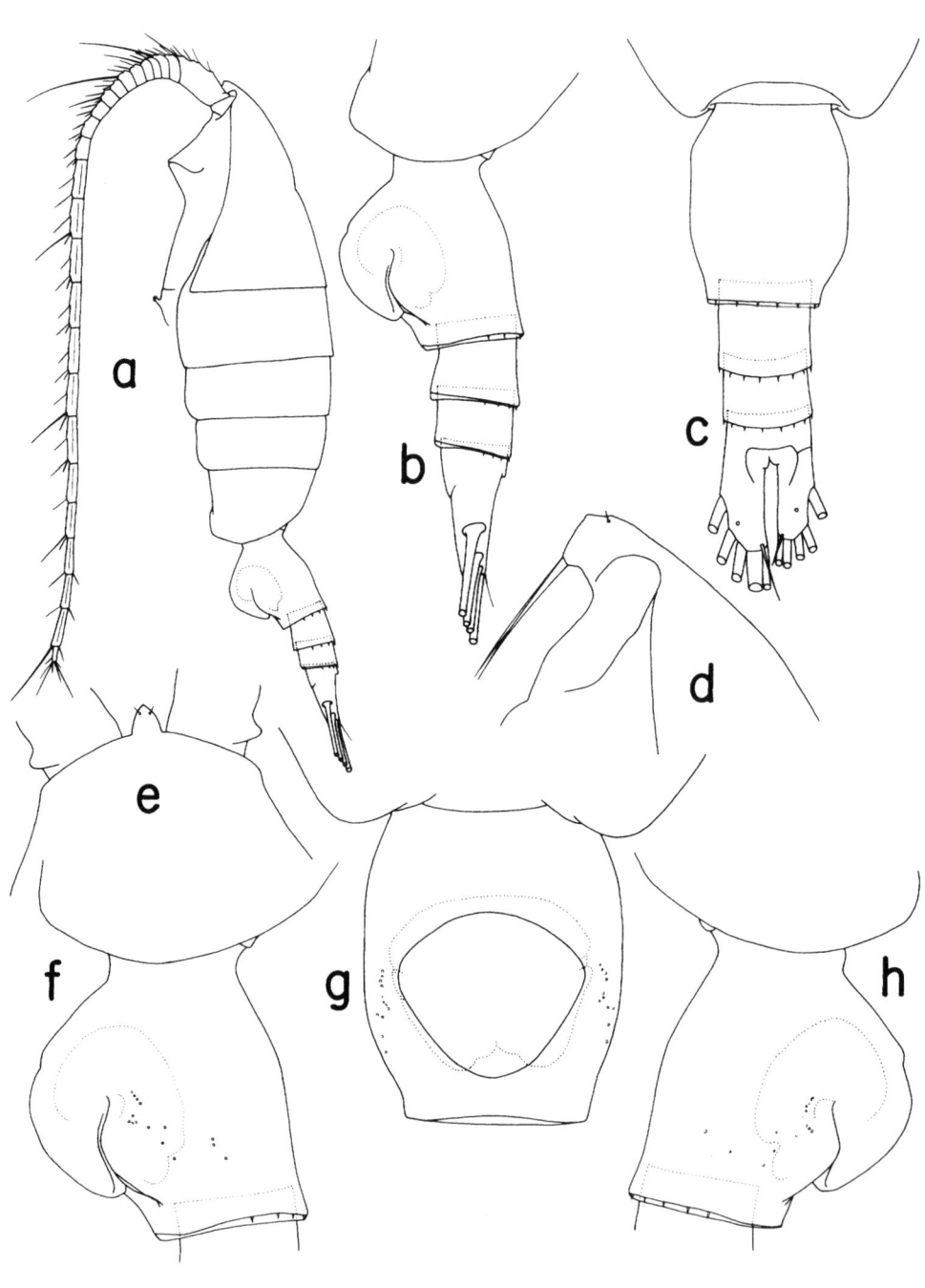

Figure 78. *Heterorhabdus prolatus*, new species, female: **a**, habitus, left; **b**, urosome, left; **c**, do, dorsal; **d**, forehead, left; **e**, do, dorsal; **f**, genital somite, left; **g**, do, ventral; **h**, do, right.

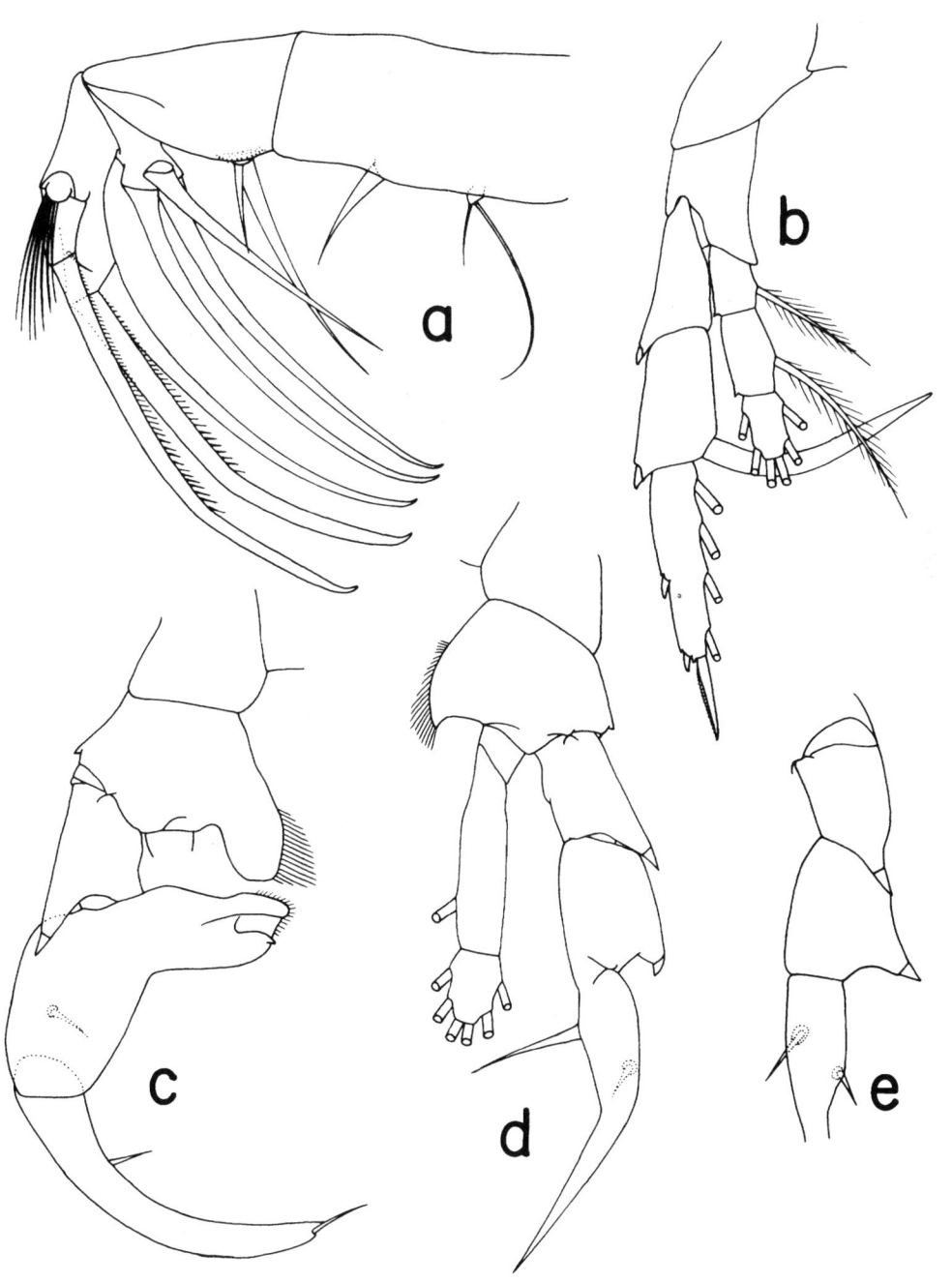

Figure 79. *Heterorhabdus prolatus*, new species, female: **a**, left maxilla, posterior; **b**, fifth leg, anterior. Male: **c**, right 5th leg with endopod omitted, anterior; **d**, left 5th leg, anterior; **e**, exopod of left 5th leg, anterior.

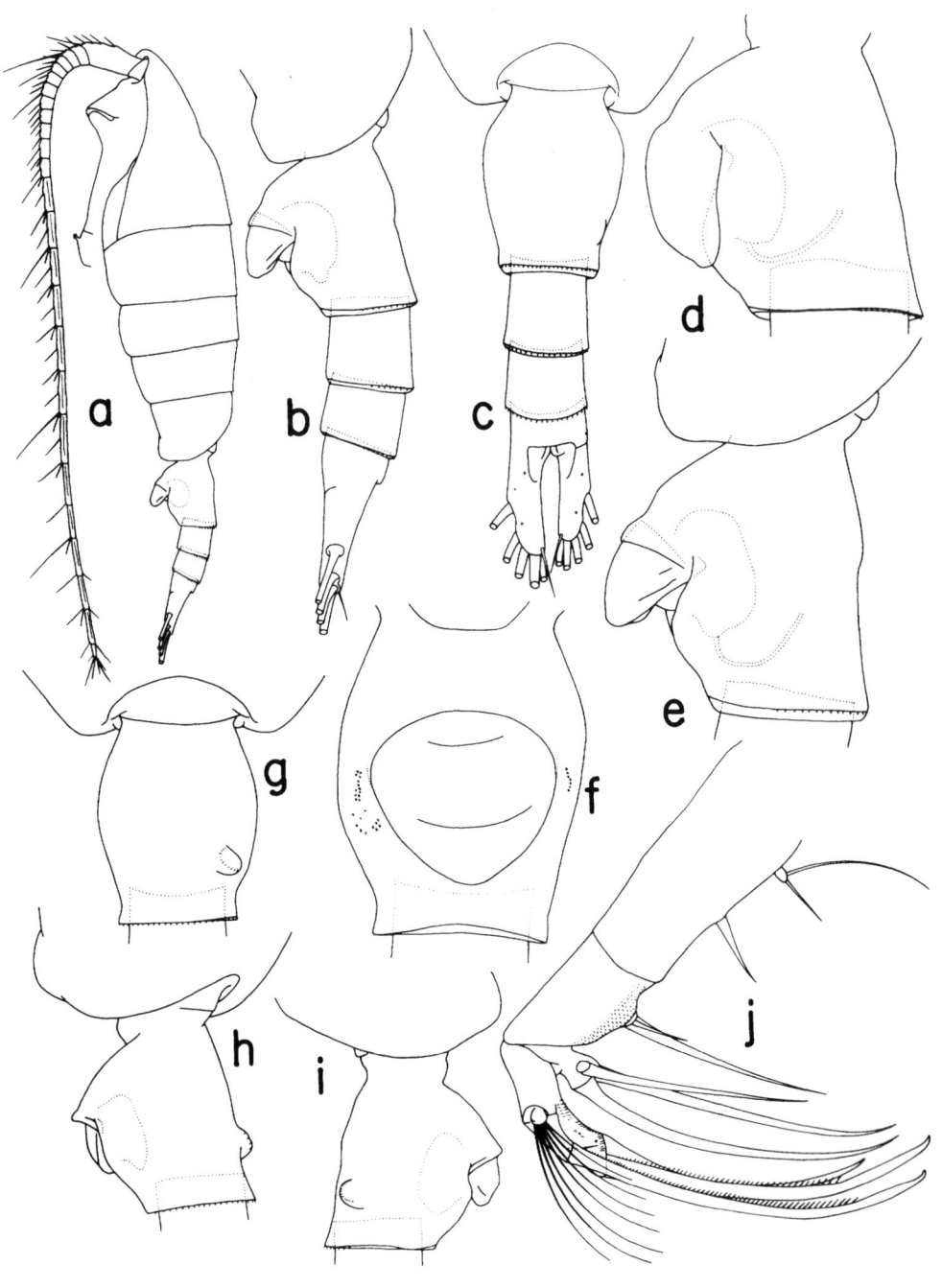

Figure 80. *Heterorhabdus fistulosus* female: **a**, habitus, left; **b**, urosome, left; **c**, do, dorsal; **d**, genital somite, left; **e**, genital somite with operculum open, left; **f**, genital somite, ventral; **g**, genital somite with a tubercular outgrowth, dorsal; **h**, do, left; **i**, do, right; **j**, left maxilla, posterior.

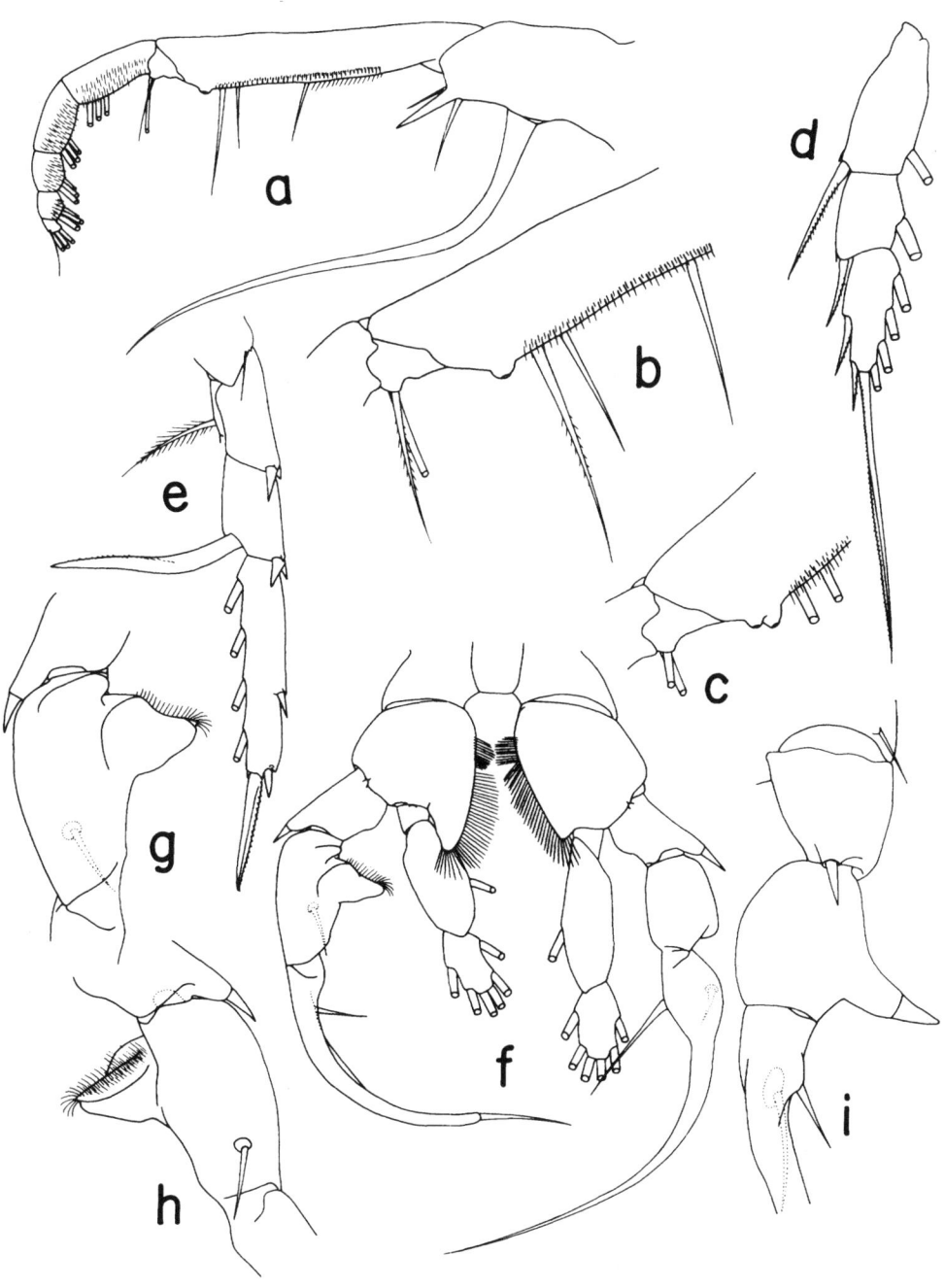

Figure 81. *Heterorhabdus fistulosus* female: **a**, right maxilliped, anterior; **b**, distal part of basis of right maxilliped, anterior; **c**, distal part of basis of right maxilliped from different specimen, anterior; **d**, exopod of 1st leg, anterior; **e**, exopod of 5th leg, posterior. Male: **f**, fifth pair of legs, anterior; **g**, second exopodal segment of right 5th leg, anterior; **h**, do, posterior; **i**, exopod of left 5th leg, anterior.

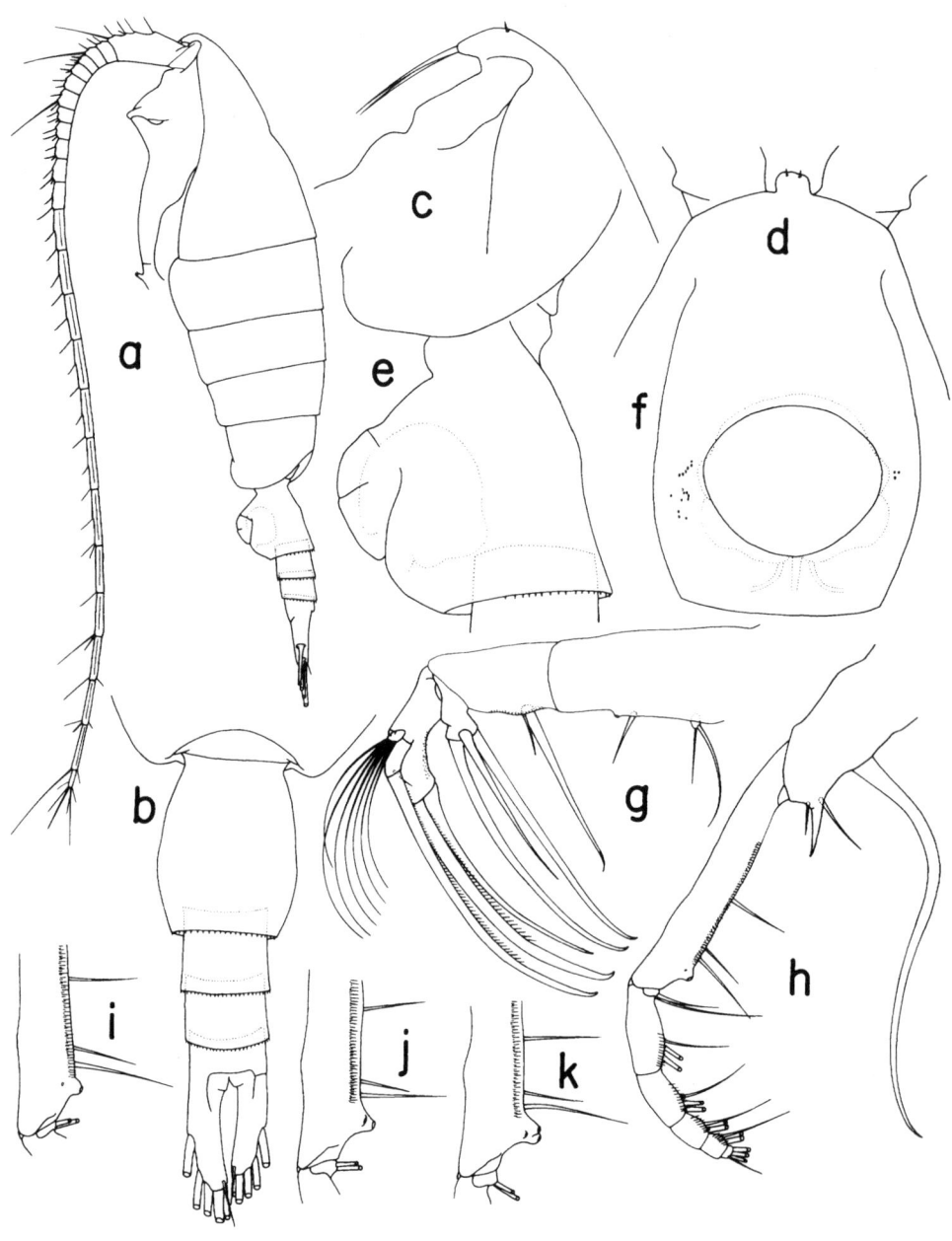

Figure 82. *Heterorhabdus egregius* female: **a**, habitus, left; **b**, urosome, dorsal; **c**, forehead, left; **d**, do, dorsal; **e**, genital somite, left; **f**, do, ventral; **g**, left maxilla, posterior; **h**, right maxilliped, anterior; **i-k**, basis of right maxilliped from 3 different specimens, anterior.

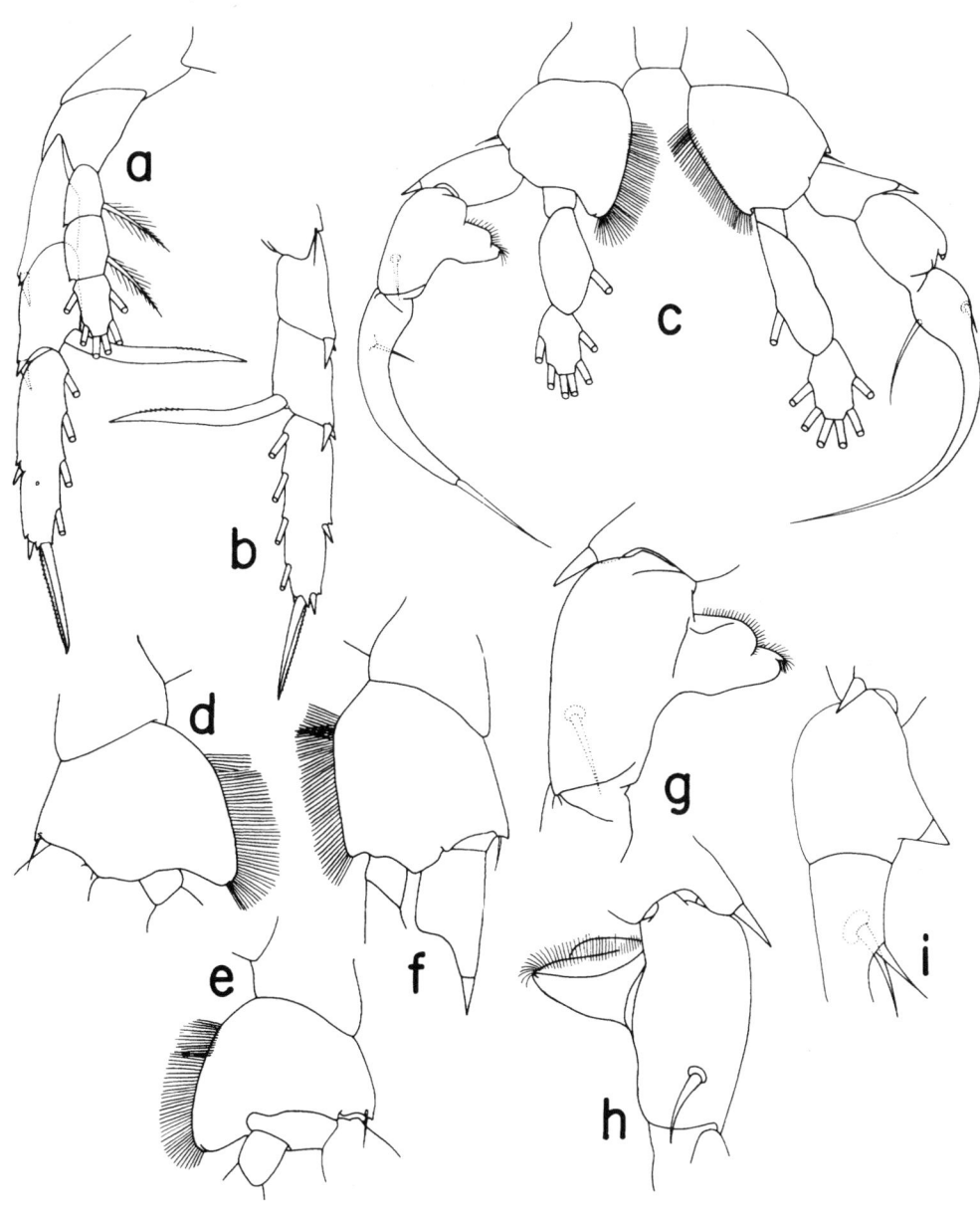

Figure 83. *Heterorhabdus egregius* female: **a**, fifth leg, anterior; **b**, exopod of 5th leg, posterior. Male: **c**, fifth pair of legs, anterior; **d**, basipod of right 5th leg, anterior; **e**, do, posterior; **f**, basipod of left 5th leg, anterior; **g**, second exopodal segment of right 5th leg, anterior; **h**, do, posterior; **i**, exopod of left 5th leg, anterior.

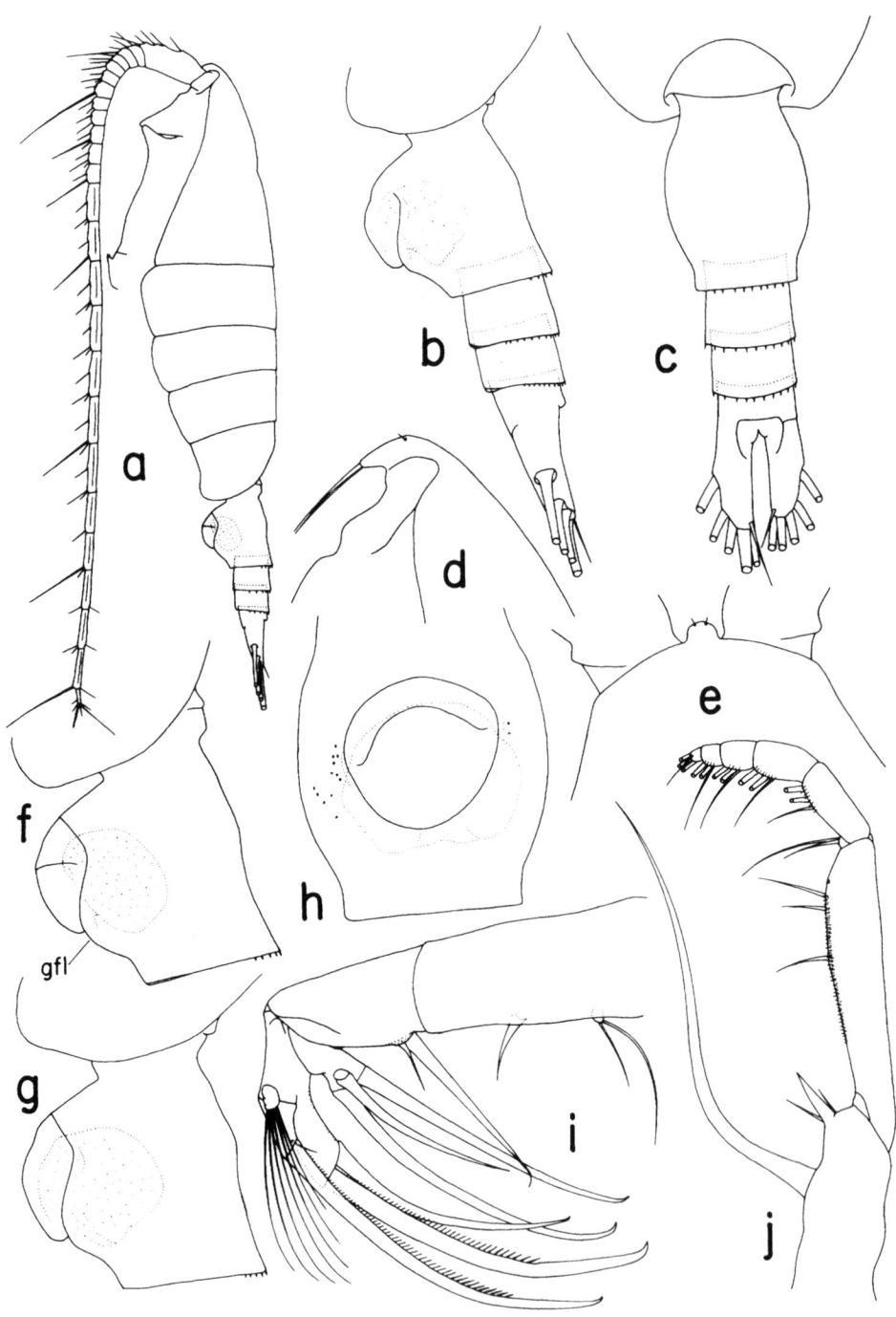

Figure 84. *Heterorhabdus habrosomus*, new species, female: **a**, habitus, left; **b**, urosome, left; **c**, do, dorsal; **d**, forehead, left; **e**, do, dorsal; **f**, genital somite, left; **g**, genital somite from different specimen, left; **h**, genital somite, ventral; **i**, left maxilla, posterior; **j**, right maxilliped, anterior. gfl = genital flange.

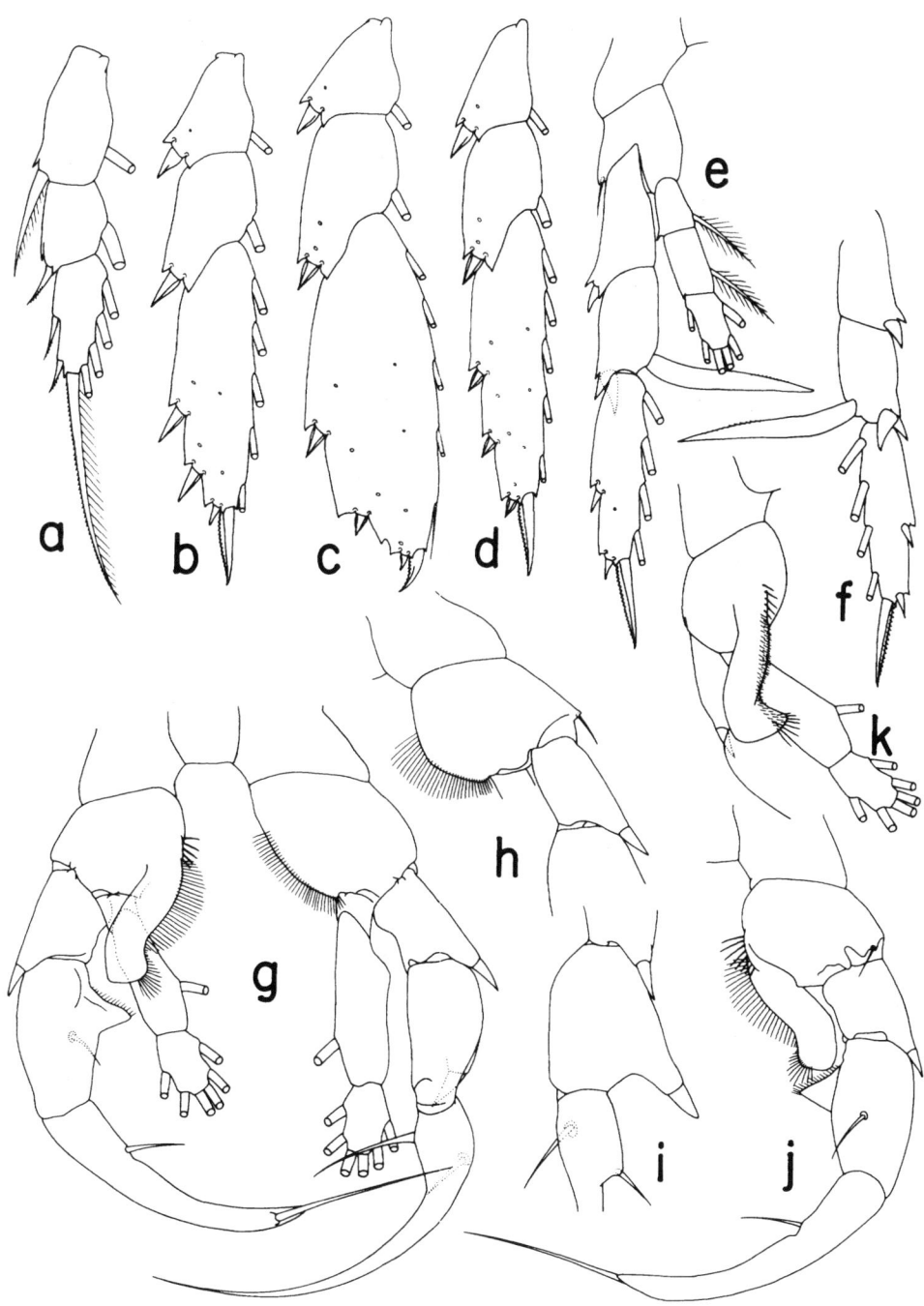

Figure 85. *Heterorhabdus habrosomus*, new species, female: **a**, exopod of 1st leg, anterior; **b**, exopod of 2nd leg, anterior; **c**, exopod of 3rd leg, anterior; **d**, exopod of 4th leg, anterior; **e**, fifth leg, anterior; **f**, exopod of 5th leg, posterior. Male: **g**, fifth pair of legs, anterior; **h**, basipod of left 5th leg, anterior; **i**, exopod of left 5th leg, anterior; **j**, right 5th leg with endopod omitted, posterior; **k**, right 5th leg with distal exopodal segments omitted, anterior, tilted counterclockwise.

237

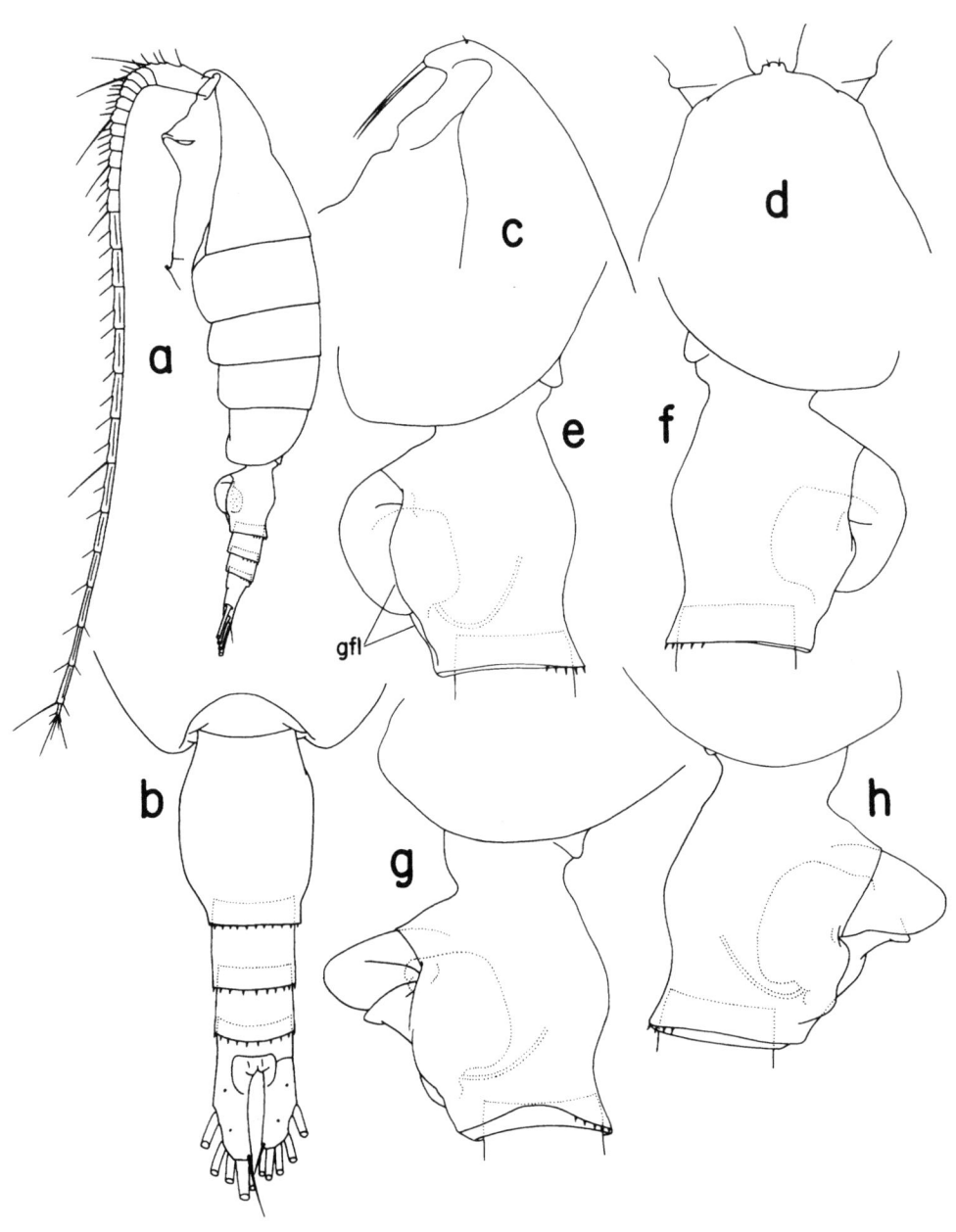

Figure 86. *Heterorhabdus prolixus*, new species, female: **a**, habitus, left; **b**, urosome, dorsal; **c**, forehead, left; **d**, do, dorsal; **e**, genital somite, left; **f**, do, right; **g**, genital somite with operculum open, left; **h**, do, right. gfl = genital flange.

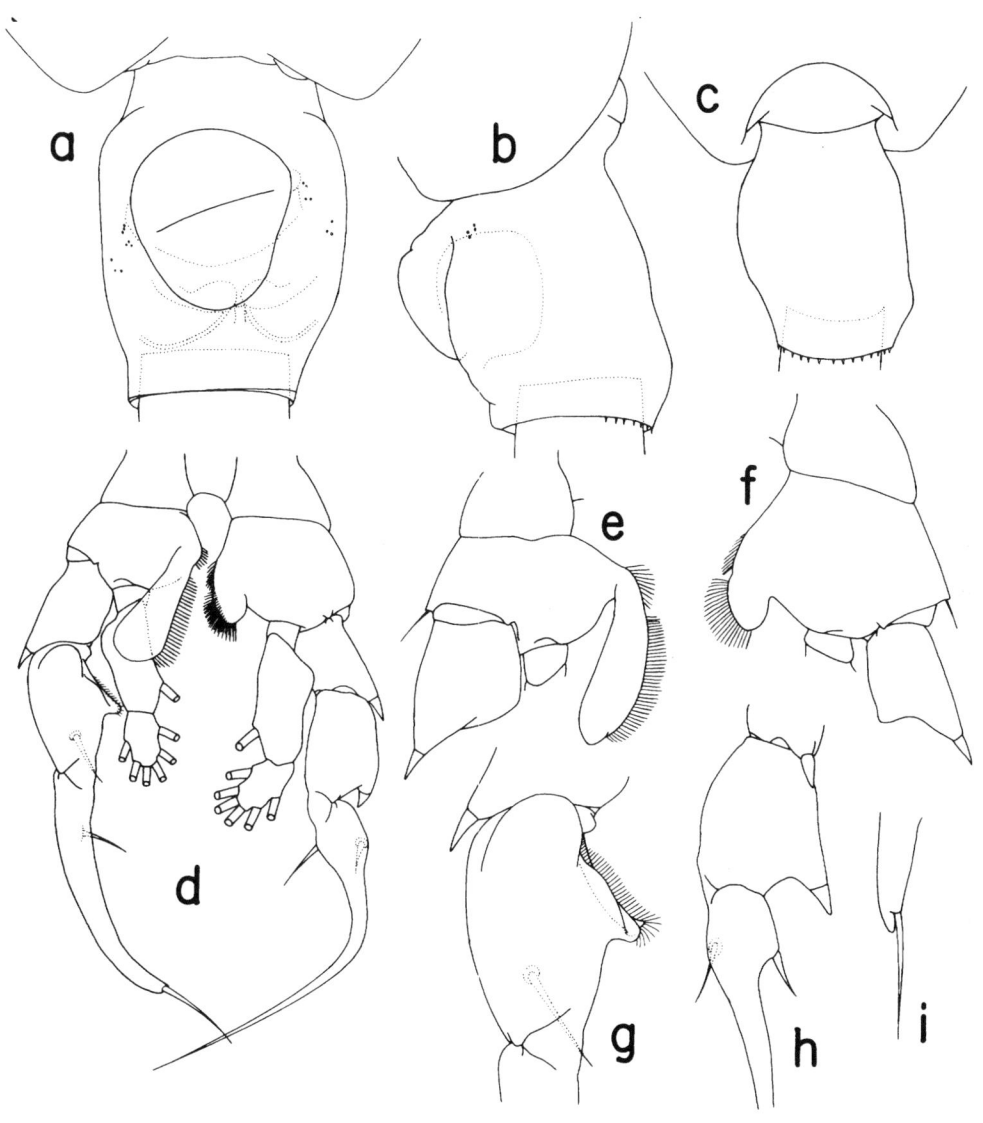

Figure 87. *Heterorhabdus prolixus*, new species, female: **a**, genital somite, ventral; **b**, abnormal genital somite, left; **c**, do, dorsal. Male: **d**, fifth pair of legs, anterior; **e**, basipod of right 5th leg, anterior; **f**, basipod of left 5th leg, anterior; **g**, second exopodal segment of right 5th leg, anterior; **h**, exopod of left 5th leg, anterior; **i**, distal end of exopod of right 5th leg, lateral.

239

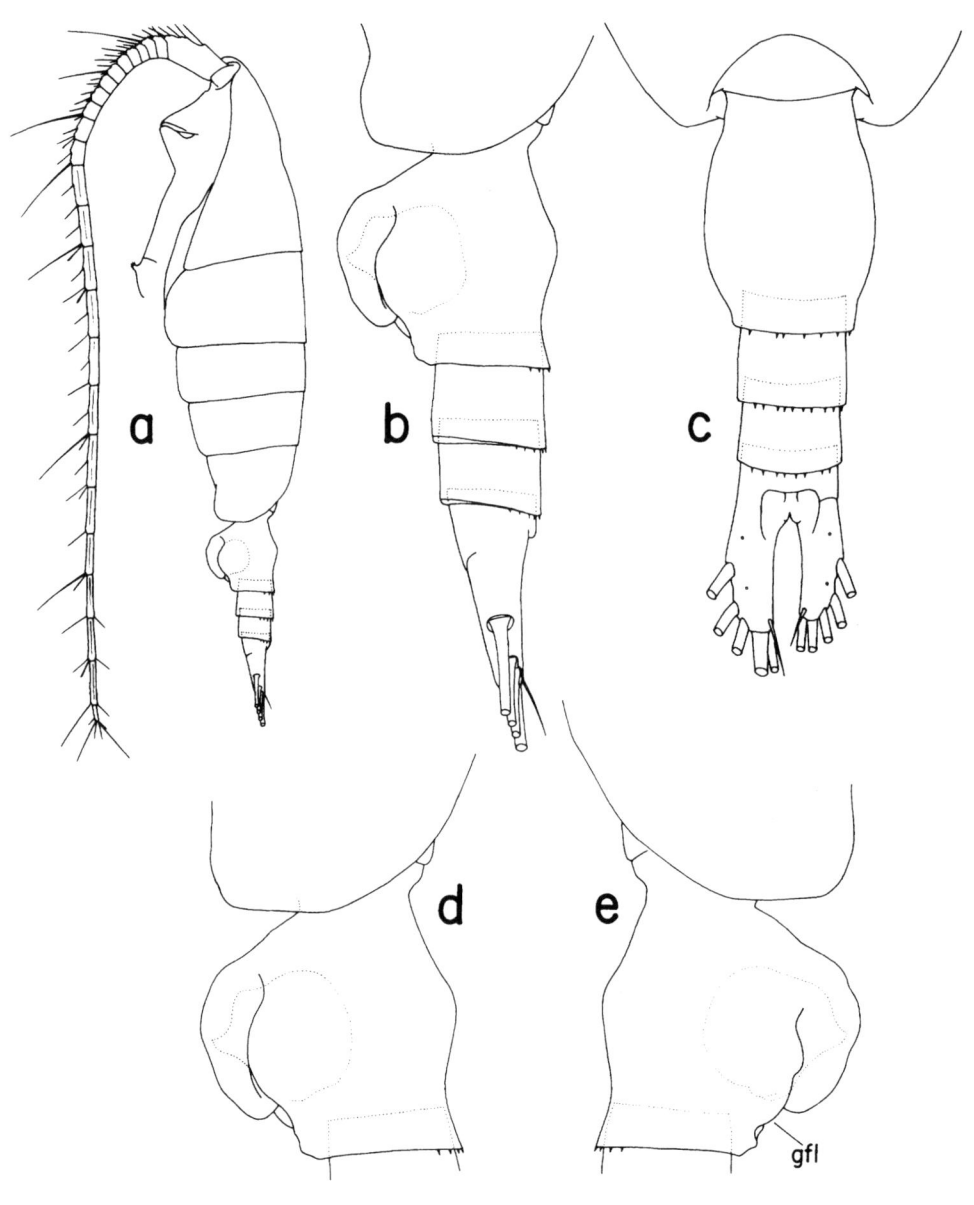

Figure 88. *Heterorhabdus clausii* female: **a**, habitus, left; **b**, urosome, left; **c**, do, dorsal; **d**, genital somite, left; **e**, do, right. gfl = genital flange.

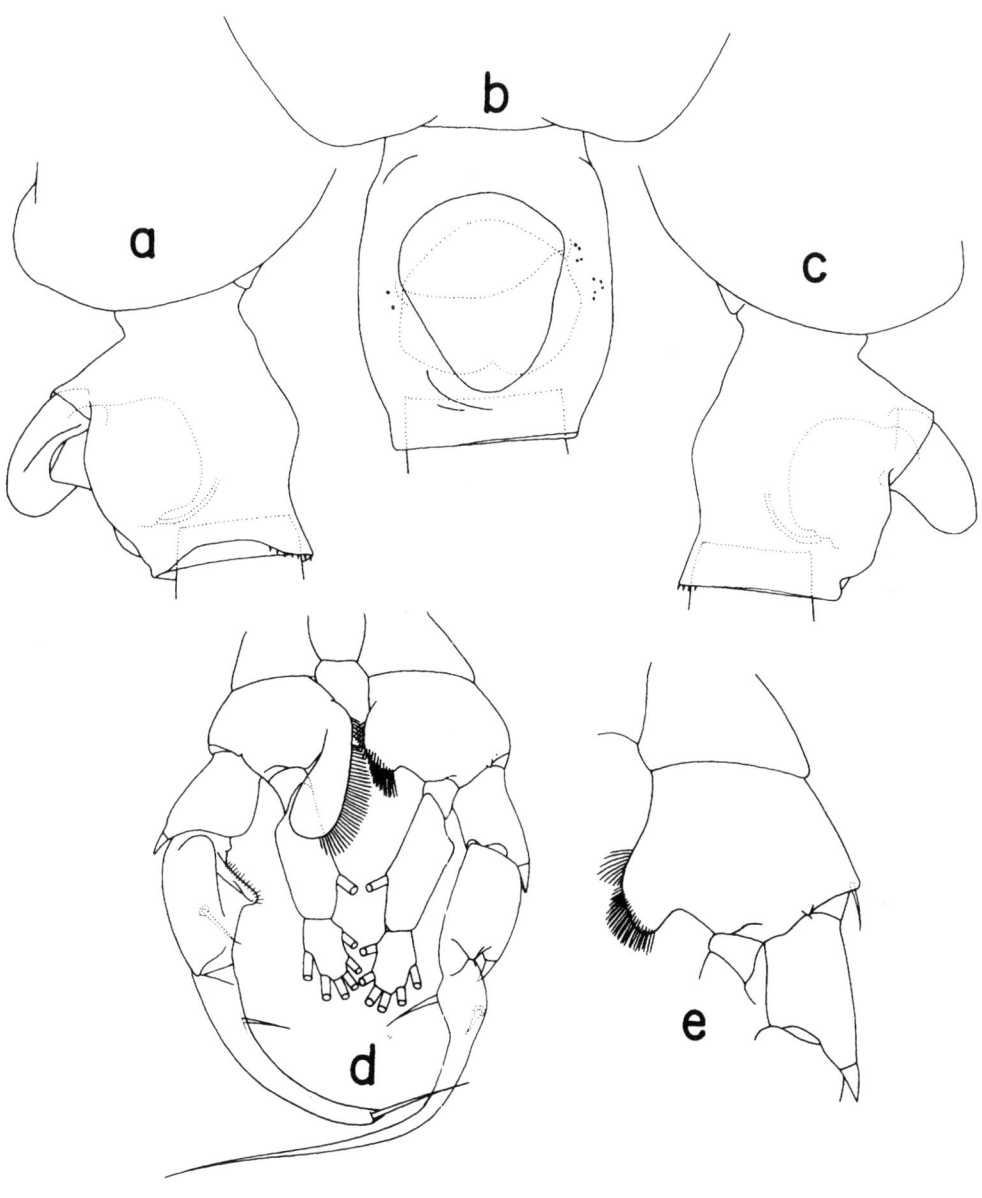

Figure 89. *Heterorhabdus clausii* female: **a**, genital somite with operculum open, left; **b**, genital somite, ventral; **c**, genital somite with operculum open, right. Male: **d**, fifth pair of legs, anterior; **e**, basipod of left 5th leg, anterior.

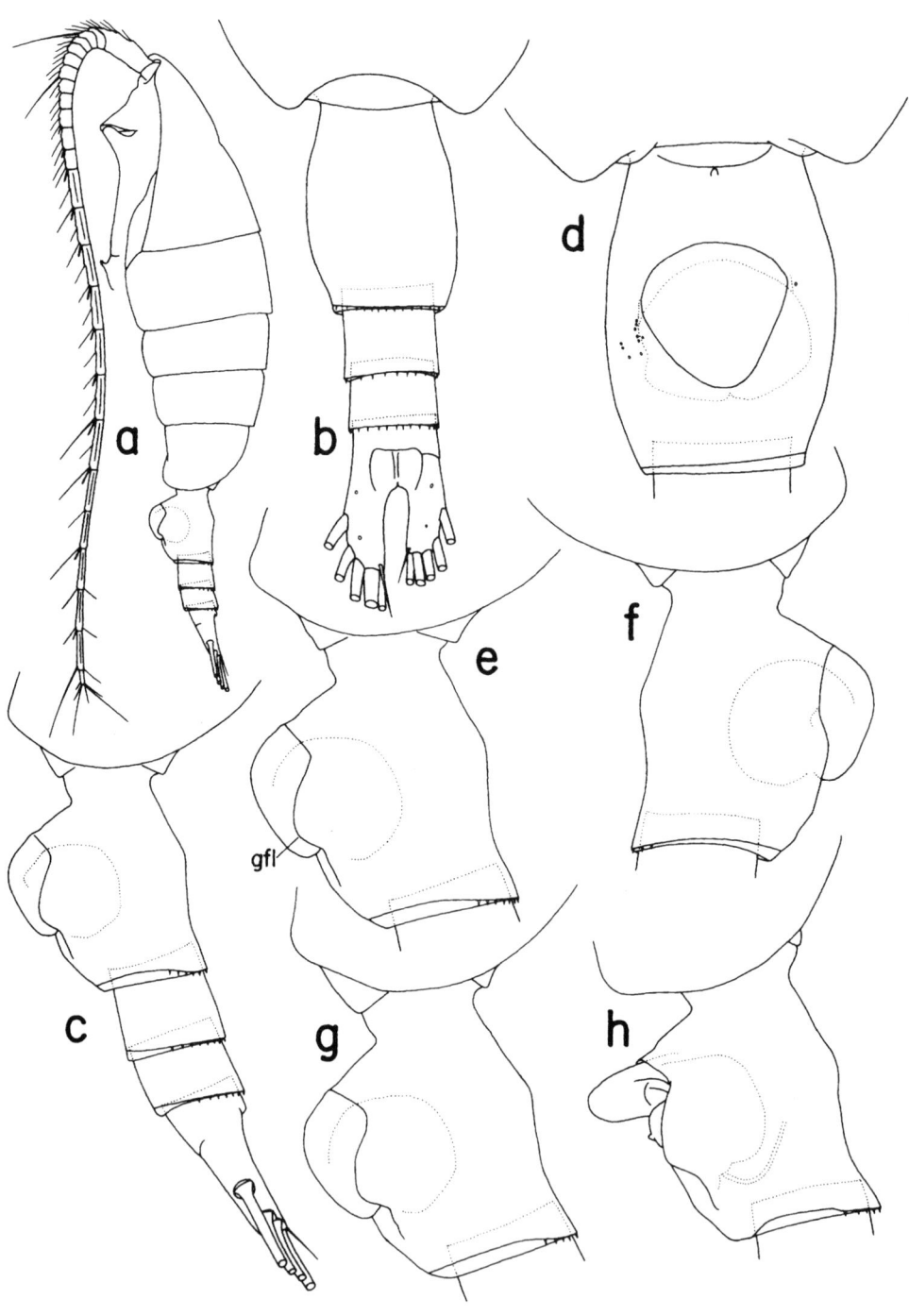

Figure 90. *Heterorhabdus tuberculus*, new species, female: **a**, habitus, left; **b**, urosome, dorsal; **c**, do, left; **d**, genital somite, ventral; **e**, do, left; **f**, do, right; **g**, do, left, tilted clockwise; **h**, genital somite with operculum open, left. gfl = genital flange.

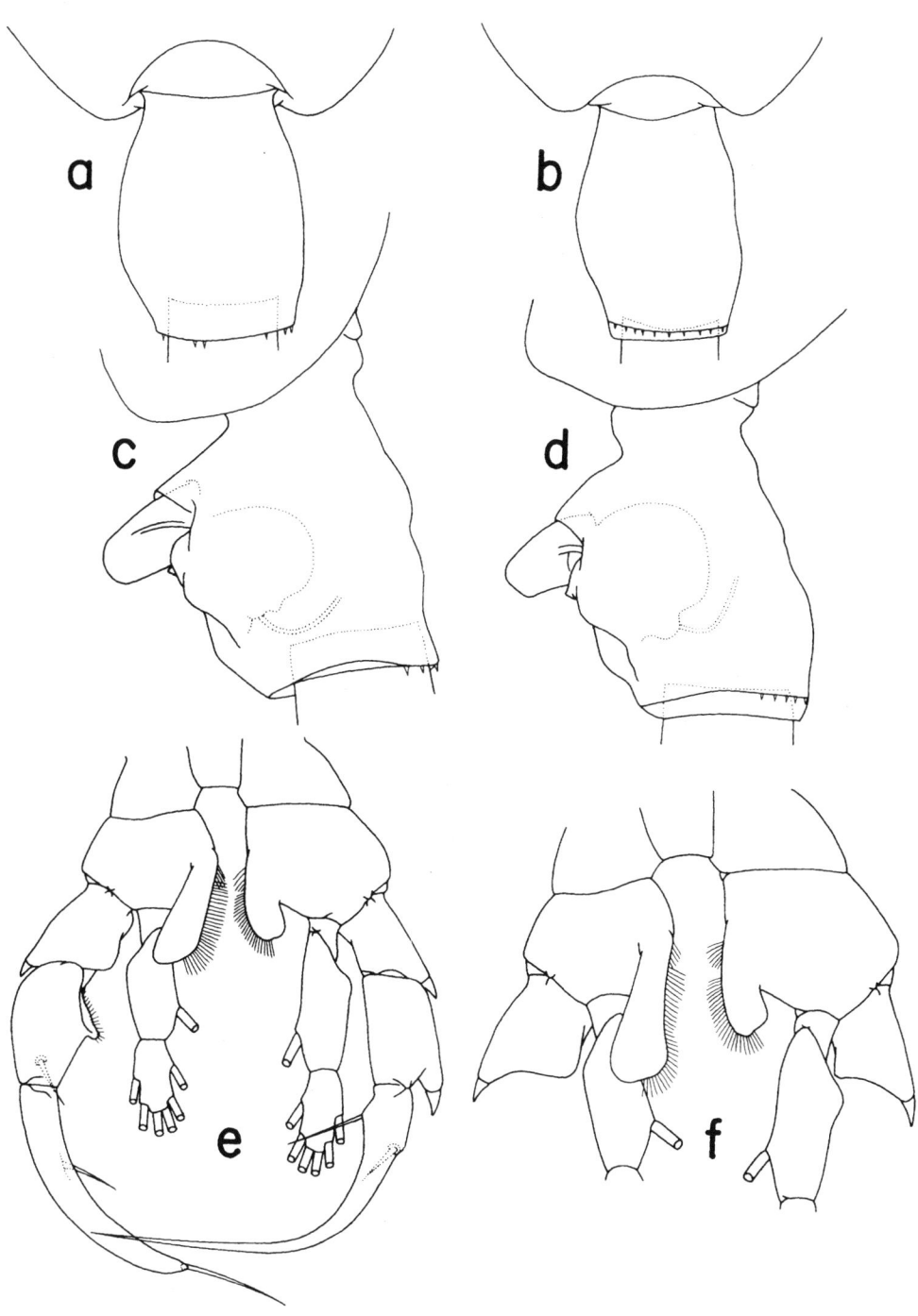

Figure 91. *Heterorhabdus tuberculus*, new species, female: **a**, genital somite, dorsal; **b**, genital somite from different specimen, dorsal; **c, d**, abnormal genital somites, left. Male: **e**, fifth pair of legs, anterior; **f**, fifth pair of legs with distal segments of rami omitted, anterior.

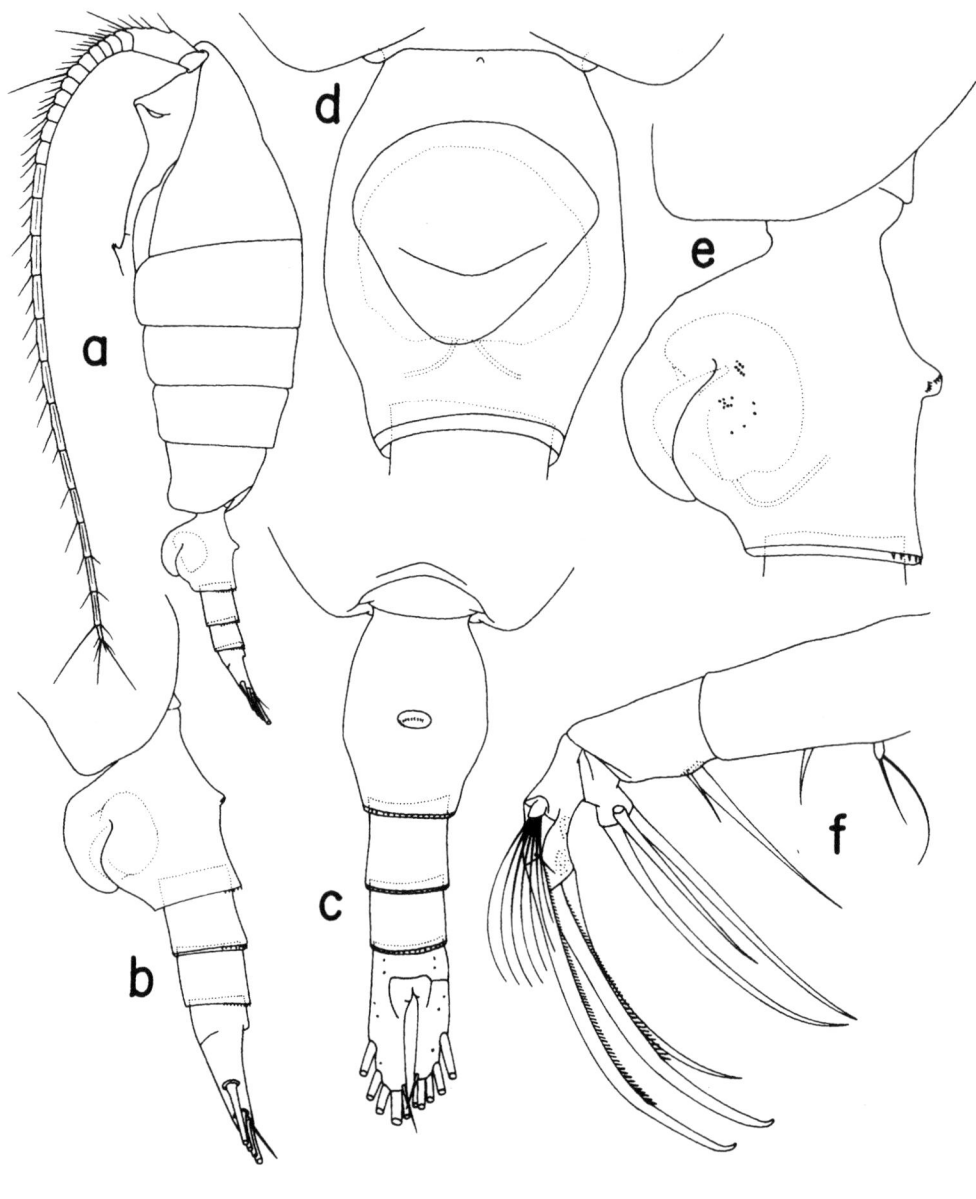

Figure 92. *Heterorhabdus pustulifer* female: **a**, habitus, left; **b**, urosome, left; **c**, do, dorsal; **d**, genital somite, ventral; **e**, do, left; **f**, left maxilla, posterior.

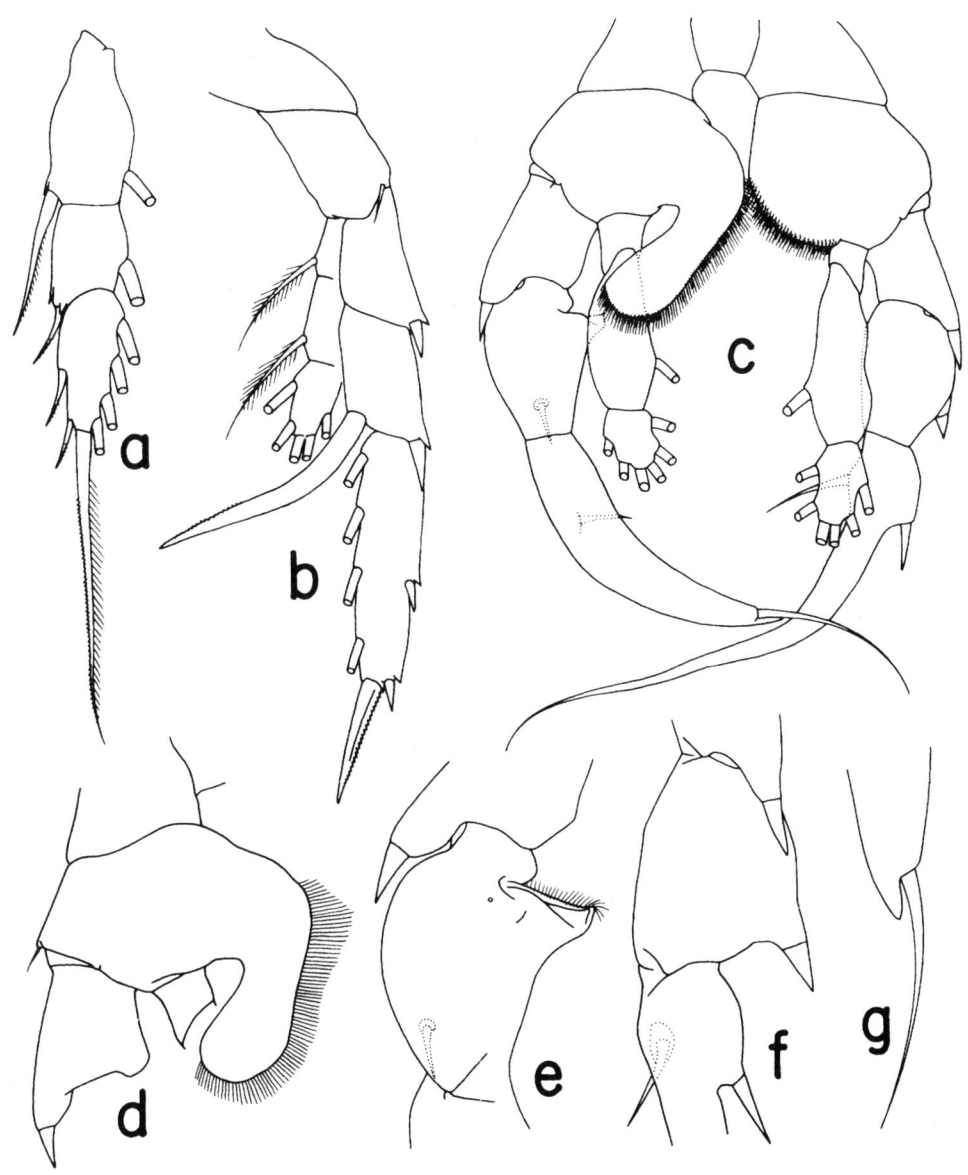

Figure 93. *Heterorhabdus pustulifer* female: **a**, exopod of 1st leg, anterior; **b**, fifth leg, posterior. Male: **c**, fifth pair of legs, anterior; **d**, basipod of right 5th leg, anterior; **e**, second exopodal segment of right 5th leg, anterior; **f**, exopod of left 5th leg, anterior; **g**, distal end of exopod of right 5th leg, lateral.

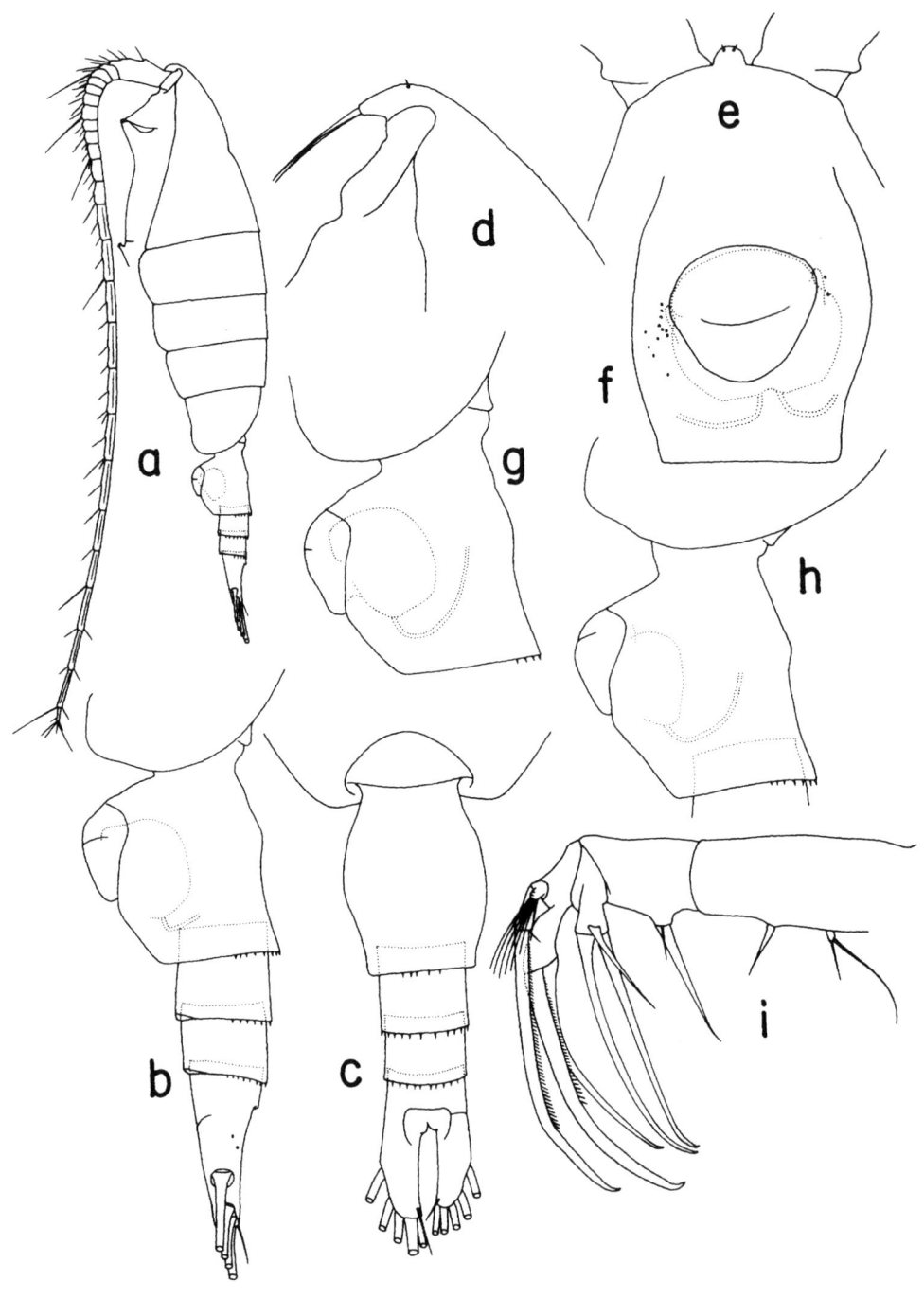

Figure 94. *Heterorhabdus oikoumenikis*, new species, female: **a**, habitus, left; **b**, urosome, left; **c**, do, dorsal; **d**, forehead, left; **e**, do, dorsal; **f**, genital somite, ventral; **g**, do, left; **h**, genital somite from different specimen, left; **i**, left maxilla, posterior.

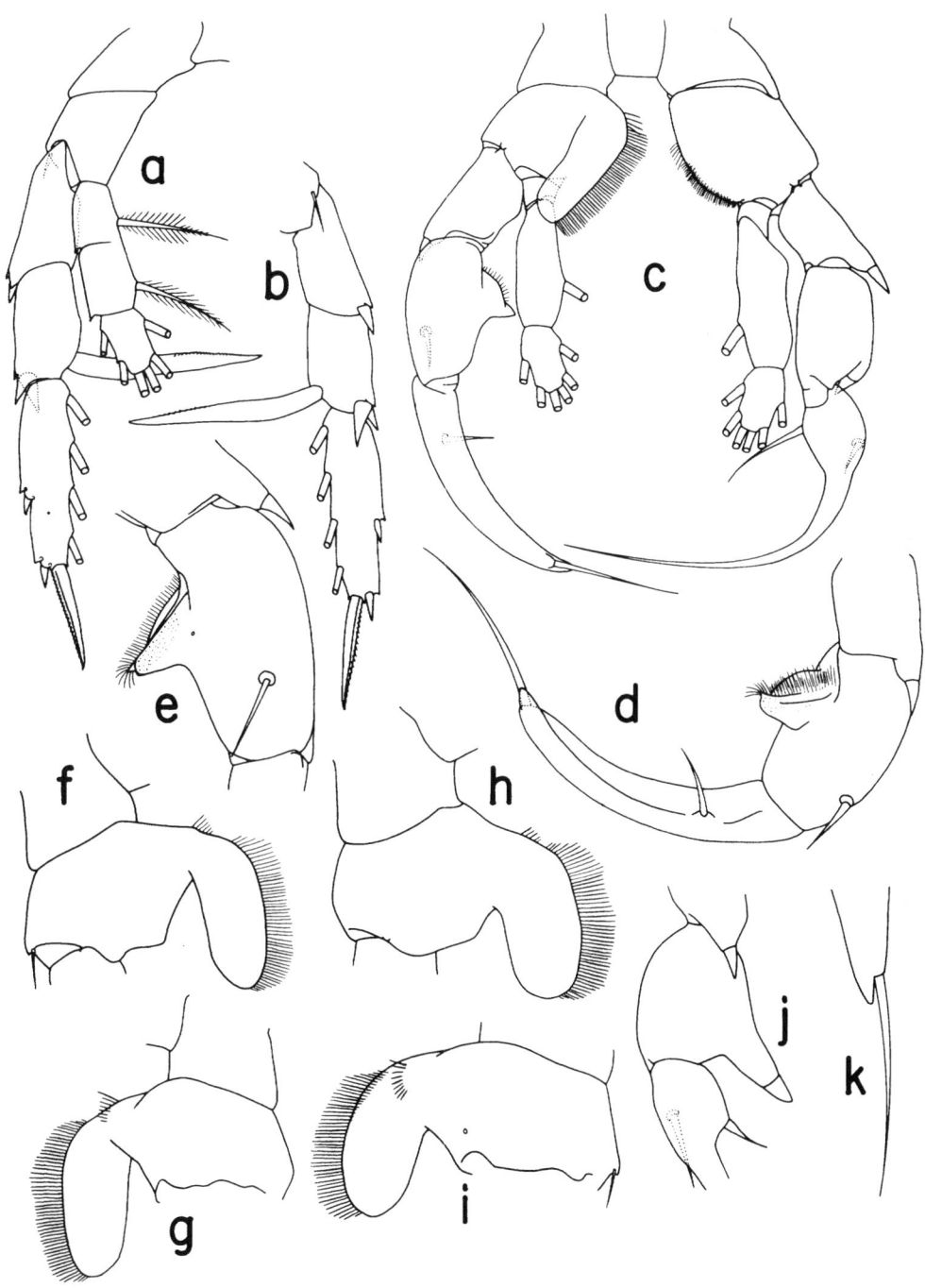

Figure 95. *Heterorhabdus oikoumenikis*, new species, female: **a**, fifth leg, anterior; **b**, exopod of 5th leg, posterior. Male: **c**, fifth pair of legs, anterior; **d**, exopod of right 5th leg, posterior; **e**, second exopodal segment of right 5th leg, posterior; **f**, basipod of right 5th leg, anterior; **g**, do, posterior; **h**, basipod of right 5th leg from different specimen, anterior; **i**, do, posterior; **j**, exopod of left 5th leg, anterior; **k**, distal end of exopod of right 5th leg, lateral.

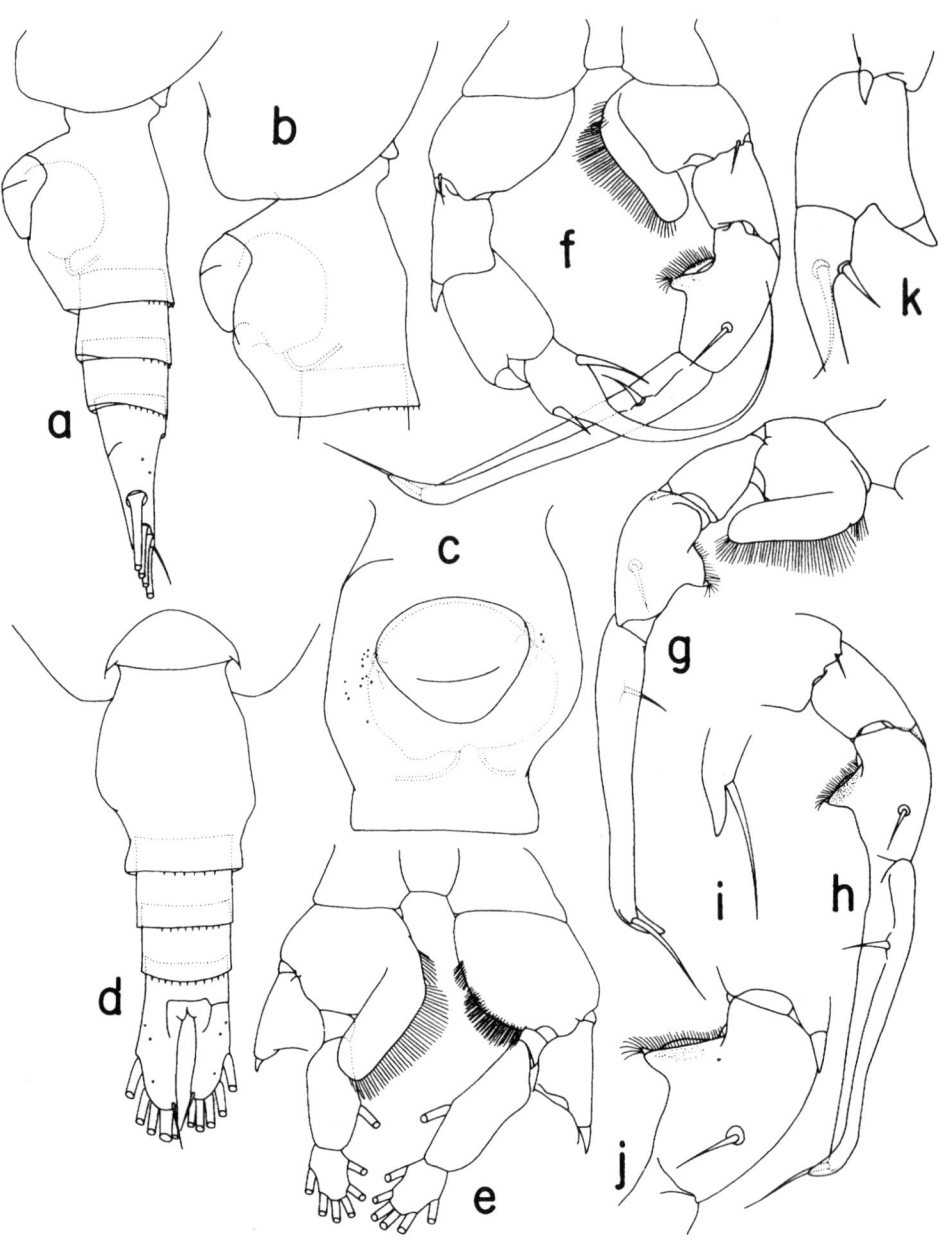

Figure 96. *Heterorhabdus longisegmentus*, new species, female: **a**, urosome, left; **b**, genital somite, left; *c*, do, ventral; **d**, urosome, dorsal. Male: **e**, fifth pair of legs with distal exopodal segments omitted, anterior; **f**, fifth pair of legs with endopods omitted, posterior; **g**, right 5th leg with endopod omitted, anterior; **h**, exopod of right 5th leg, posterior; **i**, distal end of exopod of right 5th leg, lateral; **j**, second exopodal segment of right 5th leg, posterior; **k**, exopod of left 5th leg, anterior.

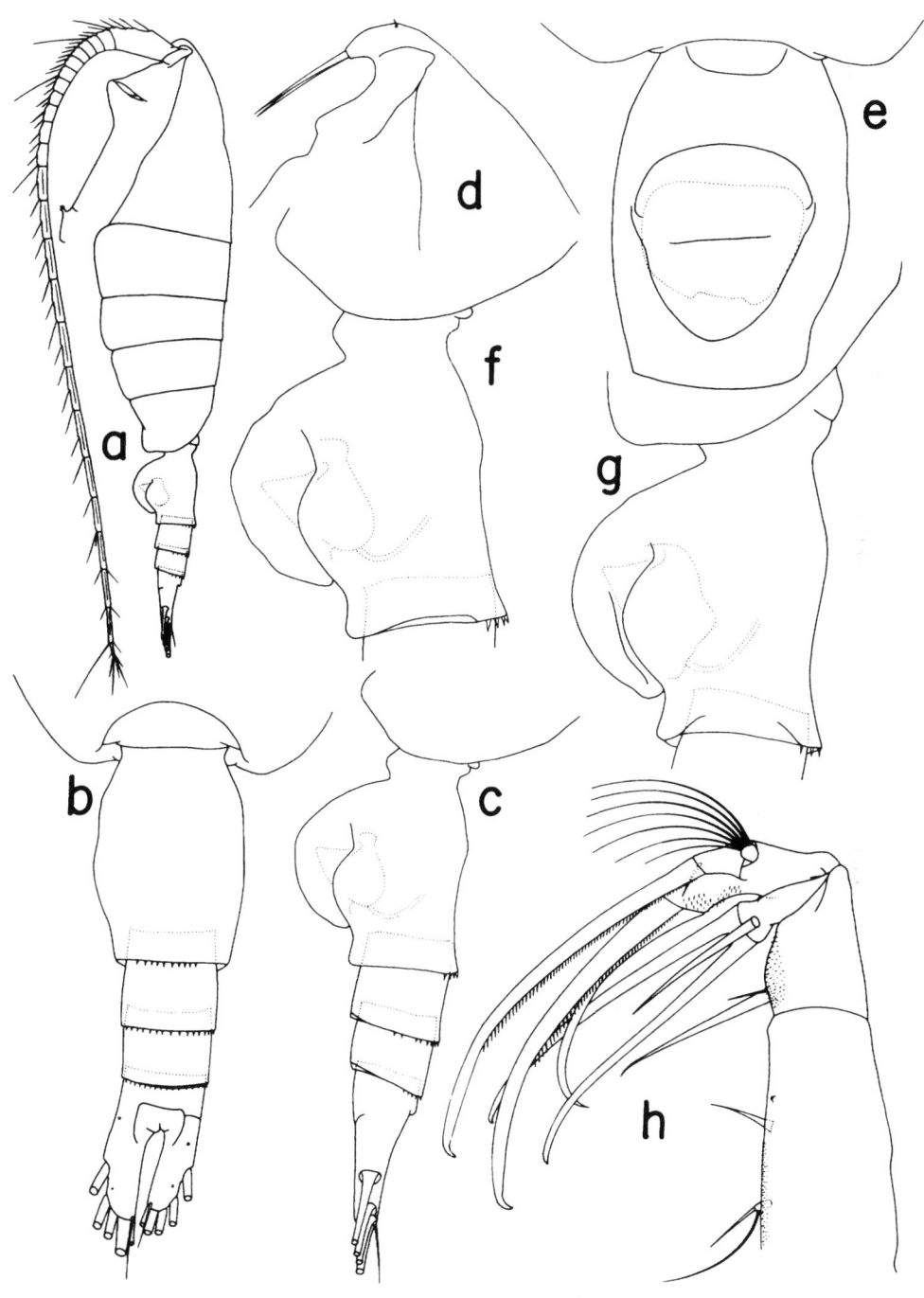

Figure 97. *Heterorhabdus americanus*, new species, female: **a**, habitus, left; **b**, urosome, dorsal; **c**, do, left; **d**, forehead, left; **e**, genital somite, ventral; **f**, do, left; **g**, genital somite from different specimen, left; **h**, left maxilla, posterior.

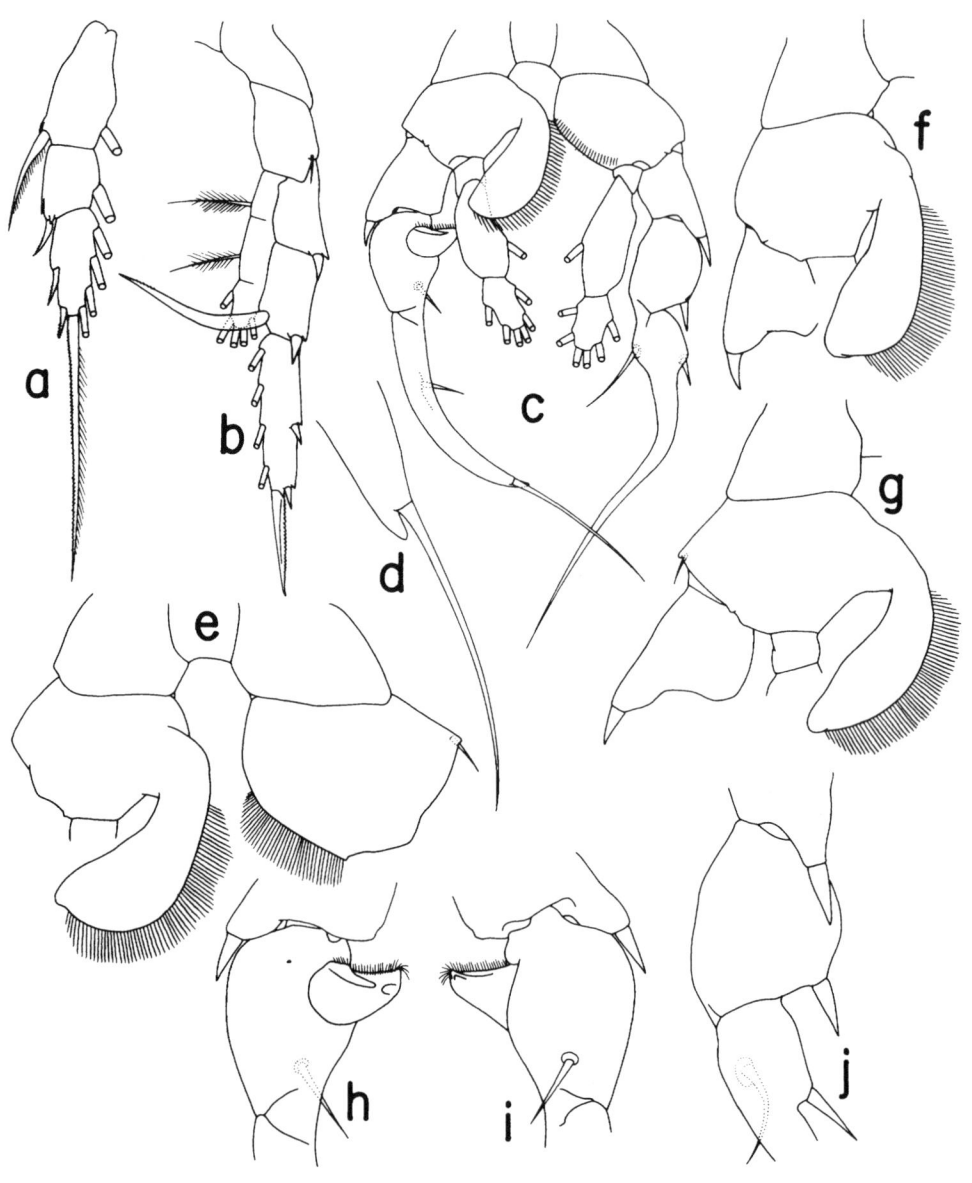

Figure 98. *Heterorhabdus americanus*, new species, female: **a**, exopod of 1st leg, anterior; **b**, fifth leg, posterior. Male: **c**, fifth pair of legs, anterior; **d**, distal end of exopod of right 5th leg, lateral; **e**, basipods of 5th legs, anterior; **f**, basipod of right 5th leg, anterior; **g**, do, tilted clockwise; **h**, second exopodal segment of right 5th leg, anterior; **i**, do, posterior; **j**, exopod of left 5th leg, anterior.

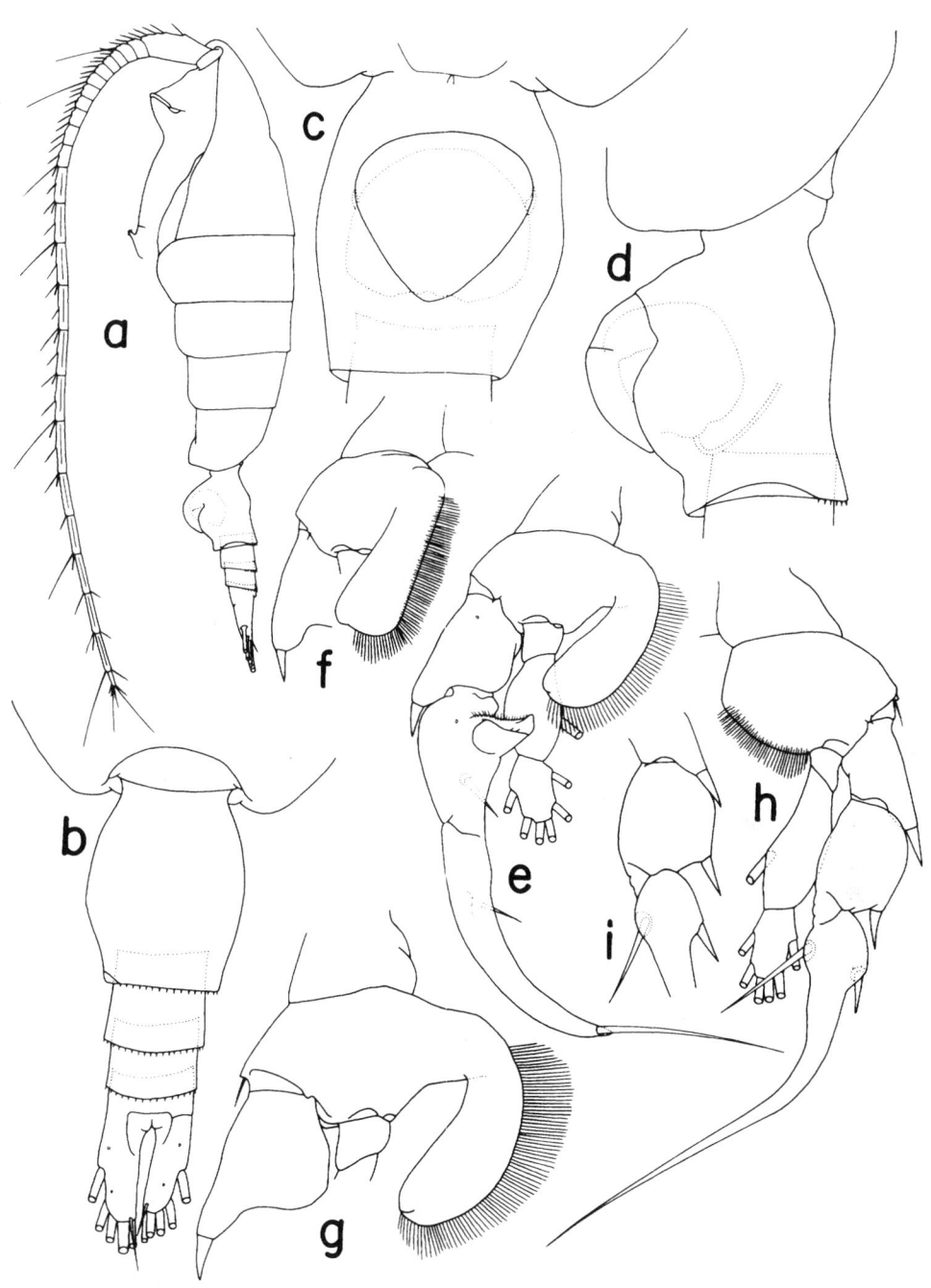

Figure 99. *Heterorhabdus cohibilis*, new species, female: **a**, habitus, left; **b**, urosome, dorsal; **c**, genital somite, ventral; **d**, do, left. Male: **e**, right 5th leg, anterior; **f**, basipod of right 5th leg, anterior; **g**, do, tilted clockwise; **h**, left 5th leg, anterior; **i**, exopod of left 5th leg, anterior.

251

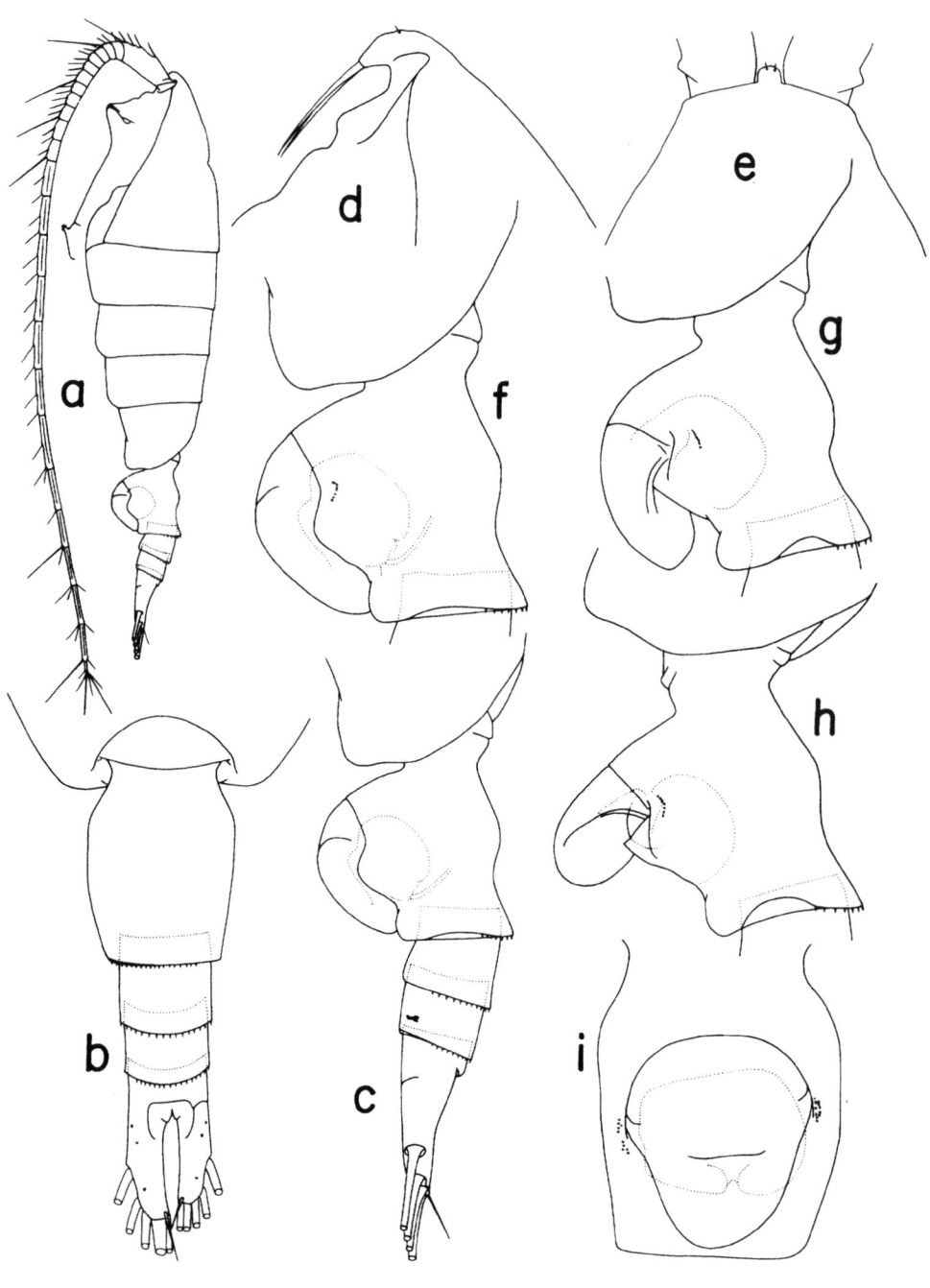

Figure 100. *Heterorhabdus spinosus* female: **a**, habitus, left; **b**, urosome, dorsal; **c**, do, left; **d**, forehead, left; **e**, do, dorsal; **f**, genital somite, left; **g**, genital somite with operculum partially open, left; **h**, genital somite with operculum fully open, left; **i**, genital somite, ventral.

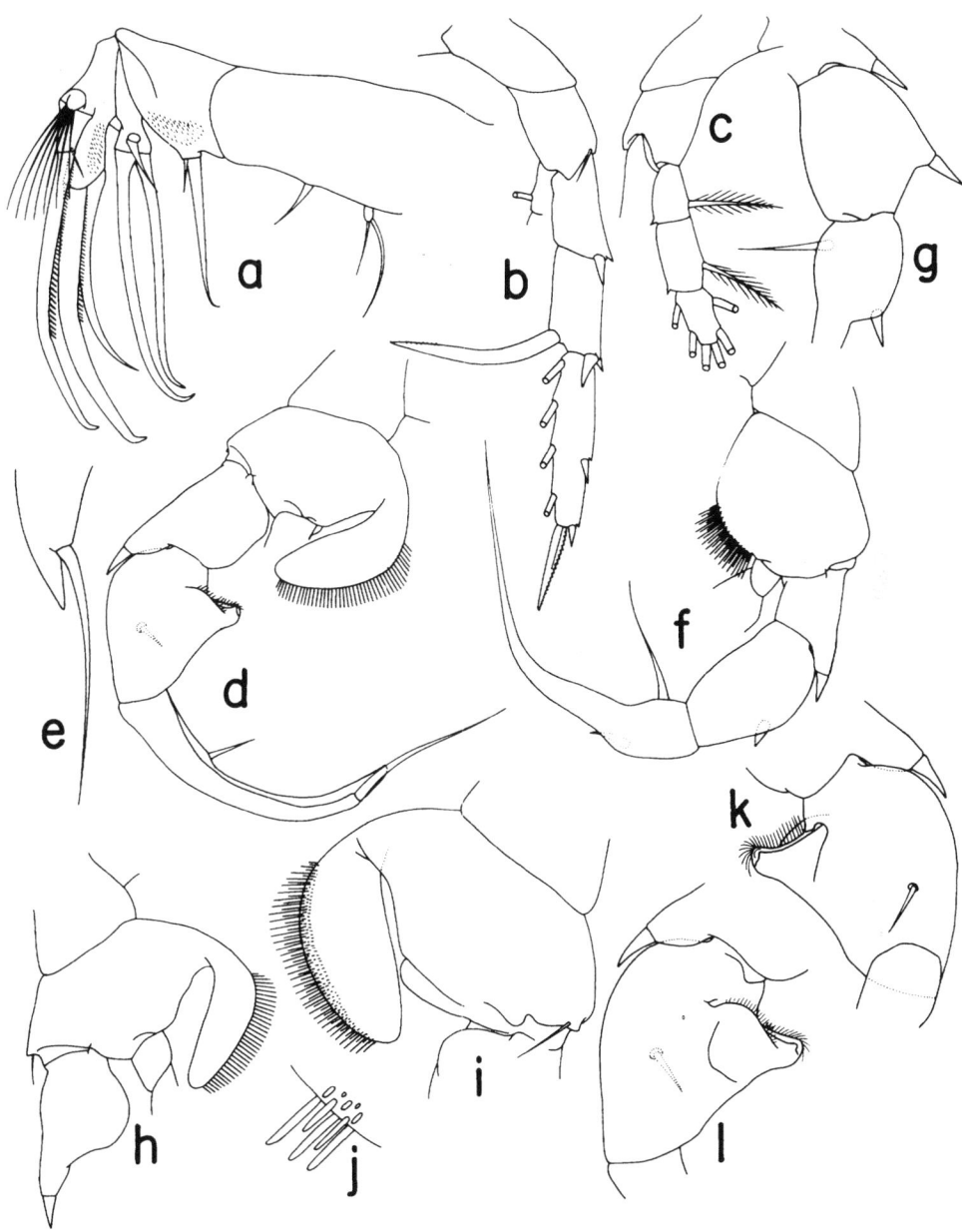

Figure 101. *Heterorhabdus spinosus* female: **a**, left maxilla, posterior; **b**, fifth leg with endopod omitted, posterior; **c**, fifth leg with exopod omitted, anterior. Male: **d**, right 5th leg with endopod omitted, anterior; **e**, distal end of exopod of right 5th leg, lateral; **f**, left 5th leg, anterior; **g**, exopod of left 5th leg, anterior; **h**, basipod of right 5th leg, anterior; **i**, do, posterior; **j**, spinules on basal lobe of right 5th leg; **k**, second exopodal segment of right 5th leg, anterior; **l**, do, posterior.

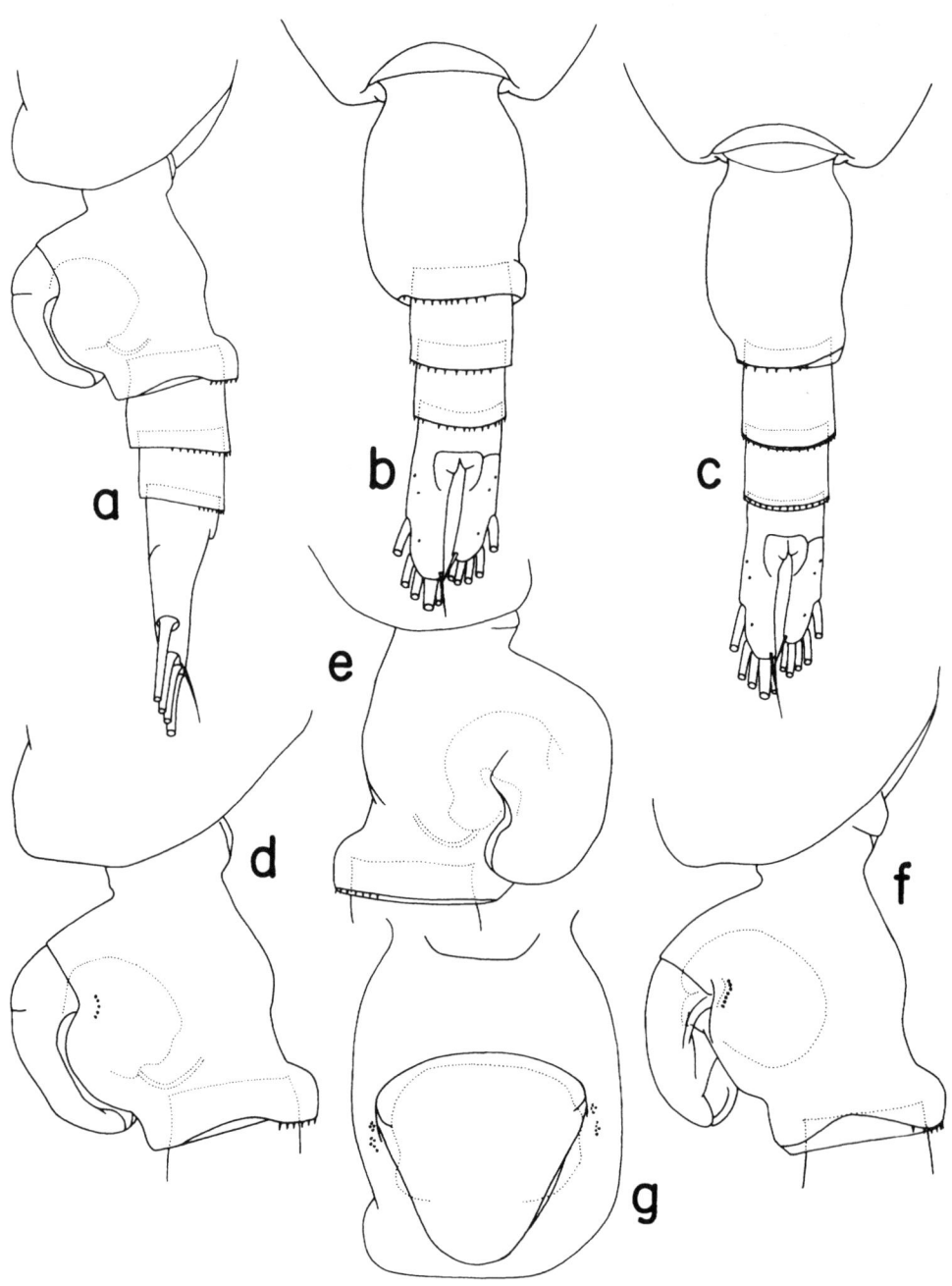

Figure 102. *Heterorhabdus paraspinosus*, new species, female: **a**, urosome, left; **b**, do, dorsal; **c**, urosome from different specimen, dorsal; **d**, genital somite, left; **e**, do, right; **f**, genital somite from different specimen, left; **g**, genital somite, ventral.

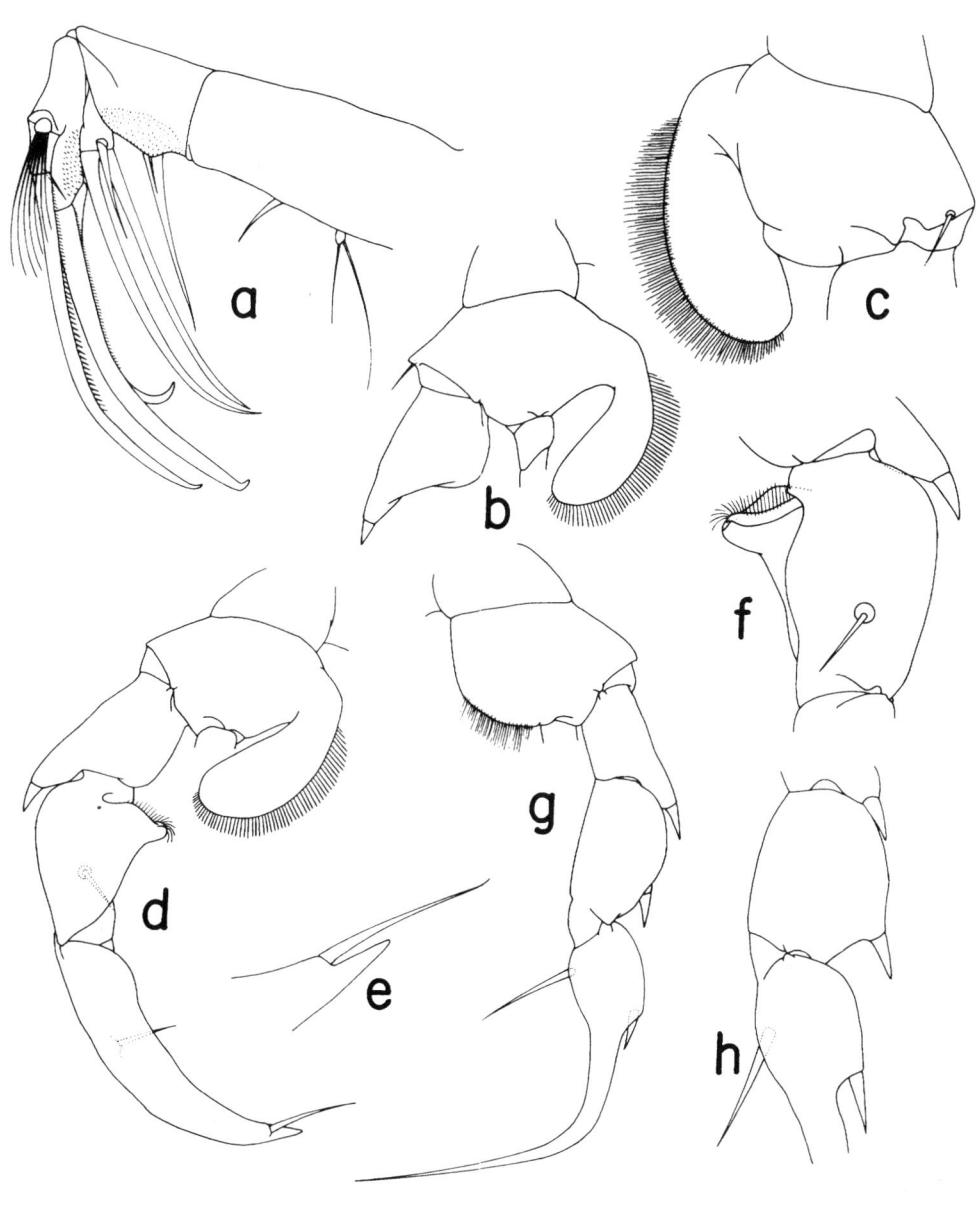

Figure 103. *Heterorhabdus paraspinosus*, new species, female: **a**, left maxilla, posterior. Male: **b**, basipod of right 5th leg, anterior; **c**, do, posterior; **d**, right 5th leg with endopod omitted, anterior; **e**, distal end of exopod of right 5th leg, lateral; **f**, second exopodal segment of right 5th leg, posterior; **g**, left 5th leg, anterior; **h**, exopod of left 5th leg, anterior.

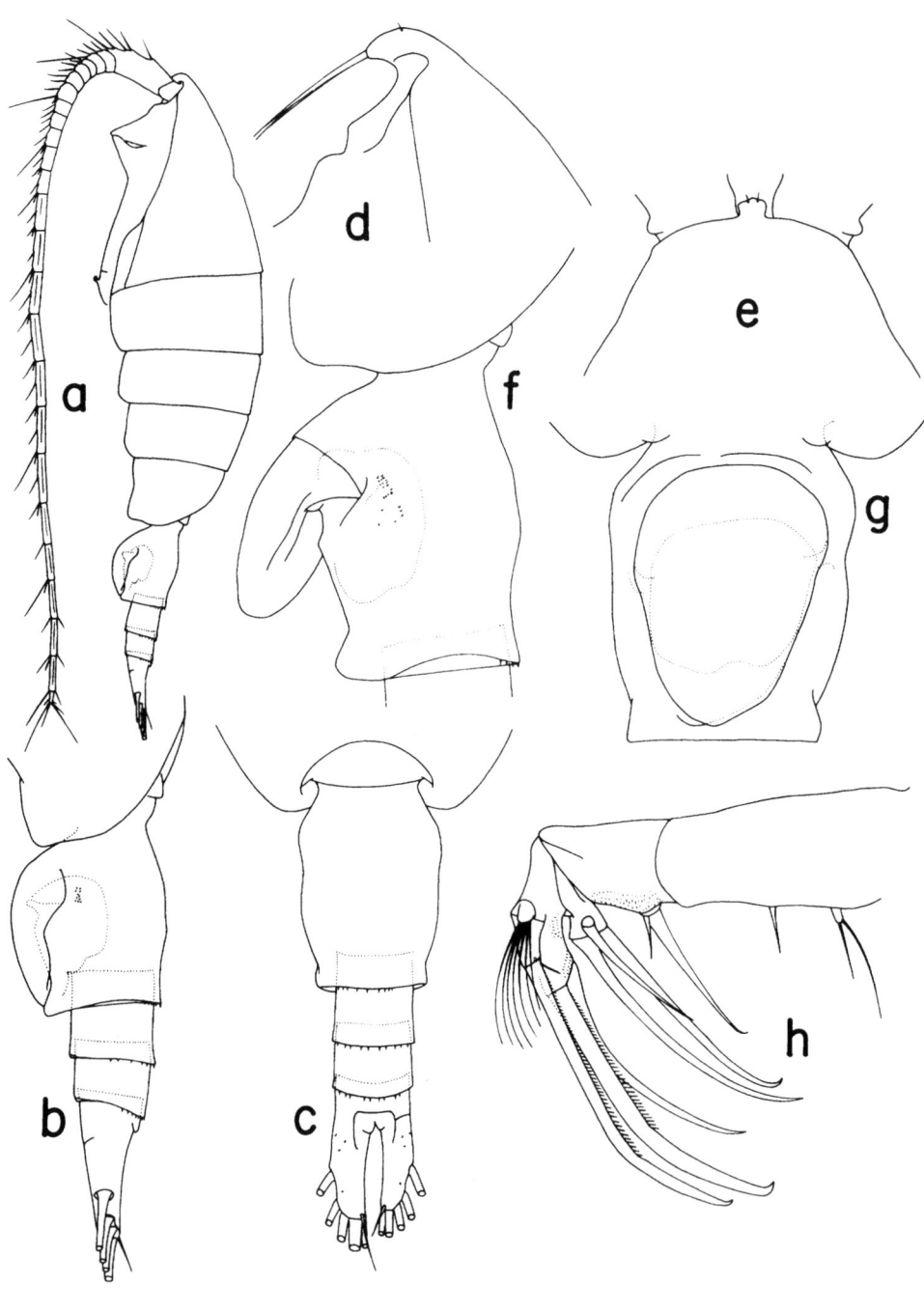

Figure 104. *Heterorhabdus austrinus* female: **a**, habitus, left; **b**, urosome, left; **c**, do, dorsal; **d**, forehead, left; **e**, do, dorsal; **f**, genital somite, left; **g**, do, ventral; **h**, left maxilla, posterior.

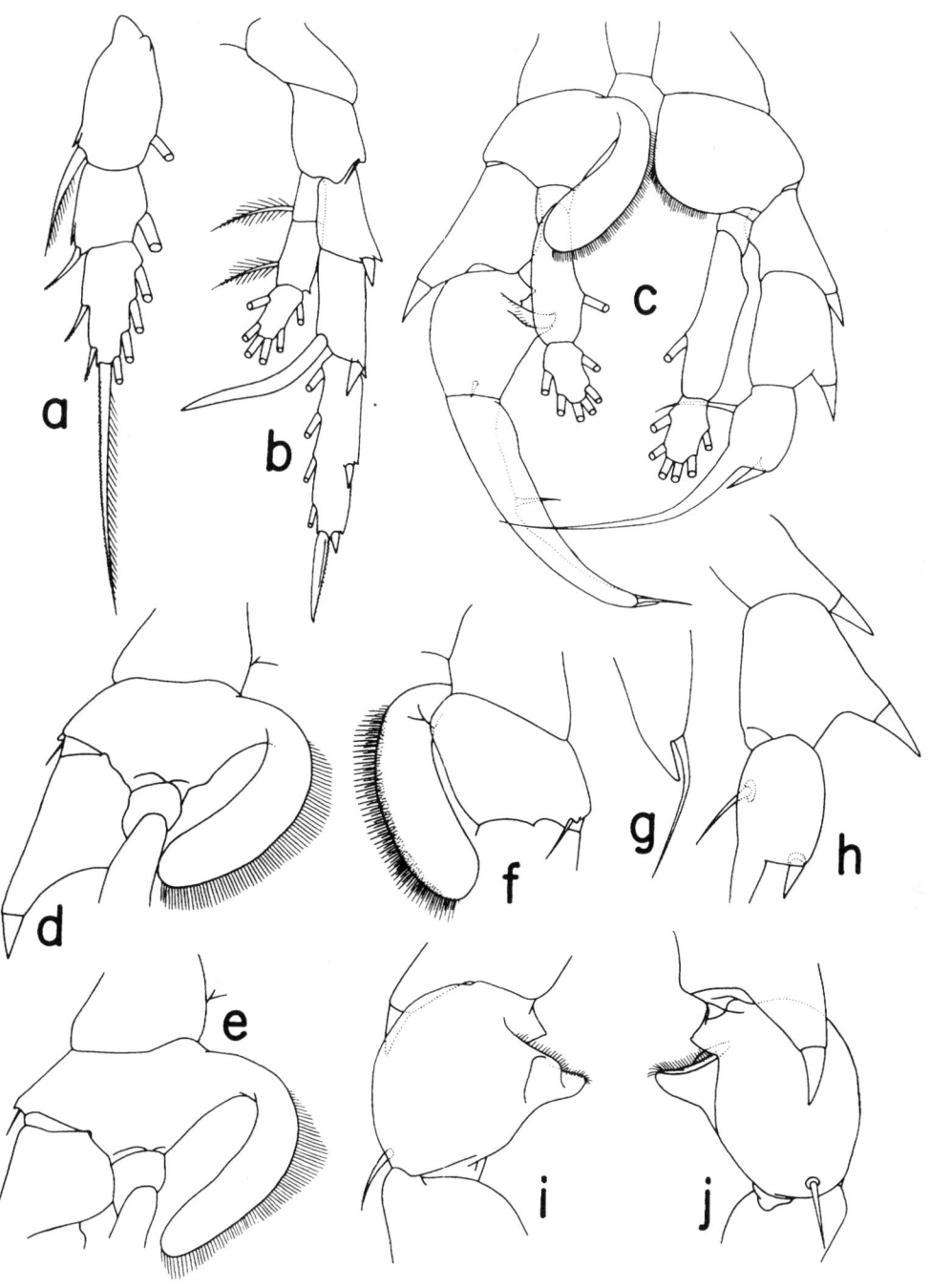

Figure 105. *Heterorhabdus austrinus* female: **a**, exopod of 1st leg, anterior; **b**, fifth leg, posterior. Male: **c**, fifth pair of legs, anterior; **d**, basipod of right 5th leg, anterior; **e**, do, tilted clockwise; **f**, do, posterior; **g**, distal end of exopod of right 5th leg, lateral; **h**, exopod of left 5th leg, anterior; **i**, second exopodal segment of right 5th leg, anterior; **j**, do, posterior.

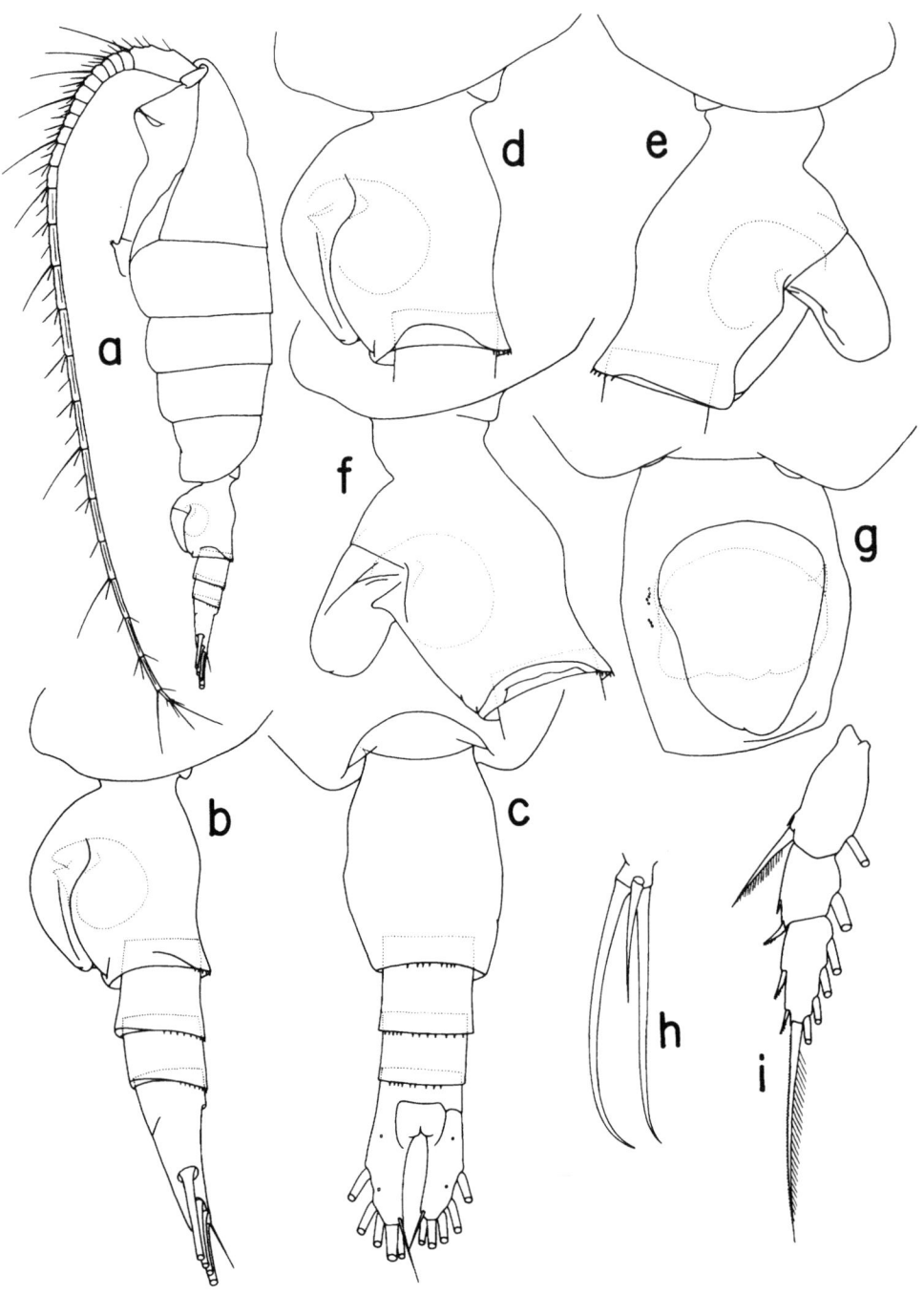

Figure 106. *Heterorhabdus abyssalis* female: **a**, habitus, left; **b**, urosome, left; **c**, do, dorsal; **d**, genital somite, left; **e**, genital somite with operculum open, right; **f**, do, left; **g**, genital somite, ventral; **h**, 4th lobe of left maxilla, posterior; **i**, exopod of 1st leg, anterior.

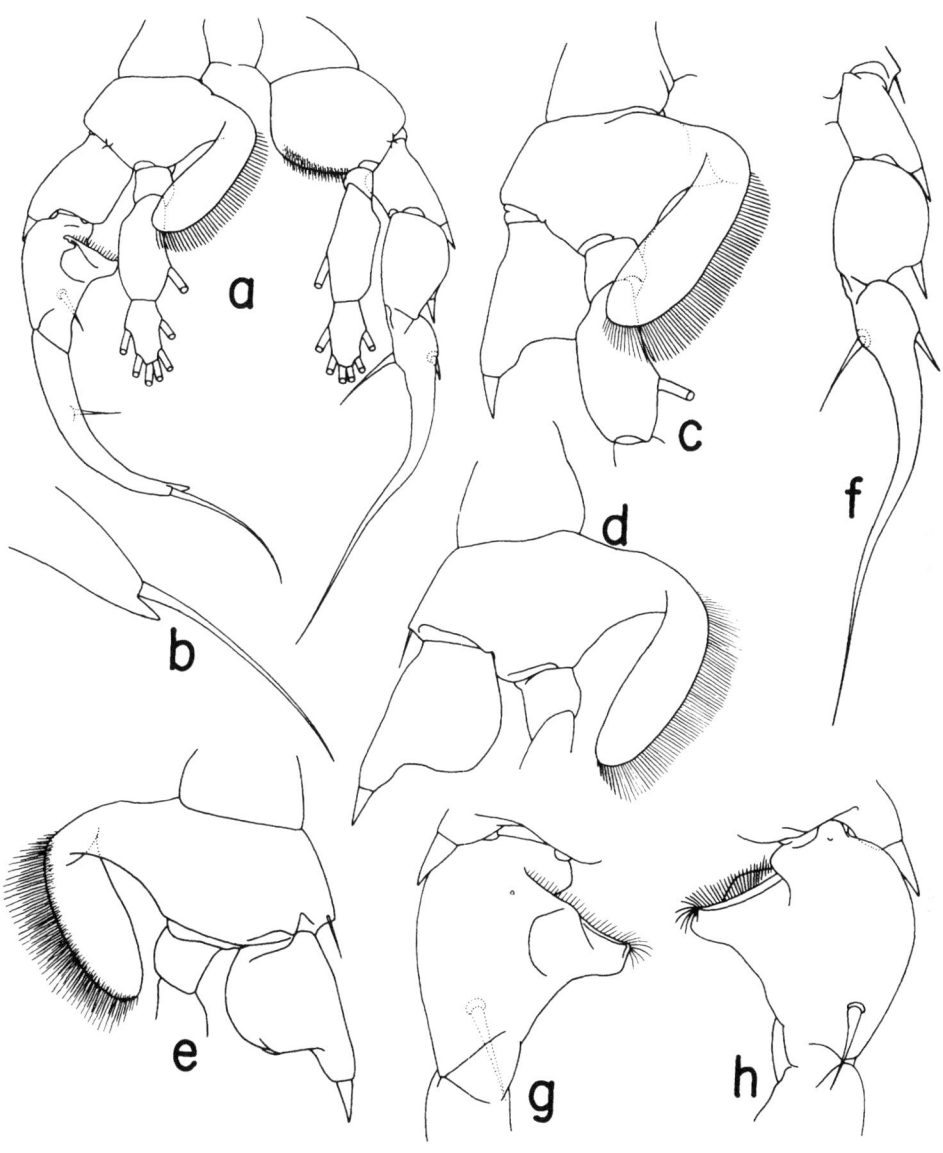

Figure 107. *Heterorhabdus abyssalis* male: **a**, fifth pair of legs, anterior; **b**, distal end of exopod of right 5th leg, lateral; **c**, basipod of right 5th leg, anterior; **d**, do, tilted clockwise; **e**, do, posterior; **f**, exopod of left 5th leg, anterior; **g**, second exopodal segment of right 5th leg, anterior; **h**, do, posterior.

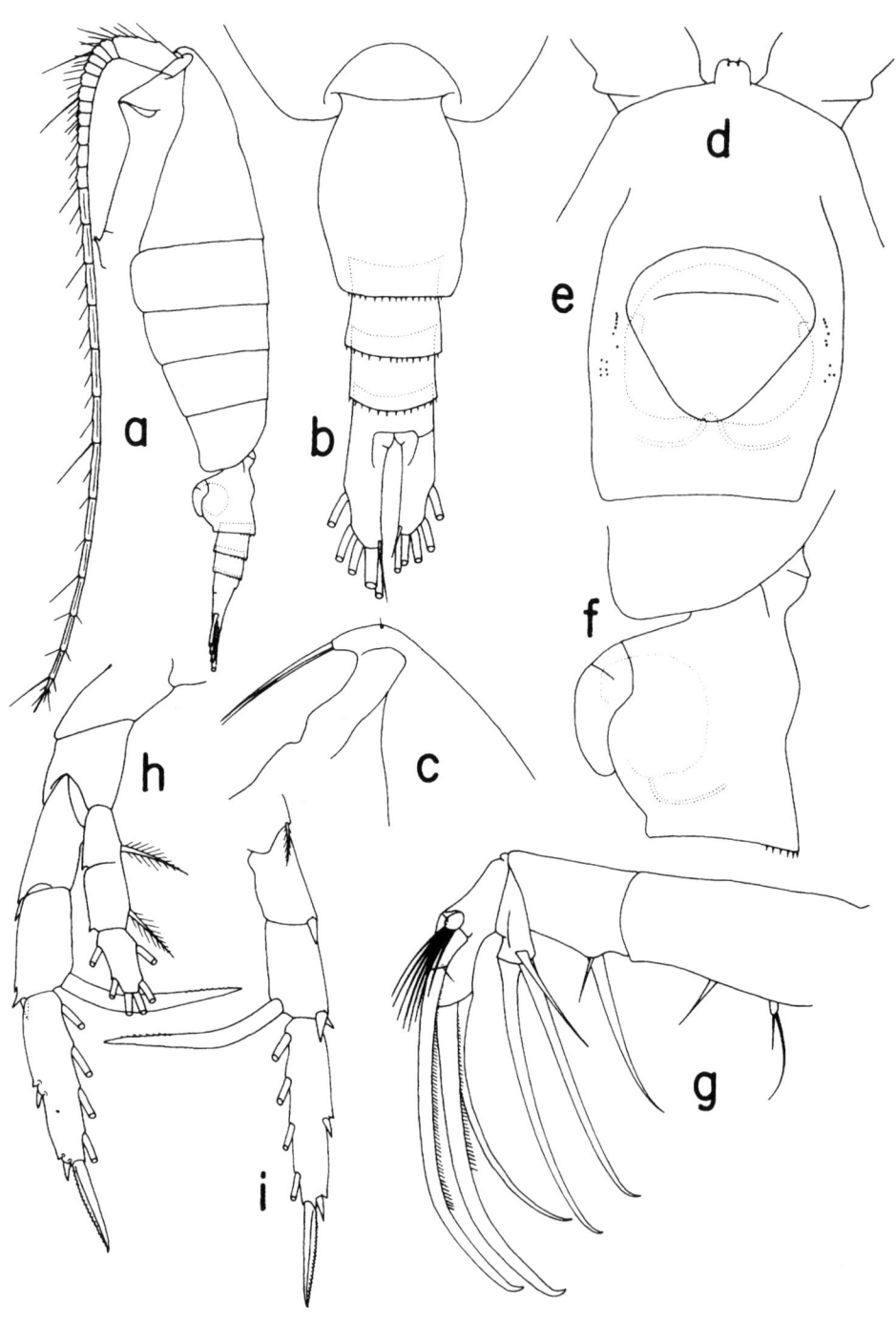

Figure 108. *Heterorhabdus pacificus* female: **a**, habitus, left; **b**, urosome, dorsal; **c**, forehead, left; **d**, do, dorsal; **e**, genital somite, ventral; **f**, do, left; **g**, left maxilla, posterior; **h**, fifth leg, anterior; **i**, exopod of 5th leg, posterior.

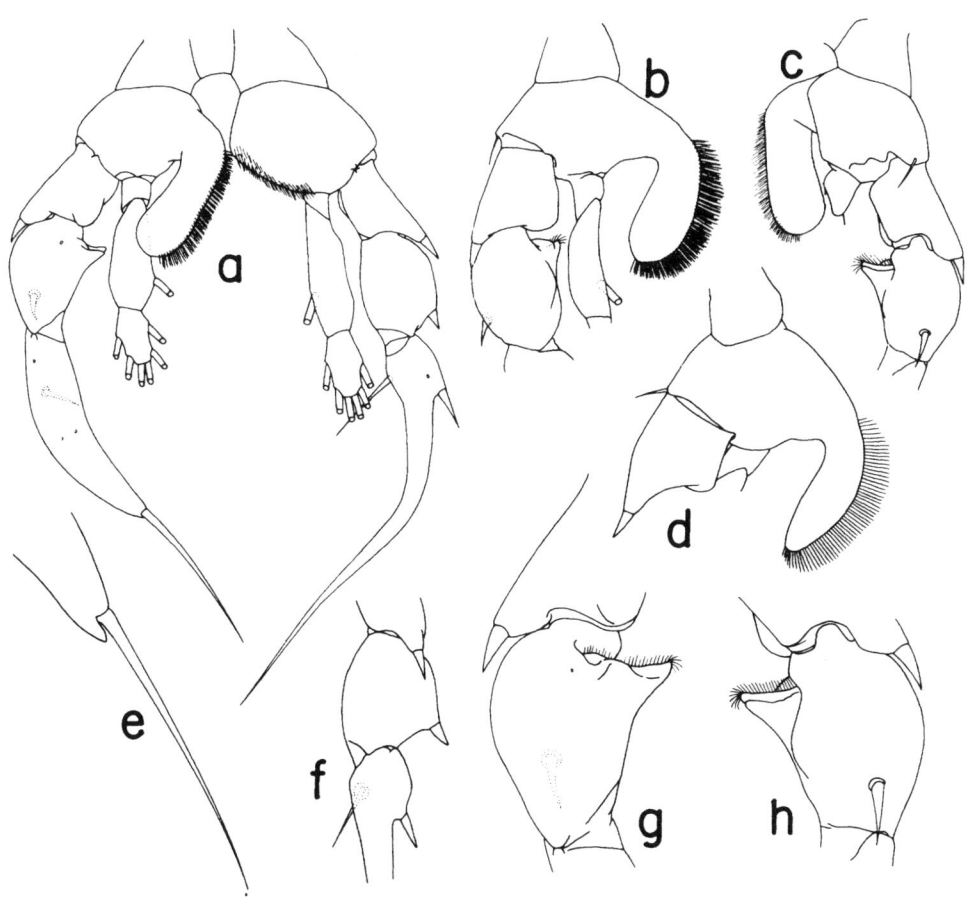

Figure 109. *Heterorhabdus pacificus* male: **a**, fifth pair of legs, anterior; **b**, right 5th leg with distal segments of rami omitted, anterior; **c**, do, posterior; **d**, basipod of right 5th leg, anterior; **e**, distal end of exopod of right 5th leg, lateral; **f**, exopod of left 5th leg, anterior; **g**, second exopodal segment of right 5th leg, anterior; **h**, do, posterior.

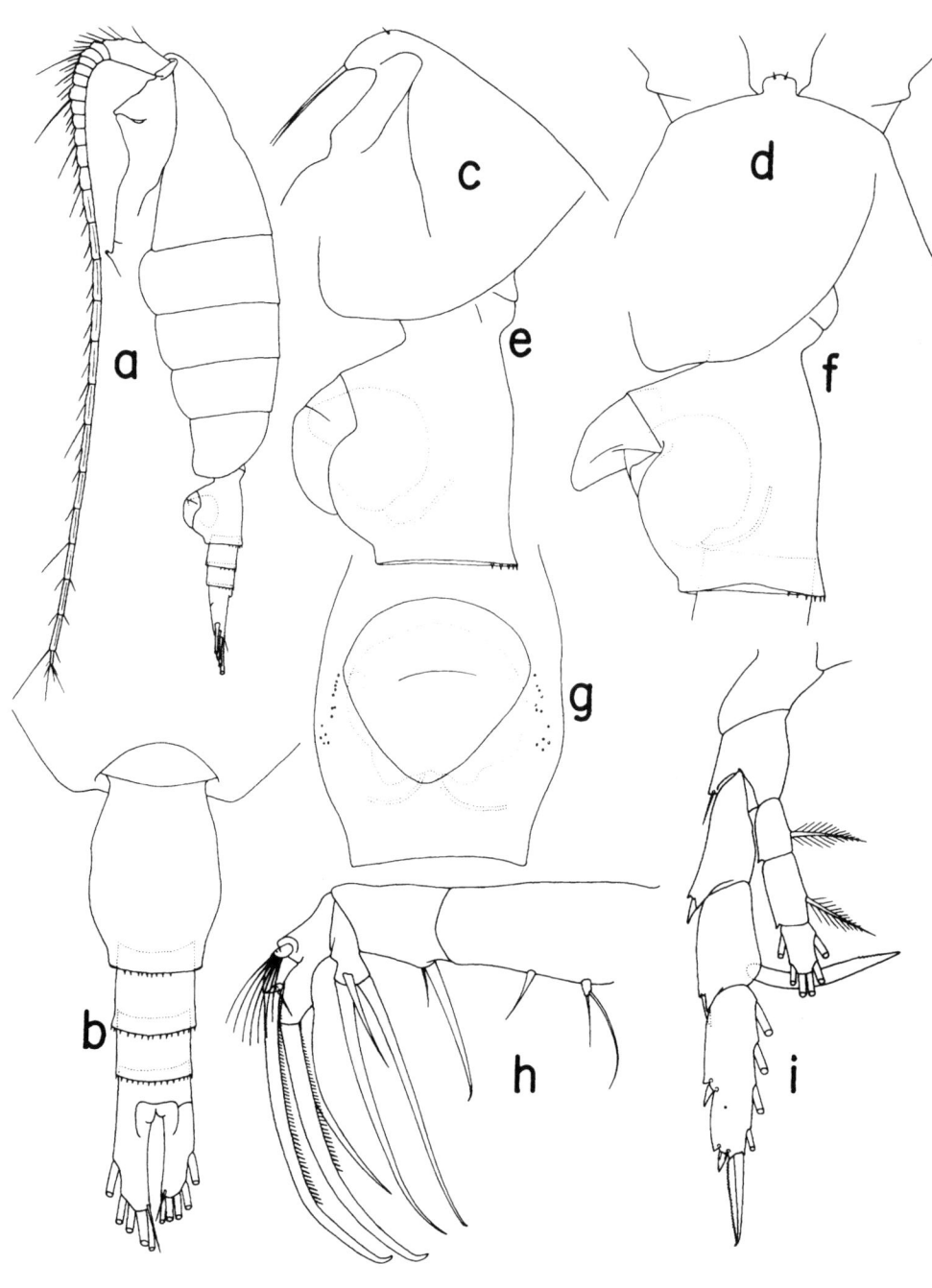

Figure 110. *Heterorhabdus confusibilis*, new species, female: **a**, habitus, left; **b**, urosome, dorsal; **c**, forehead, left; **d**, do, dorsal; **e**, genital somite, left; **f**, genital somite with operculum open, left; **g**, genital somite, ventral; **h**, left maxilla, posterior; **i**, fifth leg, anterior.

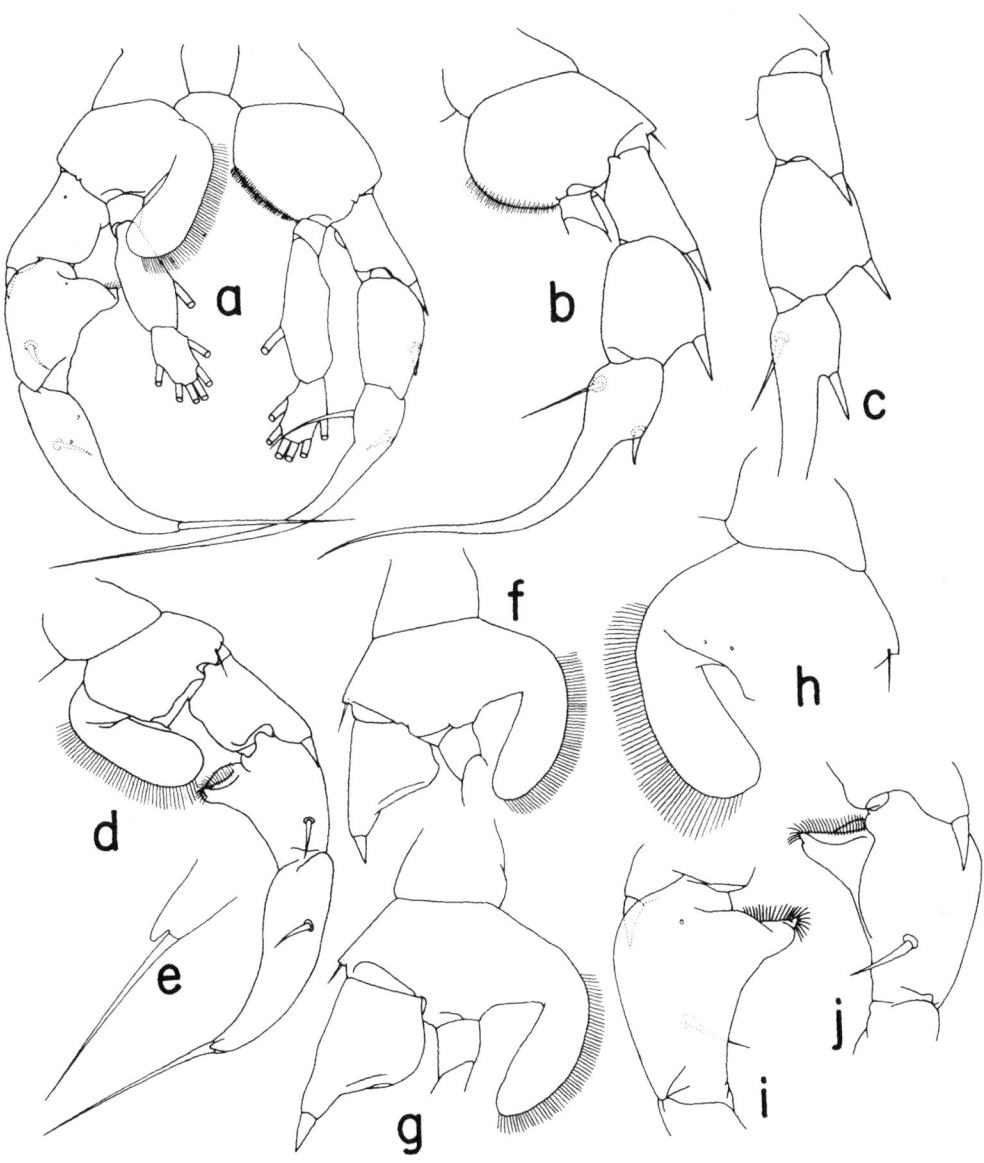

Figure 111. *Heterorhabdus confusibilis*, new species, male: **a**, fifth pair of legs, anterior; **b**, left 5th leg with endopod omitted, anterior; **c**, exopod of left 5th leg, anterior; **d**, right 5th leg with endopod omitted, posterior; **e**, distal end of exopod of right 5th leg, lateral; **f**, basipod of right 5th leg, anterior; **g**, do, slightly tilted clockwise; **h**, do, posterior; **i**, second exopodal segment of right 5th leg, anterior; **j**, do, posterior.

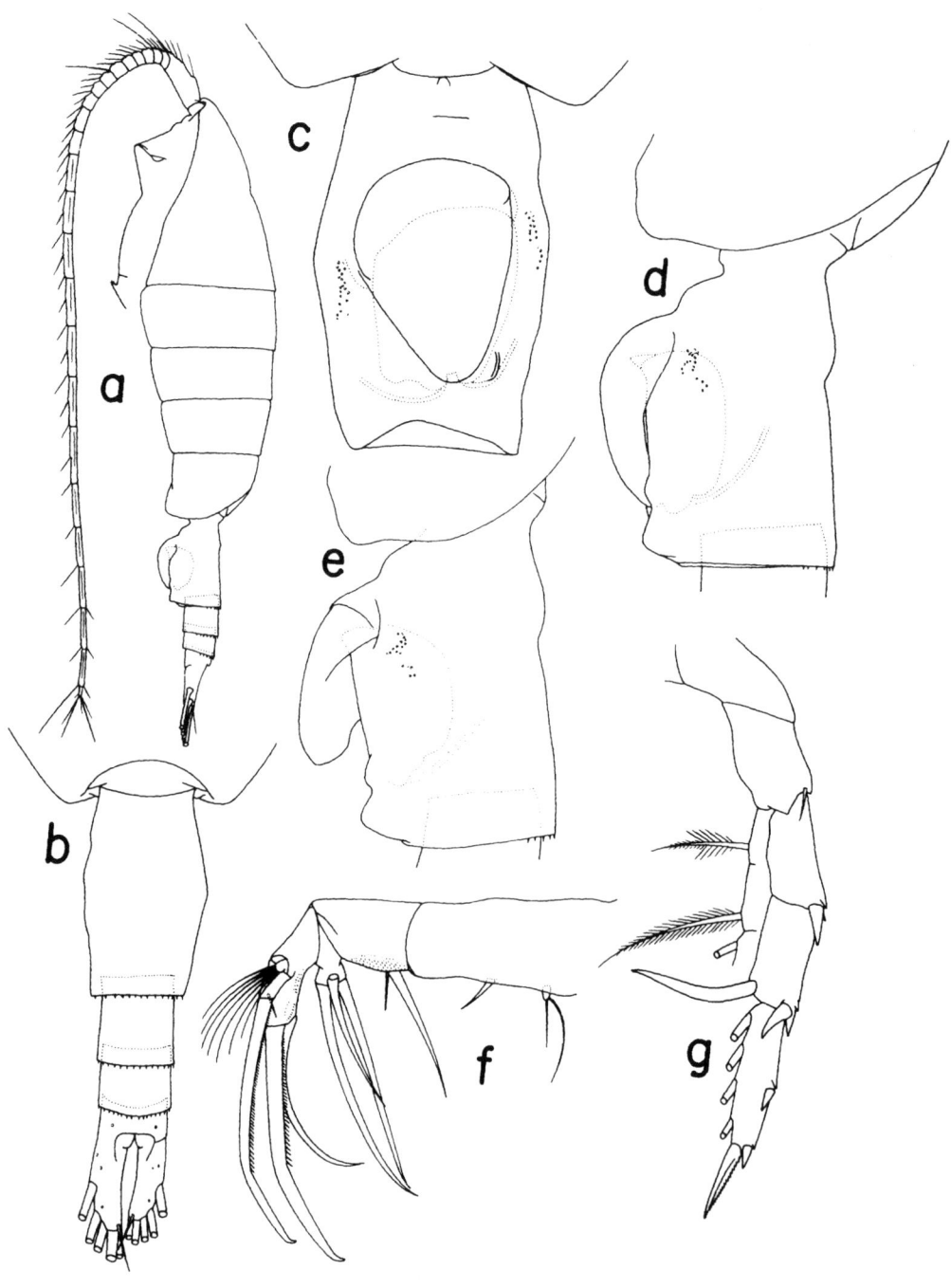

Figure 112. *Heterorhabdus tanneri* female: **a**, habitus, left; **b**, urosome, dorsal; **c**, genital somite, ventral; **d**, do, left; **e**, genital somite with operculum partially open, left; **f**, left maxilla, posterior; **g**, fifth leg, posterior.

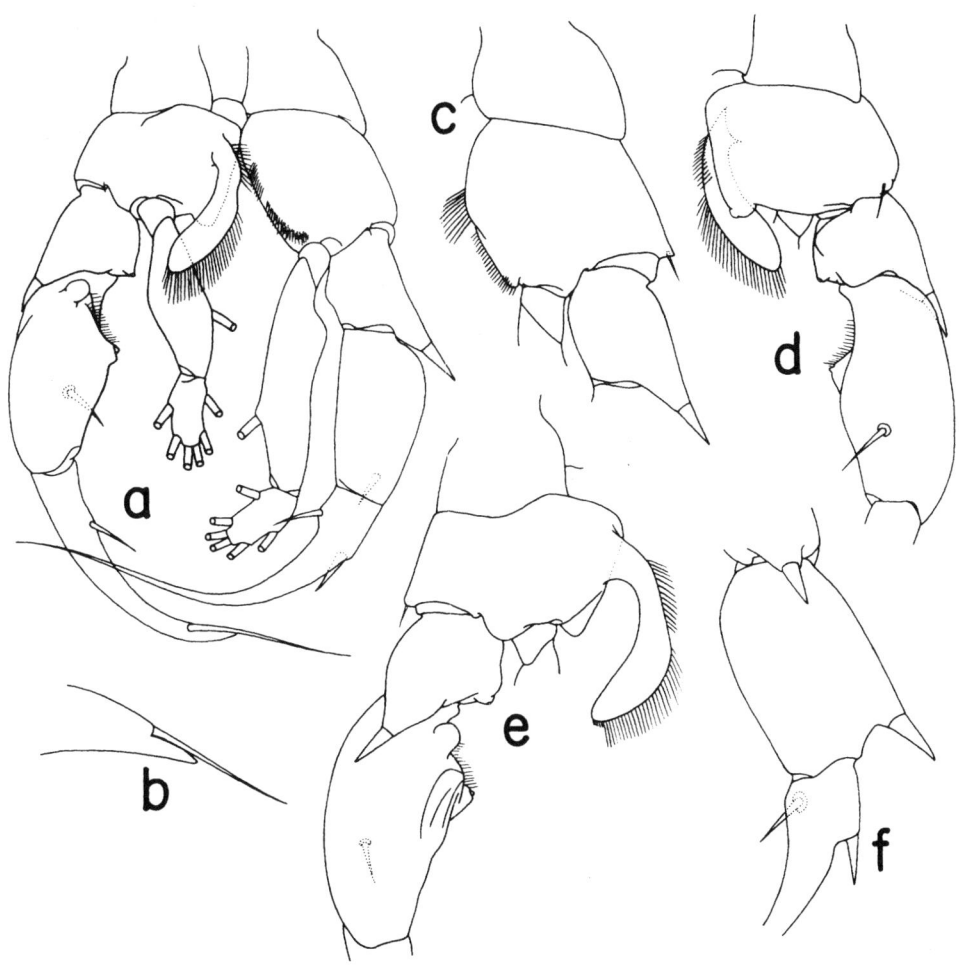

Figure 113. *Heterorhabdus tanneri* male: **a**, fifth pair of legs, anterior; **b**, distal end of exopod of right 5th leg, lateral; **c**, basipod of left 5th leg, anterior; **d**, right 5th leg with endopod omitted, posterior; **e**, do, anterior; **f**, exopod of left 5th leg, anterior.

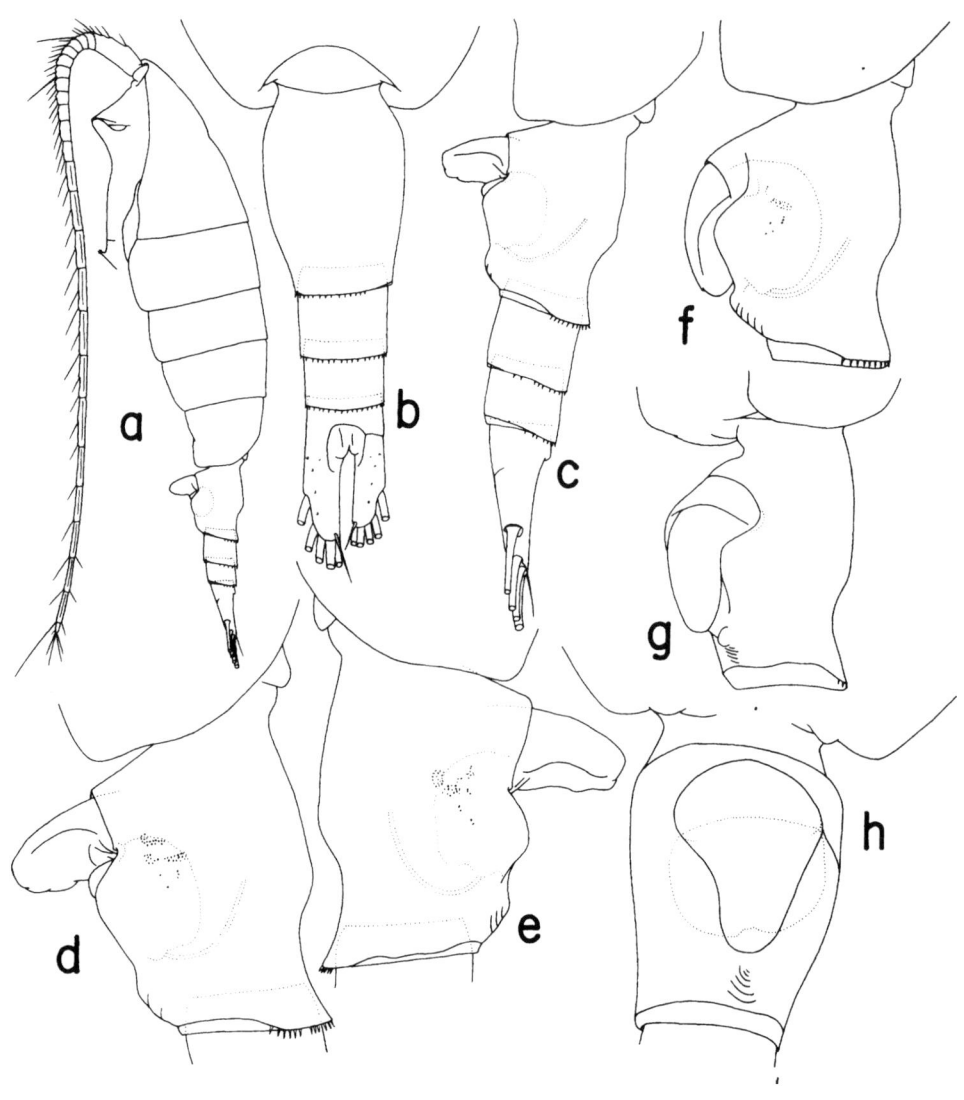

Figure 114. *Heterorhabdus norvegicus* female: **a**, habitus, left; **b**, urosome, dorsal; **c**, do, left; **d**, genital somite, left; **e**, do, right; **f**, genital somite with operculum closed, left; **g**, do, tilted clockwise; **h**, do, ventral.

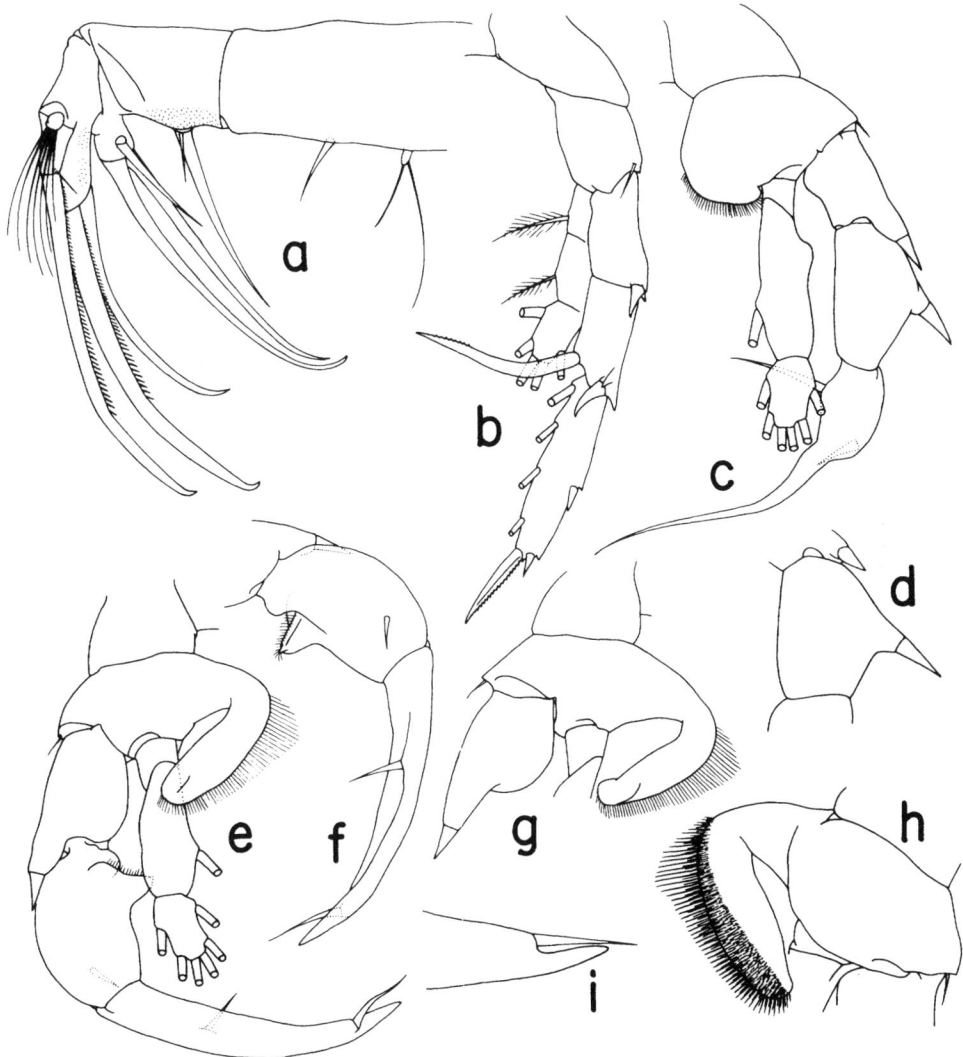

Figure 115. *Heterorhabdus norvegicus* female: **a**, left maxilla, posterior; **b**, fifth leg, posterior. Male: **c**, left 5th leg, anterior; **d**, second exopodal segment of left 5th leg, anterior; **e**, right 5th leg, anterior; **f**, exopod of right 5th leg, posterior; **g**, basipod of right 5th leg, anterior; **h**, do, posterior; **i**, distal end of exopod of right 5th leg, lateral.

Figure 116

Distribution of character states in the genera of Heterorhabdidae

Genera	Dis	Mes	Hst	Hem	Neo	Pah	Het
A. Endopod of Mx1 3-segmented (A), or 1-segmented (A')	A	A'	A'	A'	A'	A'	A'
B. Exopod of Mx1 extending to distal end (B) or far beyond distal end of endopod (B')	B	B'	B'	B'	B'	B'	B'
C. Setae of 5th and 6th lobes of Mx2 well (C) or poorly developed (C')	C	C'	C'	C'	C'	C'	C'
D. Basis of P1 with (D), or without outer seta (D')	D	D'	D'	D'	D'	D'	D'
E. Masticatory edge of Md with (E) or without (E') a group of contiguous teeth next to basal spine	E	E	E'	E'	E'	E'	E'
F. Coxa of Mxp with 1 middle seta and a group of distal setae (F') or more (F)	F	F	F'	F'	F'	F'	F'
G. P2 with 8 (G) or 7 setae (G') on 3rd endopodal segment	G	G	G'	G'	G'	G'	G'
H. Outer lobe of Mx1 with 9 setae (H) or fewer (H')	H	H	H'	H'	H'	H'	H'
I. In masticatory edge of Md, ventralmost tooth separated from the rest by a wide gap (I') or not (I).	I	I	I	I'	I'	I'	I'
J. Ventralmost tooth of Md ribbed (J') or not (J)	J	J	J	J'	J'	J'	J'
K. In masticatory edge of right Md, basal spine followed by 3 (K') or more (K) dorsal teeth	K	K	K'	K	K	K'	K'
L. Exopod of Mx1 of elongate rectangle (L'), or more or less oval (L)	L	L	L'	L	L	L'	L'
M. First segment of Mx2 elongated distal to 2nd lobe (M') or not (M)	M	M	M'	M	M	M'	M'
N. Endopod of Mx2 with 2 (N') or 6-7 setae (N)	N	N	N	N'	N'	N	N
O. Spines of 5th and 6th lobes of Mx2 serrated with large teeth (O') or not (O)	O	O	O	O'	O'	O	O
P. Left A1 of male with segments 22 and 23 fused (P') or separate (P)	P	P	P	P'	P'	P	P
Q. First segment of Mx2 with 2+1 seta (Q') or more setae (Q)	Q	Q	Q	Q	Q	Q'	Q'
R. Mxp coxa with middle seta elongated (R') or not (R)	R	R	R	R	R	R'	R'
S. Mx2 with 5th lobe reduced (S') or normally developed (S))	S	S	S	S	S'	S	S
T. Mxp coxa with large spine (T') or without (T)	T	T	T	T	T	T	T'

Dis = *Disseta*, Mes = *Mesorhabdus*, Hst = *Heterostylites*, Hem = *Hemirhabdus*, Neo = *Neorhabdus*, Pah = *Paraheterorhabdus*, Het = *Heterorhabdus*.

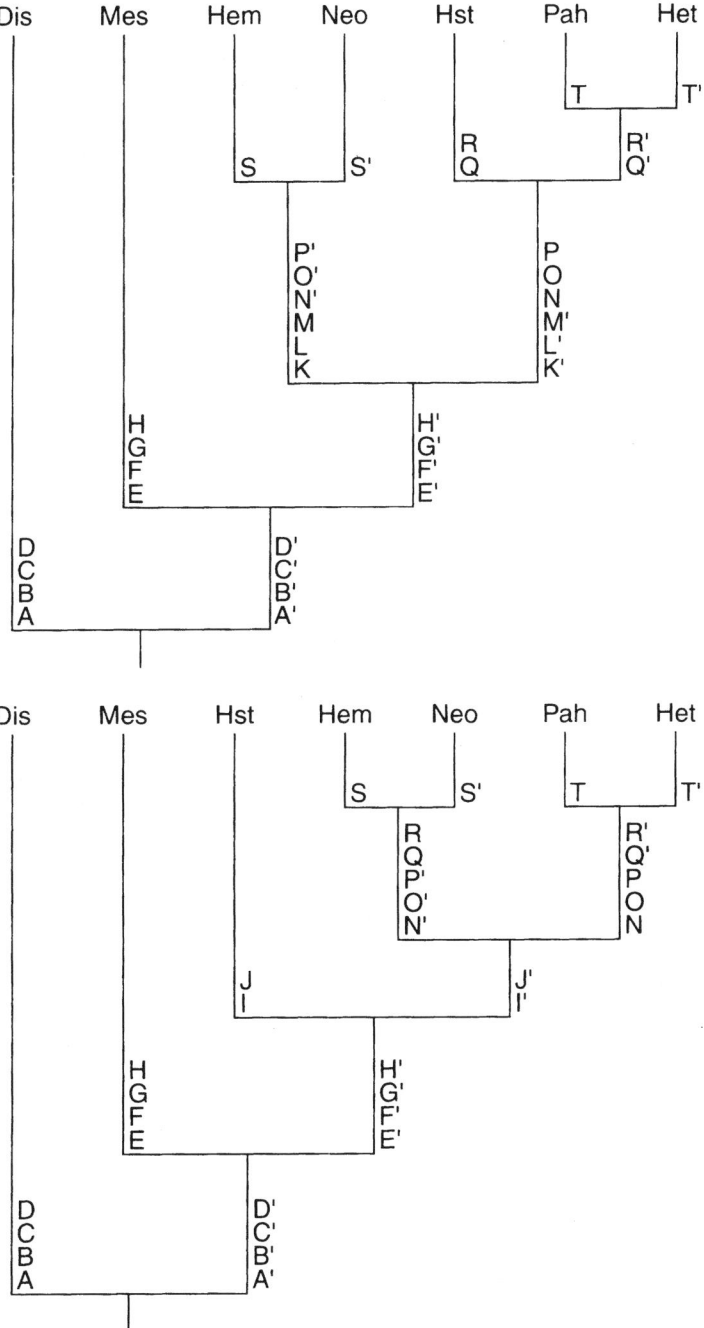

Figure 116. Cladograms showing phylogenetic relationships among the genera of the family Heterorhabdidae. Based on the 20 pairs of character states selected, two cladograms are possible.